Manfred Peschel · Werner Mende

The Predator-Prey Model:
Do We Live in a Volterra World?

# The Predator-Prey Model: Do We Live in a Volterra World?

by Manfred Peschel and Werner Mende

with 113 figures
and 11 tables

Akademie-Verlag Berlin
1986

Authors:
Prof. Dr. Manfred Peschel
Forschungsbereich Mathematik und Informatik
der Akademie der Wissenschaften der DDR
Dipl.-Math. Werner Mende
Institut für Geographie und Geoökologie
der Akademie der Wissenschaften der DDR

Erschienen im Akademie-Verlag Berlin, DDR-1086 Berlin, Leipziger Straße 3—4

Lizenznummer: 202 · 100/417/85
Printed in the German Democratic Republic
Gesamtherstellung: VEB Druckerei „Thomas Müntzer", 5820 Bad Langensalza
Lektor: Dipl.-Phys. Gisela Lagowitz
Einbandgestaltung: Ralf Michaelis
LSV 1095
Bestellnummer: 7634510 (6895)
04800

# Preface

The study of growth phenomena is now a dominant problem in many scientific disciplines. In macroeconomies the growth of the economy of a state is studied as a determining factor for the dynamics of national income; in the reliability of constructions, the growth of cracks plays an important role; in agriculture and the wood industry, the most important processes depend on the growth of animals and plants; in ecology research problems deal with the increase of emissions into the air, water and soil; and growth processes are, of course, of great importance in microbiology and chemistry.

Essentially, we frequently confront the opinion that we should try to control the growth rates $d \ln x/dt$ of the growth indicators $x$. In this context, rate-coupling is considered to be the basic mechanism for all such interactions. In this monograph, we study problems of growth and structure building from the point of view of ecology.

We frequently meet in the complex systems (in which we study growth phenomena), basic structures in the forms of chains and cycles. In these chains and cycles, usually the functional interaction of consecutive modules is determined by rate coupling, which means that the logarithmic derivative is something like a functional elementary operation in such systems.

Using an analogy from architecture of ecological systems, we propose a Structure Design Principle. This principle yields the result that, with the help of a base module "logarithmic derivative", a large set of ordinary differential equations can be transformed into a unified description by the Lotka-Volterra equations containing, besides the original state-variables, some additional state-variables. This seems to be a remarkable step forward in the systematisation of growth processes.

Important in this context is the option that we meet in objective reality only rarely uniform proportional growth; in contrary, we encounter very often explosive growth or extinction processes in natural growth phenomena.

By imbedding the Lotka-Volterra equations into the more general case of multinomial differential equations, we get a huge class of equivalence transformations admitting structural transformations for Lotka-Volterra representations. The introduction of the so-called Riccati transformation is very suitable here. It is based on the purpose to remove all autocatalytic effects from the primary Lotka-Volterra description, meaning to get rid of the diagonal elements of the interaction coefficients matrix.

The Riccati representation offers some advantages:

- The determination of closed analytical solutions, if such solutions exist.
- A key for determination of bifurcations in simulation experiments.

- To get rid of parasitic poles of the additional state-variables.
- A base for hardware realisation of function-generators, inverters and coordinate transformators.
- The possibility for a hardware processor for ordinary differential equations.

The transformation of systems of ordinary differential equations into Lotka-Volterra representations leads to interesting possibilities for the study of the qualitative properties of nonlinear systems. A necessary condition for the existence of returning motions (especially for limit cycles) is derived, connected with a certain shift-cone construction. General autonomous Lotka-Volterra equations can be interpreted as differential games, and a certain insight gained into the mechanism to decompose a set of species into permanent predators and permanent prey.

There is a connection between growth measured by the biomass and the corresponding resource flow demanded by the system for further growth. This connection exposes clearly the important role of explosive growth and extinction phenomena. The replicator form of the Lotka-Volterra equations is a good basis for the study of structure-building processes. From the dynamics of Lotka-Volterra equations we found useful cluster-building procedures which can be used for the explanation of the building of biotopes in ecological systems.

It is worth mentioning that the questions of growth, qualitative behaviour and structure-building cannot be considered completely separately from the dialectics of "determinism and indeterminism". This monograph does not study these questions in detail. We restrict ourselves to robustness studies, specifically robustness of systems behaviour against random parameter disturbations with bounded amplitudes.

This English version of the monograph is based on a German edition published in 1983 [110]. However, the content of this English version of the book has been enlarged considerably. We introduced new discrete chain concepts to win local models for time-series or to generate new function generators. We give now a transparent picture of the qualitative properties of Exponential Chains in connection with graph theory, the theory of finite automata and coding. In a natural way, we meet a vector-optimization task for the optimal choice of models using chain approximations.

The normal form theory has been enriched by many new results, especially the problem of a robust vector space basis has been studied in detail. The importance of theRiccati representation is proved here more convincingly than it has been previously. First applications of the Lotka-Volterra approach have been made in the theory of linear systems, leading especially to new possibilities for system simulation, and also for hardware solutions (for example, a special processor for the integration of ordinary differential equations and a proposal for a hyperlogistic follow-controller, resp. regulator).

Many concrete computer simulations are included, accompanied by BASIC programs which can be run on any personal computer. New arguments have been found for the fundamental role of hyperlogistic growth models. The hyperlogistic growth model is used in case studies, dealing, for example, with the growth of the world population, the prognosis for the growth of world energy production, and the growth of certain European cities in the past.

We heartily thank some of our colleagues for valuable remarks and for informations from their work, which we were allowed to include into this monograph. We want to thank especially Dr. Jansen, Dr. Voigt, Dr. Albrecht, Dr. Kriegel, Dr. Straube, Dr. Grote and Prof. Schilling from ZKI (Zentralinstitut für Kybernetik und Informationsprozesse der Akademie der Wissenschaften der DDR), Dr. Richardt from Institut für Nachrichtentechnik, Berlin, and Prof. Schneider, Dr. Feistel, Dr. Schneider and other regular attendants of the Volterra-Seminar, and Dr. Hofbauer from the University of Vienna. Of great help to us were many problem discussions in the UNESCO working group on Systems Theory and Systems Analysis, especially with Prof. Schuster, Prof. Sigmund and Prof. Allen. We had a good and successful cooperation with Dr. Breitenecker from the Technical University of Vienna. We have to thank further some interested scientists in IIASA (International Institute for Applied Systems Analysis in Laxenburg, Austria), especially Prof. Wierzbicki, Dr. Grauer, Dr. Loeser and Prof. Marchetti, and the leader of the Laboratory "Mathematical Ecology" in the Computer Centre of the Academy of Sciences of USSR, Prof. Svireshev. We thank the SINCLAIR Company for providing us with some suitable hardware tools for performing our computer experiments.

We are thankful to the editors from Springer-Verlag Wien-New York and Akademie-Verlag, Berlin, especially Dr. Weingärtner (Springer-Verlag), the Stylus Scientific & Technical Communication Consultants, Inc. and G. Lagowitz (Akademie-Verlag) for all their activities to secure a quick publication in cooperation.

Berlin, January 1985

M. Peschel
W. Mende

# Contents

# 1. Phenomenon Evolution: Growth and Structure-Building

## 1.1. *The Dialectics Between Continuous Growth and Discontinuous Structure-Building and Structural Change*

Evolution and growth are the most impressive phenomena in the world. Every year when nature wakes up anew, we observe the growth of the plants and the animals and the awakening of a large variety of species. We always see life and death against the background of a rich manifold of evolutionary processes. There is a tremendous number of natural processes around us. Our world not only *is*, it is continuously growing; its being is nothing but a slow growth.

There are two different classes of changes in time. In the first class we observe changes where the direction of time can be inverted without an essential influence. If we invert the direction of the motions of the planets, or if we let a car run from $A$ to $B$ instead of from $B$ to $A$, we do not alter our understanding of the world. In the same manner we can change the time direction in any deterministic mechanical motion.

To the second class, there belong such changes in time to which a unique direction in space in the sense of a trajectory does not correspond. For example the eruption of a vulcano, the break of a cup into pieces, the development of an animal from a germ cell, or the evolution of the biosphere around the empty earth are processes of the second kind.

It is usual to call the processes of the first class *reversible* and of the second class *irreversible*. In the strictest sense of thermodynamics, reversible processes do not exist. Even the motions of planets are accompanied by dissipative, energy-distributing irreversible processes (for example, the friction that is connected with the tides). We do not demand that a globally reversible process should invert all physical processes on which it is based, but only that it should revert the resulting trajectory. Even this, though, would be absurd for the second class of processes, as can be visually demonstrated by showing a motion picture reel backwards. At first glance, we can distinguish in the class of irreversible processes two different subclasses.

In the cases of the eruption of a vulcano and the breaking of a cup, no ordered structures are built up. On the other hand, the result fo the development of a germ cell is a complete organism, and is an ordered structure of the highest perfection. The evolution of the biosphere is an organisation process which has already continued for billions of years, and has led to an overwhelming accumulation of ordered structures on the earth. This second subclass of irreversible processes, the so-called structure-accumulating processes, consists of growth and evolutionary processes.

Hereby, growth means not only the increase of an appropriate parameter, of a growth indicator, but, always at the same time, also a process of inner organisation being reflected only by that corresponding parameter. We are not willing to restrict growth only to biotic processes, but we speak more generally of growth as the result of an aim-oriented organisation process in different systems.

Frequently, growth processes are based on an inner program. Organisation and structure-accumulating processes for which such inner program either does not exist or has only a small influence are called evolutionary processes. Of course, growth processes accompanying evolution processes play an important role.

We meet growth and evolution processes in general in the microcosmos and the macrocosmos, in all timescales and ranges. Growth and evolution are present everywhere and are the most astonishing and impressive phenomena in our world.

In reality, no system is isolated or completely autonomous; it is always imbedded into an environment in which it must compete with other systems. The environment can have a stimulating, but also a restricting, influence on a considered system. We must take into account that it is usually impossible to speak of environment in a global sense. Normally, every system is situated on a certain level of a hierarchy of systems. In this hierarchy, the environment in the narrow sense designates only the next higher level. In order to best understand systems behaviour, we must base our study on two opposite viewpoints.

There is, on one hand, the consideration from below to above (bottom-up), from the microscopic to the macroscopic level; on the other hand, we can consider the opposite direction, from above to below (top-down).

The idea of evolution is as old as mankind. All philosophical systems of the past had to formulate a specific point of view in considering evolution. In the natural sciences, essential progress in understanding evolution is connected with the names of Darwin in biology and of Boltzmann and Clausius, Prigogine, Eigen and Schuster in physical and chemical evolution. There is no doubt that the term "evolution" designates phenomena of high complexity, in which some dialectic polarities give important contributions like statics/dynamics, continuous/discontinuous, whole/part, randomness/necessity, and structure/function. In our opinion the dialectic connection of growth and structure-building plays an important role.

In this section, we shall stress the special role of the dialectic of growth and structure-building, because this dialectic is necessarily connected with the modern understanding of evolution. The growth dynamics of some species is, on one hand, a source of structure-building (function), but on the other hand, growth in its dynamics is only possible given the background of special structures (structure). Therefore, the connection between growth and structure-building is a partial aspect of the dialectic of structure/function.

In the following section, we will point out how growth and structure-building condition each other, and how, on the basis of the dynamic unity of both sides of this dialectic polarity, concrete evolution is progressing.

### *Structure as a Precondition of Growth*

Growth manifests itself in the dynamic behaviour of growth indicators. It can be expressed by the so-called *relative growth rate*. $x$ may be a growth indicator of a species (biomass, matter concentration, population density etc.). Then, the relative growth rate is given by the expression

$$\mathrm{d}x/\mathrm{d}t : x ,$$

with the velocity of change of the indicator $x : v = \mathrm{d}x/\mathrm{d}t$. Uniform proportional growth is then a growth with a constant growth rate

$$\mathrm{d}x/\mathrm{d}t : x = K .$$

The solution of this condition is an exponential function.

Exponential growth is theoretically characterized as the growth of a large number of uncoupled and self-reproducing elements which are fed by an unrestricted resource, and the refuse of which is put into a sink of unlimited capacity. Every element grows separately, without influencing its neighbouring elements or its environment.

Therefore, in general, real growth cannot be determined by uniform proportional growth, but is ruled by laws of growth which are characterized by the given coupling of structure and interaction. Thus, the character of real growth processes must depend upon the coupling structure, on the cooperative interactions between the elements with each other and with the concrete environment.

### *Growth Dynamics and Dynamic Interaction as a Supposition of Structure-Building*

We postulate that quantitative growth processes are determined by an interaction given by the Lotka-Volterra equations [1]

$$\mathrm{d}x_i/\mathrm{d}t = x_i(\sum G_{ij}x_j) \quad \text{with} \quad x_i \geqq 0 .$$

This whole monograph aims to provide proof of this opinion.

It is known that this law holds widely in ecology, when the variables $x_i$ are the population densities of species $i = 1, 2, \ldots, n$. It is also very well known that, in an ecosystem with interacting species, clusters of certain species are lumped to form communities (biotypes). This usually gives the ecosystem as a whole a higher stability. We refer for example to Allee [112], who, in the 1940's, had already gathered much empirical material, demonstrating the stabilizing effect of animal and plant aggregates and species cooperation (commensalism, mutualism, symbiosis etc.).

The decomposition of the whole set of species in a region into communities means nothing but that the considered ecosystem has a structure. What are the conditions for such a stability increasing structure-building? Is this structure-building process a consequence of the quantitative interactions between species given by the Lotka-Volterra equations or are new, specific first laws necessary to explain the structure-building?

The structure of the interaction graph, and therefore the structure of the ecosystem by different stable communities recruiting from a given reservoir of species, can be understood from the quantitative connections given by the Lotka-Volterra equations under the supposition that we extend the notion of *biological species* by other species, habitats, requisites and other extensive milieu conditions.

## *1.2. The Stability of Growth Against Disturbances*

The dialectic of randomness/necessity has a great influence on growth and structure-building. Usually, people think that the occurrence of a random influence leads to adverse consequences on the system's behaviour, to unwanted deviations from the objectives, to fluctuations which destabilize the motion, etc. However, this understanding of the role of random factors oversimplifies the real properties. In general, a growth step is defined as the transfer of a system from a lower level to the next higher level, as expressed by the values of a growth indicator. The system starts from a certain initial state on the lower level. In this stage, the influence of randomness can be very high, meaning that the whole future can depend entirely upon fluctuations of the initial values. Even in the case of meeting only a small random influence, the consequences of that influence in the future can be remarkable. When the system has started and is on its way to the higher level, however, the situation changes.

In this phase, the random effects usually increase, meaning that we observe a relatively large variance of the fluctuations. Here, however, we frequently meet a contraintuitive situation: although the variance increases, the influence on the system's path decreases because of a high stability of the growth process against disturbances in the phase of an extensive evolution. This growth phase is usually accompanied by cooperative effects, an intensive building up of new structures which are well-fitted to stimulate future growth.

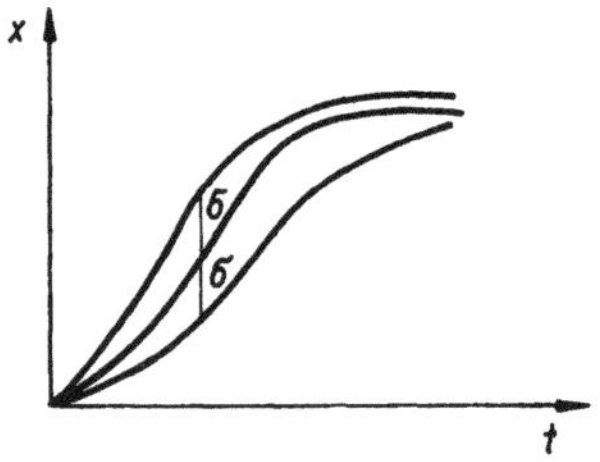


**Fig. 1** Sausage model of an EVOLON transition

This is a very important fact. In order to achieve a realistic model of a growth process, we must demand robustness against random disturbances below a certain level. The initial sensitivity to random inputs is analogous to a random seeking process, or to the generation of mutations.

As we approach the end of an evolutionary step, saturation of the growth occurs, and the growth velocity decreases. The structures which were built up for the purpose of growth are no longer necessary. The system begins to decompose, the existing structures will be gradually destroyed and their components used by the system as an inner resource. Parallel to the destruction of structures, the stability against disturbances diminishes. As the system decomposes into elements, the consequences of random influences become more and more important. This, though, is a useful effect, because a system decomposed into elements offers new possibilities to build up structures by recombination of its elements and the increased influence of random factors represents broad possibilities for experiments with the germs of new structures. Therefore, the situation at the end of such an evolutionary step is very similar to that at its beginning.

This is natural because usually the end of one evolutionary step is the beginning of the next. We restrict ourselves here to discussion of the main features of interaction between randomness/necessity and growth/structure-building during an evolutionary step. Further on, we shall return to this topic in more quantitative terms. One result of this consideration is of great importance for all our future studies: we must look for types of motions which are robust against disturbances of a realistic class (for example, amplitude-bounded random disturbances).

## *1.3. Are Structures Preformed Possibilities or Are They Built Up by Mutation and Selection?*

If we consider any system as being on an evolutionary path, then it must have a certain structure, determined values of the growth indicators, and a function given by its properties. The problem is what will occur in the future.

How does the system change its structure? On which factors does the structural change of the system depend?

After Darwin, the main method for a system to change its structure consists of the application of two operators: mutation and selection. We considered the influence of these operators on the growth already in Section 1.2. The simplest way to understand mutation is through a game where we throw at random a die, meaning that we perform steps from a certain resource of steps, each with the same probability.

Having chosen such a possible step, we must check the validity of the choice. For example, we can perform a step and compare its consequences with our given objectives or restrictions. That means that to evaluate the benefit of a step we need always to have a test for a sense of the "fitness of life". If we apply this approach to structure-building, we must define the steps or the resources of feasible structural changes (introduction of new arches into a given structure graph, or adjoining nodes to the graph, yields a spectrum of incremental changes of the structure).

In nature, these samples of new structures are realized and checked against the environment (the whole system of all restrictions on the unfolding of the system). Those structures which fit best (for example, biological structures) are chosen and can continue their existence. Unfortunately this clear picture is in general too simple. We cannot speak of free mutation if we have to take restrictions on the mutation process into account. Such restrictions can be the boundary of the set of feasible steps or the intensity of a step (for example, step width). These possibilities can be predetermined or can themselves be outputs of an experiment, of a mutation operator working within a whole hierarchy of mutations. Also, in the selection process we can meet many dynamic, predetermined or possible restrictions.

In general, we are confronted with a situation formulated after Hegel [113] in the sentence:

> "The possibilities are determined, but the possible steps within the restrictions of these preformulated possibilities are randomly generated and selected".

We conclude that, in general, we meet a hierarchy of structures, in which the structures on one level determine the restrictions and the possibilities for the next lower level. Within the spectrum of degrees of freedom left under these restrictions, a relatively free mutation and selection is possible.

This is not a static picture, but a living one because, on every level, we must deal with growth, the main task of a structure being to guarantee the best conditions for future growth. Therefore, the growth dynamics on every level produce a dynamic influence on the structure-building process, giving the whole system of restrictions a dynamic character.

## *1.4. The Paradigm of Exponential or Uniform Proportional Growth*

In connection with continuous growth processes, very often the slogan *uniform proportional growth* is used. First, we shall explain how this slogan must be understood. Afterwards, we will analyse critically this notion from the ecological point of view.

A system which, during every fixed time period, receives a fixed portion of food can accumulate the corresponding effects, in the ideal case, like an integrator; that means that, in proportional to the food ration $y$, the value of a corresponding growth indicator $x$ (for example, the storage of fat) increases according to the following growth law:

$$\mathrm{d}x/\mathrm{d}t = Ky'.$$

The initial point in time shall, by definition, be $t = 0$. For the special case of a constant food ration $y = y(0) = y_0$ we thus get a linear growth law:

$$x(t) = Ky_0t\,.$$

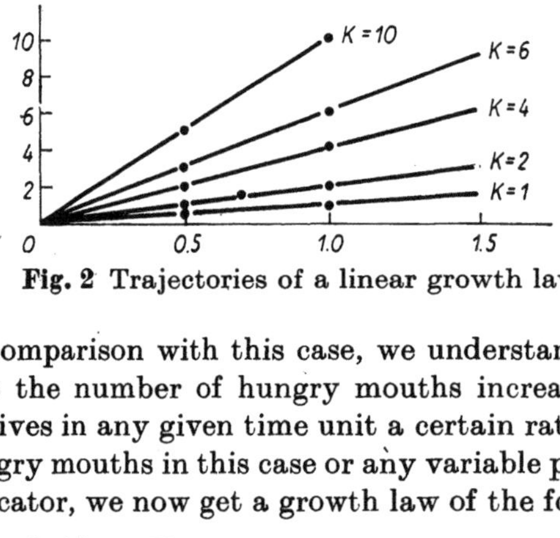


**Fig. 2** Trajectories of a linear growth law

In comparison with this case, we understand uniform proportional growth to mean that the number of hungry mouths increases over time, and every hungry mouth receives in any given time unit a certain ration of food, $y_0$. If we take the number of hungry mouths in this case or any variable proportional to this number as the growth indicator, we now get a growth law of the following form

$$\mathrm{d}x/\mathrm{d}t = Kxy_0\,.$$

In this case, the overall demand for food is now proportional to $x$. We understand uniform proportional growth therefore as a so-called exponential growth law, frequently called the *law of natural growth*, perhaps because it is so unnatural

$$x(t) = x(0)\,\mathrm{e}^{Ky_0t}\,.$$

Historically, the exponential law served as an argument for Malthus [4], who stated that the population on earth increases in a geometrical progression, meaning exponentially, while, at the same time, the possible global food supply increases only in an arithmetic progression, meaning linearly. Therefore, the discrepancy between supply and demand for food for the ever-increasing world population will grow very fast, and already for this reason, mankind cannot avoid future catastrophes. The considerations in this monograph later on will show how far the Malthus's ideas must be corrected. First we bring some examples, where we in fact find uniform proportional growth to understand better under which conditions it can arise.

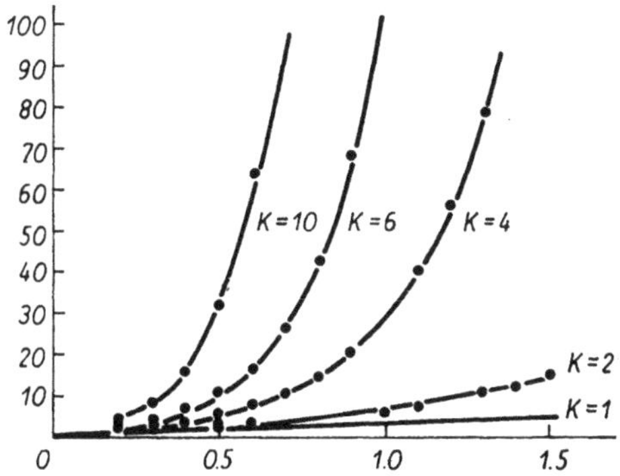


**Fig. 3** Trajectories of the exponential growth law

*1. Compound interest of a capital in the bank*

If we have in the bank an initial capital sum $x$, with a yearly interest of $p\%$ where $p = y_0$, then every dollar (hungry mouth) increases during every year by the factor $(1 + p)$; therefore, the initial capital $x(0)$ has changed after a year as follows

$$x(t) = x(0)\,(1 + p)\,.$$

If it is possible to get interest $n$ times per year, we must divide the quota $p$ into $n$ steps, each of the amount $p/n$. We get the following equation for the change of the capital:

$$x(1) = x(0)\,(1 + p/n)^n\,.$$

For unlimited growing number $n$ of the time-points of compound interest, we get in the limit for $n \to \infty$ the law

$$x(1) = x(0)\,\mathrm{e}^p\,.$$

If we do not restrict the consideration to one year only, but take into account a time interval $(0, t)$, we again meet the exponential law

$$x(t) = x(0)\,\mathrm{e}^{pt}\,.$$

This growth curve obviously obeys the following differential equation:

$$\mathrm{d}x/\mathrm{d}t = px\,.$$

An active capital investor is not satisfied with the result from simple compound interest. Therefore, he frequently transfers his capital to industry (stock investments), in order to achieve a monetary return of an amount corresponding to an increase of the interest $p$ proportional to the invested capital $x$, perhaps even with a very small proportionality factor

$$p = (p_0 + \varkappa x) \quad \text{leading to} \quad \mathrm{d}x/\mathrm{d}t = (p_0 + \varkappa x)\,x\,.$$

This phenomenon-based law leads to large deviations from an exponential law. It assumes strong coupled structures between bank capital, production capital and the market, and it necessarily leads to explosion. It produces a driving force which tries to reach an infinite amount of capital $x$ in a finite time.

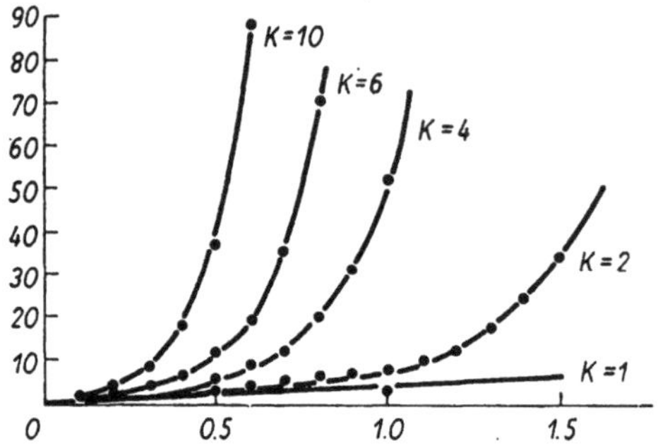


**Fig. 4** Trajectories of an explosive growth law

*2. The law of radioactive decay*

The radiation activity of a radioactive specimen with the initial activity $x(0)$ decreases, as is very well known, according to the exponential law

$$x(t) = x(0)\, e^{-pt}$$

because the number of newly decaying atoms depends proportionally upon the number of atoms actually present in the undecayed initial state. Both processes, the increase of capital and radioactive decay, have the common property of treating a number of isolated elementary systems in the same manner: there is a common factor, which is greater than 1 in the case of compound interest (reward) and less than 1 in the case of radioactive decay (punishment) which gives multiplied with the "mass" $x$ of the given system the whole increment in time. Obviously, uniform proportional growth is related to isolated growth processes with no stimulating or inhibiting coupling structures, where every element is treated in the same way.

*3. The growth of a bacterial culture in a chemostate*

Under constant environmental conditions, after a phase of adaptation, bacteria begin to divide. This so-called log phase growth continues as long as the bacteria do not disturb each other with demands for room or food. In this log phase, each bacterium grows individually with its inner growth rate $K$ (a typical case is $K = 2/\text{h}$), taking no notice of its neighbours. Therefore, the growth process is completely determined by the individual reproductive processes within each bacterium. This means that exponential growth is growth of a sufficiently large number of self-reproducing uncoupled elements.

*4. Cell division processes in the embryonic development*

In embryonic development the cells divide completely independently from each other. Therefore, the number of cells increases after the exponential law $x(0) = 2^t$ where $t$ is the number of division periods. However, according to the latest observations [43], this exponential law holds only up to the 8th division period. Afterwards, coordination of the division processes of the different cells can already be observed as coming from a centre within the blastula. Beginning at this time, we can no longer speak of exponential growth. The consequence is a subexponential growth controlled by a certain genetic program. The considerations and examples in this section lead to the idea that always when partial processes which contribute their part to the overall growth are no longer independent from each other but show sufficiently strong coupling, the global growth process cannot be described as exponential growth. On the contrary, the coupling links lead to a stimulation or an inhibitation of the growth rate. It is reasonable to assume that the coupling structure produces a growth rate $dx/dt : x$. The value of the growth rate is no longer constant, but dependent on the value $x$ of the growth indicator itself, or on other influencing variables.

## *1.5. Empirical Findings Proving the Existence of Hyperbolic and Parabolic Growth and Extinction Phenomena*

Highly aggregated growth indicators of large homogenous systems show significant deviations from exponential growth if we consider sufficiently large times and sufficiently large systems. The relative growth rates manifest a time-dependent trend representing the coupling features between the elements.

For uncoupled exponential growth processes (exponentially driven and exponentially saturated), the introduction of growth-damping coupling is connected in a very simple way with the space at the disposal of the system; perhaps $N$ is the number of places for existence (logis). Then the relative growth rate will be linearly proportional to the number of places $(N - x)$ not yet occupied:

$$dx/dt = Kx(N - x) \, .$$

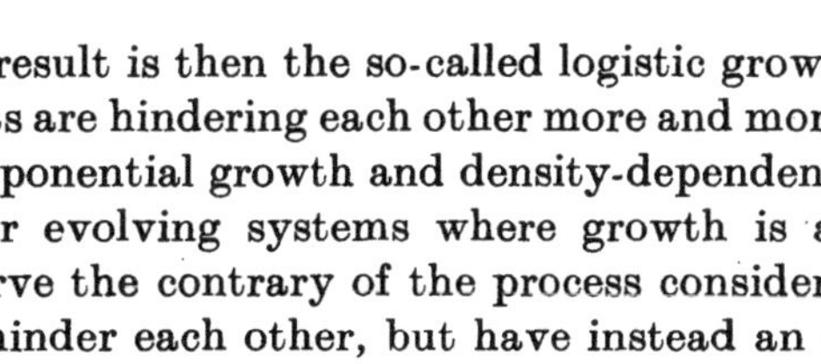


**Fig. 5** Trajectories of the logistic growth law

The result is then the so-called logistic growth law. With growing density $x$, the ele ments are hindering each other more and more. Therefore, logistic growth is composed of exponential growth and density-dependent exponential saturation.

For evolving systems where growth is accompanied by structure-building, we observe the contrary of the process considered above. In this case, the elements do not hinder each other, but have instead an enhancing effect on each other. Growth here is not negatively, but positively density correlated. A very wellknown example of this effect is population growth. From demography we know that we find the largest growth rates in big cities (at least in the initial phases of such an evolution we find a positive density correlation). Also, the growth of world population shows this property. World population on the whole is a very complex coupled system; its growth, measured by the number of individuals living at the same time, depends upon a tremendous number of coupling factors such as food, resources, region, fertility and climate, as well as such factors as production, society, scientific/technical level, communication, culture, etc.

The question is whether, despite this variety of influencing factors, it is possible to recognize general and sufficiently simple growth laws which describe reliably the growth of the number of individuals over very long periods of time. Let us consider the number of human beings $P(t)$ living at the same time $t$ and draw the corresponding

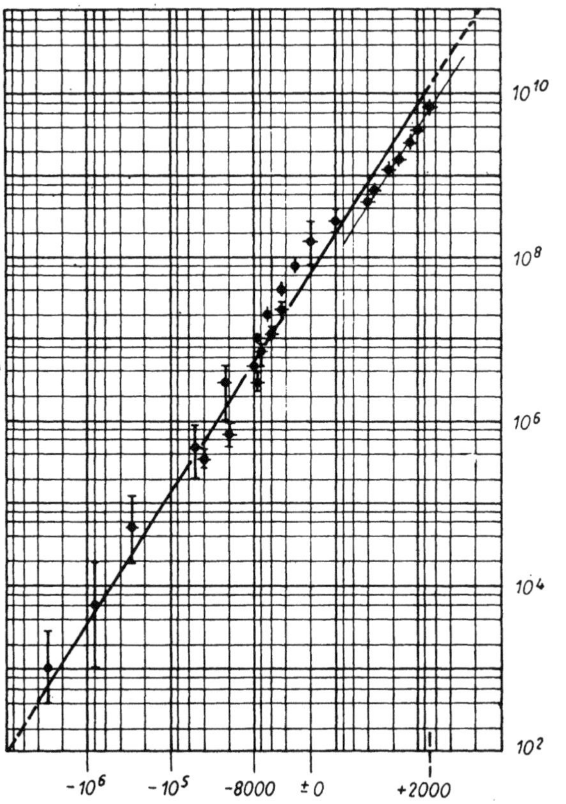


**Fig. 6** The growth of world population in double-logarithmic representation

curve into a double-logarithmic net (Mende [78]). Some billions of years ago the genus Homo originated from a tribe of hominids with some hundreds of thousands of individuals. This number is characteristic of a primitive species of primates forming population. Since that time, we may observe a continuous increase of the growth rate as a long-term trend over a time period of about 2 million years. In this time, the overall number of people has increased by a factor of several millions. Today, the growth rate of world population is about 10,000 times larger than the long-term median value of the growth rate during the earlest days of mankind.

In this long-term perspective, we can see a relatively uniform increase of the growth rates with astonishingly small deviations about the straight line in Fig. 6 correspondng to a hyperbola with the equation

$$P(t) = 1.1 \cdot 10^{12}/(2050 - t)^{1.3} .$$

This is remarkable, because mankind went through a history with many major changes, such as the Ice Ages, natural catastrophes, epidemics, wars and hunger. These all are influences responsible for important local variations which lasted a relatively short time. Besides these we had long waves around the long-term trend, such as the Neolithic revolution and the stagnation in the Middle Ages. Obviously, the overall trend is robust against all these events.

In this long-term view, we cannot speak of an explosion of population in a certain time. On the contrary, we have to accept the continuous acceleration of the growth of mankind during our whole history: world population is not exploding only now in the present time, but already has been for millions of years.

This tremendous and relatively uniform expansion of mankind is surely no random effect, but the expression of a very deep systems law. Increasing growth rates are characteristic of undamped growth processes in self-organising, structure-accumulating, meaning evolving systems. The hyperbolic growth of mankind reflects the successful fighting against degradation and the ever-increasing challenges of the system "mankind" by its environment, and is based on the cooperative behaviour of the individuals within the society.

The working activities of the people infer social links; this increased cooperativity leads to more and more successful appropriation of natural resources. Even taking into account relatively large error intervals for the paleodemographic data, the fact of increasing growth rates is beyond doubt even on a much smaller time scale.

An exponential process projected back into the past would yield very quickly population numbers which are much too small. If we assume a long-term median value of the yearly growth rate of about 0.1% corresponding to a doubling time of 700 years, we would after 32 bisection steps (corresponding to about 23,000 years), reduce todays population of 4 billion people to only one person.

We now bring additional examples for global hyperbolic growth processes. The corresponding data are not so reliable, but exponential growth can surely also be excluded in these cases.

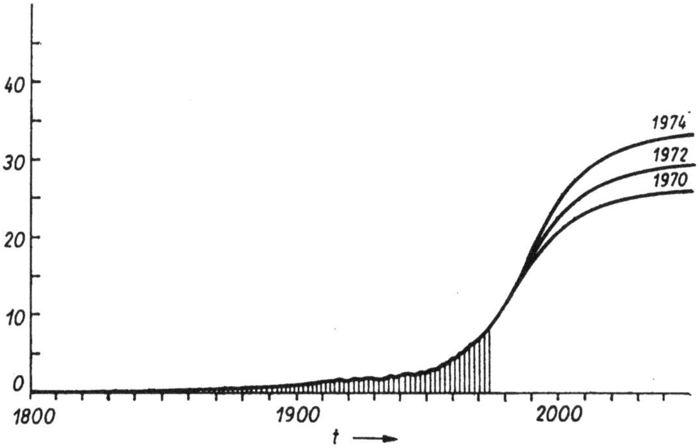


**Fig. 7** Primary energy consumption of the world after data from IIASA in TW

| | | |
|---|---|---|
| Accuracy of time measurement | $\Delta t/t = 10^{24}/(2010 - t)^8$ , | $k = 1.12$ |
| Yearly publications in mathematics | $n = 9.5 \cdot 10^9/(2030 - t)^3$ , | $k = 1.33$ |
| Energy production | $E = 5 \cdot 10^{17}/(2070 - t)^{1.9}$ , | $k = 1.52$ |
| World population | $P = 1.1 \cdot 10^{12}/(2050 - t)^{1.3}$ , | $k = 1.77$ |
| Brain volume of primates | $K = 3 \cdot 10^9/(2 \cdot 10^6 - t)^{1.0}$ , | $k = 2.0$ |
| Output of sun (Present $t = 4.5$) | $L/L_0 = 3.83/(11.5 - t)^{0.69}$ , | $k = 2.45$ |
| $O_2$-level in atmosphere | $O = 3 \cdot 10^{-3}/(2 \cdot 10^6 - t)^{1.3}$ , | $k = 1.77$ |

The hyperbolic growth curves obey the following autocatalytic differential equation:

$$\mathrm{d}x/\mathrm{d}t = Kx^k , \qquad k > 1 .$$

A rough look at these results shows that the exponent in the hyperbolae will decrease, or the autocatalytic exponent $k$ will increase with the increasing complexity of the corresponding system. Of course it is not allowed to project the above processes into the far future without changing the type of motion, because, in most of these processes (without the example of the sun's output), saturation mechanisms have already begun to work.

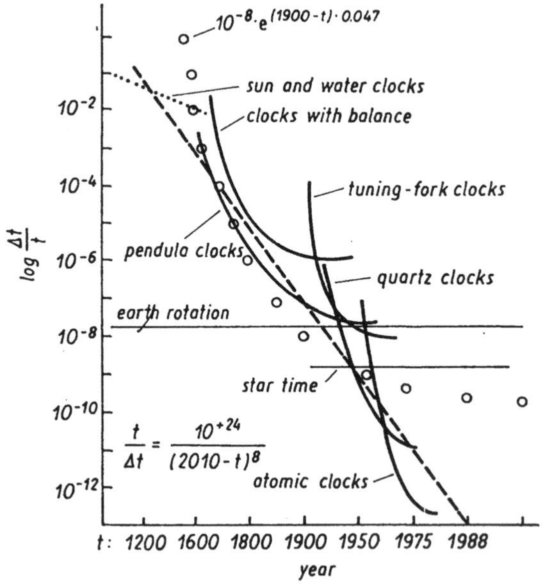


**Fig. 8** Precision of time measurement
(W. Mende (1975), after a diagram of Spoljanskij (1974))

Quite different phenomena occur if the future aim of the growth of a system is predetermined by an inner program. Such growth processes are met in the autogenesis of organisms. Here, the aim of achieving an adult organism is already fixed by a genetic program. Such growth processes arise also in connection with ecological successions, which move under unchanged environmental conditions to their climax stage. This

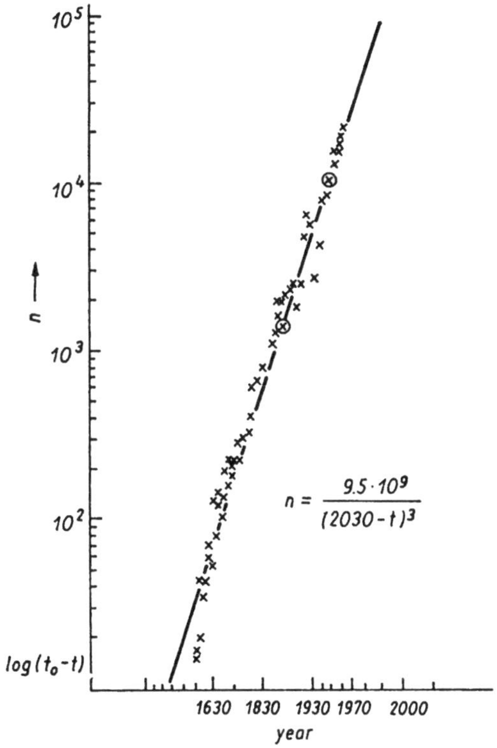


**Fig. 9** Number of new mathematical publications per year in 1550—1970

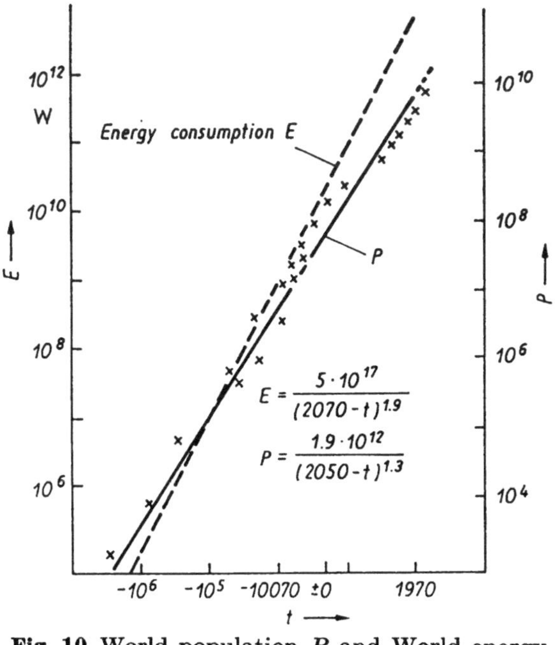


**Fig. 10** World population $P$ and World energy consumption $E$ in W for the time $-2.10^6$ to 2000 years

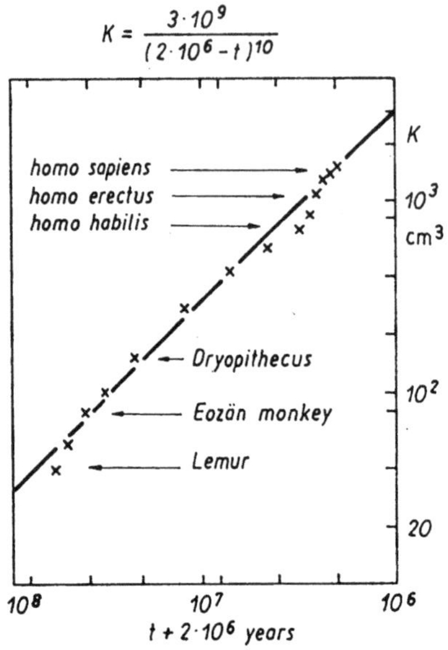


**Fig. 11** Evolution of primates measured by the growth of brain capacity (W. Mende (1975), after data of Stebbins (1968) and Heberer (1974))

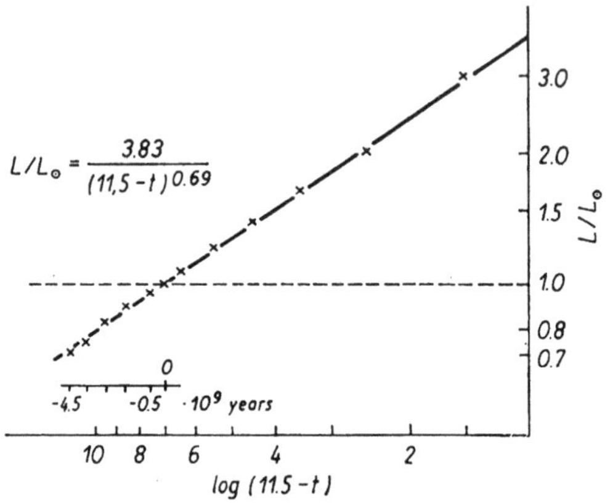


**Fig. 12** Evolution of the energy output of sun in the time interval $-4.5 \cdot 10^9$ to $6 \cdot 10^9$ years

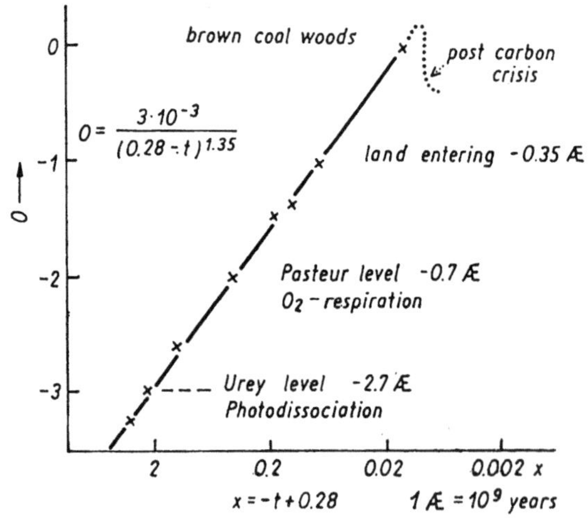


**Fig. 13** Atmospheric oxygen level (hyperbolic trend) (W. Mende (1977), data from Lauterbach and Schidlowski)

stage is completely predetermined by the inventory of the corresponding species and the physical environment.

In growth processes of this kind, we very often meet parabolic processes characterized by decreasing growth rates induced by the control program. In such a case,

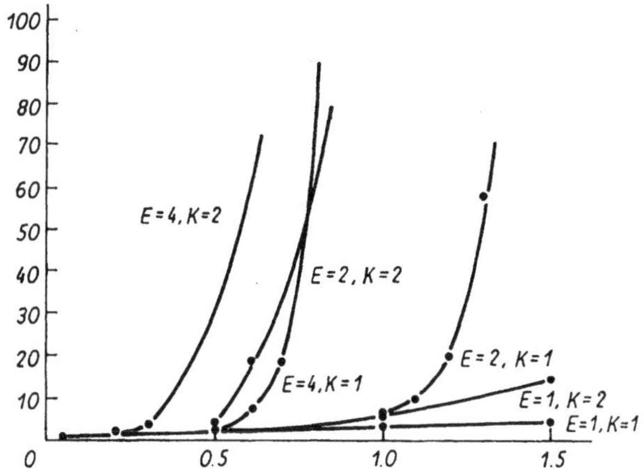


**Fig. 14** Autocatalytic hyperbolic growth

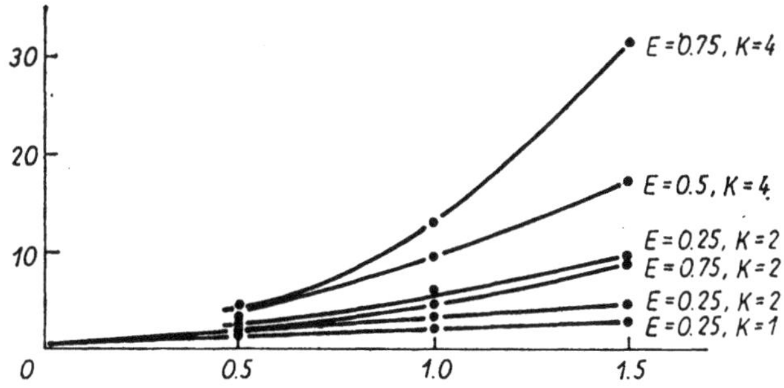


**Fig. 15** Autocatalytic parabolic growth

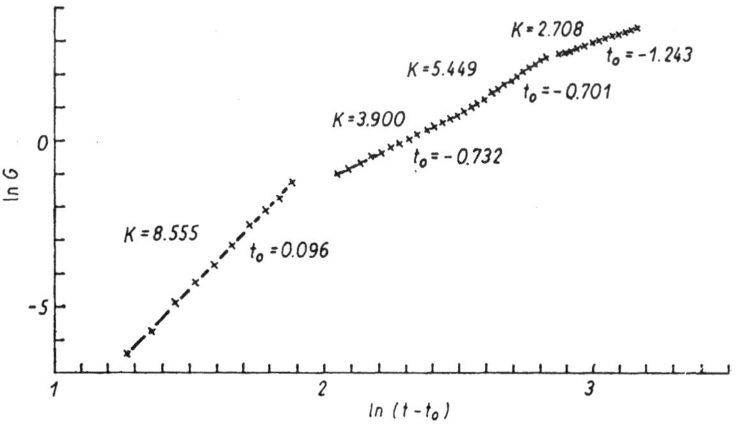


**Fig. 16** Growth pushes of chick embryos

usually an autocatalytic law

$$dx/dt = Kx^k \quad \text{with} \quad k < 1$$

holds.

Genetic regulation or the laws of ecological evolution usually yield a damping of the growth. A well-known example of this is the parabolic growth law of Schmalhausen, which provides us with a good approximation of embryonic growth. We can show this with chick embryos for data from Gerškova [65]. In full agreement with the conclusions of the author, we get three different growth pushes.

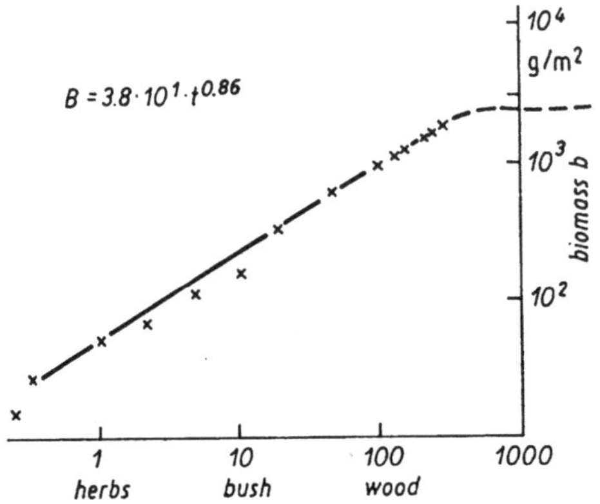


**Fig. 17** Reforesting of fallow land by a mixed wood succession, expressed by the biomass $B$ in g/m²

For an ecological succession we use data from Whittaker [68]. Also in this case, we get a good approximation by a parabolic growth curve. These examples show that a characteristic difference in the growth dynamic between hyperbolic and parabolic growth can be guessed depending upon whether the corresponding process is self-organising and structure-accumulating or controlled by an inner program.

## *1.6. Properties of Hyperbolic and Parabolic Growth, Saturation and Extinction*

The simplest way to describe autocatalytic effects is with the help of the following ordinary differential equation

$$\mathrm{d}x/\mathrm{d}t = Kx^k$$

with the growth rate $\mathrm{d}x/\mathrm{d}t : x = Kx^{k-1}$. We can use $m = k - 1$ as a certain measure for the complexity of the corresponding growth process. For $k > 1$, we get hyperbolic, and for $k < 1$ we get parabolic growth. Both growth types are structurally stable in the sense that small deviations of the exponent $k$ cannot change the behavioural type.

The case $k = 1$, the case of exponential growth, separates these types of motion from each other, and is, in regard to the parameter $k$, structurally unstable. Every disturbance, even a very small one, leads the exponential growth either to hyperbolic or to parabolic growth. Obviously exponential growth separates hyperbolic and parabolic growth types with the following consequence:

An exponential growth process can always be well approximated by a hyperbolic or a parabolic growth process with an exponent $k$ approximately equal to 1. Conversely however, it is impossible to make a good approximation of a hyperbola or a parabola from an exponential growth function.

The autocatalytic differential equation apparently has the following integrals:

$$x(t) = \varkappa/(t_g - t)^{1/(k-1)} \quad \text{for} \quad k > 1 .$$

This curve has the form of a hyperbola, therefore, we speak of hyperbolic growth in this case. Essentially, infinity $\infty$ is, in this case, approached in a finite time.

$$x(t) = \varkappa(t_g + t)^{1/(1-k)} \quad \text{for} \quad k < 1 .$$

The monotonous approach of infinity $\infty$, however requires infinite time. For the corresponding parameters, we get the following expressions

$$\varkappa = 1/(|k - 1|\, K)^{1/(k-1)} \quad \text{and} \quad t_g = 1/(x_0^{k-1}\, |k - 1|\, K) .$$

The simplest case to model an autocatalytic saturation uses the following ordinary differential equation

$$\mathrm{d}x/\mathrm{d}t = K(B - x)^k \quad \text{with} \quad K > 0!$$

After the substitution

$$\xi(t) = B - x(t)$$

$\xi(t)$ describes an autocatalytic extinction process obeying the following differential equation

$$\mathrm{d}\xi/\mathrm{d}t = -K\xi^k .$$

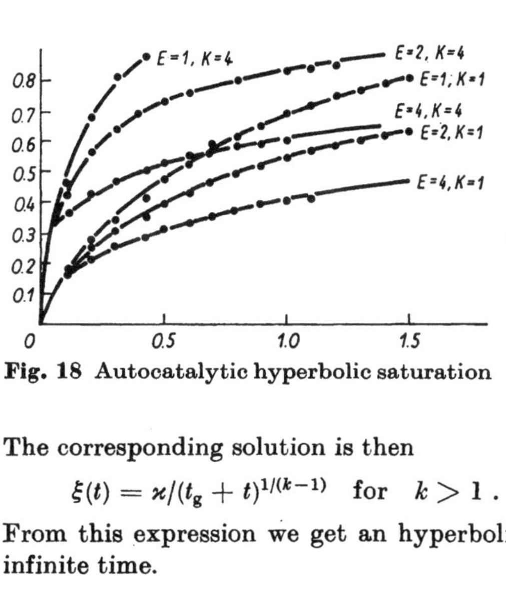


Fig. 18 Autocatalytic hyperbolic saturation

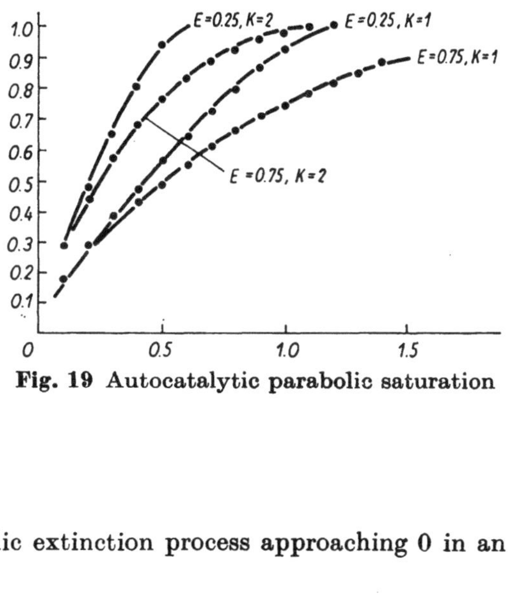


Fig. 19 Autocatalytic parabolic saturation

The corresponding solution is then

$$\xi(t) = \varkappa/(t_g + t)^{1/(k-1)} \quad \text{for} \quad k > 1 .$$

From this expression we get an hyperbolic extinction process approaching 0 in an infinite time.

$$\xi(t) = \varkappa(t_g - t)^{1/(1-k)} \quad \text{for} \quad k < 1 ,$$

this is a parabolic extinction process approaching 0 in a finite time. Of course, driven or damped processes do not occur in this pure way. Normally we meet the dialectics of driving and damping, that meaning a superposition of forces which, on one hand, stimulates the growth with forces which, on the other hand, produce a damping influence. This superposition leads to $s$-formed transition processes (evolution step) for transferring the system from a given level to the next higher level of evolution. In general, some of these evolutionary steps follow, one after the other and form what we call a *staircase of evolution.*

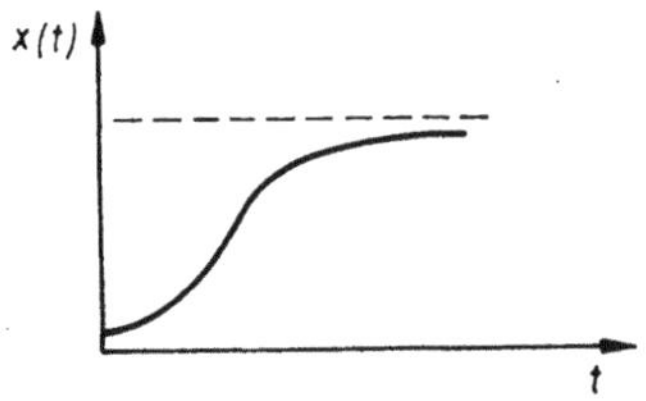


**Fig. 20** *s*-formed transition: an evolutionary step

We get important hints for the identification of the character of the growth processes if we consider the $p$-factor amplification times. For our autocatalytic power growth process, we get the result

$$x(t_1) = x_1 \qquad x(t_2) = px_1 \quad \text{and} \quad x(t_3) = p^2 x_1 ,$$

$$1/p^{k-1} = \frac{t_3 - t_2}{t_2 - t_1}, \qquad p > 1 .$$

This means the $p$-factor amplification times are shortened for $k > 1$ and increased for $k < 1$, always in a geometric progression, and they stay constant for $k = 1$. For the saturation process, we get the reverse result. This geometric progression explains that hyperbolic growth or parabolic saturation are stopped in a finite time because the shortening of the $p$-factor amplification times necessarily leads to the occurrence of a limit point on the time-axis in a finite time $t_g$ because of the convergence of the sum of a geometric series based on the increments considered above.

For a relation

$$\mathrm{d}x/\mathrm{d}t = Kx^k$$

the growth increment depends only upon the mass $x(t)$ which is present at a certain time $t$. It is, however, reasonable to assume that the corresponding coupling mechanism is independent of the mass $x$. This could be interpreted as a homogenous interaction structure between a large number of elements within the global system. The growth mechanism is realised by a large number of relatively independent elementary processes and therefore possesses a certain redundancy. If we would remove a small number of elements from the system, we do not modify the existing growth mechanism. Obviously such a coupling between elements is functionally a weak interaction (coupling redundancy).

This is opposite to the behaviour for example of an electronic device, which usually has many elements of individual importance. Frequently, dropping one such element out leads to a failure of the whole device (strong coupling).

Homogenous weakly coupled large systems are usually realised in chemical reactors and in all chemical reactions which are contained within a large volume (This is because of the high value of Avogadro's number $10^{23}$). Such large systems are also typical in sufficiently large populations or other subsystems of an ecosystem. In such cases, the individuals are only representatives of a general functional relation. If a huntsman shoots some foxes or some rabbits in a forest system, the system usually continues to function in the same way as it did before. Such homogenously coupled large systems, being robust against partitioning to a certain degree, we call *soft systems*. Under this redundancy condition, it is no longer astonishing that complex interactions in ecosystems can be summarized and represented simply only by stimulating or damping influences on growth rates.

## 1.7. The Structure of Complex Ecological Systems. The Downflow Trails of Energy. The Special Role of Chains of Rate-Coupled Systems (Exponential Towers)

Firstly we will explain the notion of *rate-coupling.*

We call any interaction rate-coupled in which the rate of the controlled variable $\mathrm{d}x/\mathrm{d}t : x$ is influenced by a control variable $x'$ according to a law of the form

$$\mathrm{d}x/\mathrm{d}t = xf(x') .$$

Rate-coupling in chains is bound to the existence of a linear structure, consecutively linked rate-coupled systems, described by equations of the form

$$\mathrm{d}x_i/\mathrm{d}t = x_i f(x_{i+1}) , \qquad i = 0, 1, 2, \dots .$$

In the simplest case, the functions $f(x_{i+1})$ are linear, meaning that the growth rate of the controlled variable is linearly proportional to the state-variable of the next higher level in the chain

$$\mathrm{d}x_i/\mathrm{d}t = K_i x_i x_{i+1} , \qquad i = 0, 1, \dots .$$

Such a behaviour can be observed in food chains, in chains of enzymatic reactions, in the growth processes of organs within living beings, in energy cascades. The coupling is multilinear if the growth rate $\mathrm{d}x_i/\mathrm{d}t : x_i$ of level $i$ depends on a linear superposition of some inputs. We will discuss this case in detail later on. For the moment, we restrict ourselves to the case in which we have only one input on every level.

The Exponential Tower

$$\mathrm{d}x_i/\mathrm{d}t = K_i x_i x_{i+1}$$

is characteristic for all downflowing trails of energy. Let us interpret this in connection with our earth. The lowest level may be the population of microorganisms in the soil. The corresponding growth rate $\mathrm{d}x_0/\mathrm{d}t : x_0$ depends at first glance on the food input $x_1$ produced as garbage from plants and animals. The growth rate of this garbage $\mathrm{d}x_1/\mathrm{d}t : x_1$ depends on the energy level $x_2$ contained in the biosphere as a whole. The growth rate $\mathrm{d}x_2/\mathrm{d}t : x_2$ of the biosphere depends on the energy state on the surface of the sun $x_3$.

There is also a direct sunlight energy input on the medium level, meaning that we may meet parallel chains between the sun and the microorganisms in the soil and the sun and the organisms on the medium level, but this does not change our argument.

If the downflow trail of energy in a system is large enough, we meet structure-building phenomena. Structure is built on the different levels of energy waste (distribution). This leads to a certain energy storage in the structures built on the different levels. Necessarily, we observe between these levels functional relations given by rate-coupling properties. Therefore, the structure, on one hand, hinders the degeneration of energy. It is also a resistance for the free distribution of energy expressed by increased entropy. On the other hand, existing structures and stimulated growth by these structures are entropy machines of the highest efficiency; through growth, they thoroughly stimulate the consumption of energy. Thus, we meet this dialectic contradiction on the downflow trails of energy.

More generally, in chains of rate-coupled systems, we frequently have to deal with the following situation described by the equation

$$dx_i/dt = K_i x_i x_{i+1} - L_i x_i x_{i-1} - a_i x_i , \qquad i = 0, 1, \ldots .$$

On the $i$th level, the growth rate of the corresponding state-variable $x_i$ is proportionally controlled (driven for $K_i > 0$ and damped for $K_i < 0$) by the state-variable $x_{i+1}$ of the next higher level. The same function as $x_{i+1}$ for $x_i$ has the state-variable $x_i$ to fulfill in respect to $x_{i-1}$. This can lead to a certain loss on the level $i$ of the amount $L_i x_i x_{i-1}$ in its influence on the growth rate of $x_i$. Besides these two effects, perhaps a certain garbage unit $a_i x_i$ arises which is not usable within the connection of the given chain. Of course, it is possible that such a garbage unit can be used as a resource elsewhere in the global system to which the considered chain belongs as a substructure.

We speak of a *hypercycle* [9] if a finite chain of rate-coupled systems is closed to a circuit by a feedback loop from the bottom level up to the top, meaning that the following closing conditions hold:

$$x_{N+1} = x_0 \quad \text{and} \quad x_{-1} = x_N .$$

It is called a hypercycle because of the bilinear connection

$$dx_i/dt = K_i x_i x_{i+1}$$

in which each element already has its own feedback. Fig. 21 shows a chain of rate-coupled systems together with a hypercycle of third order.

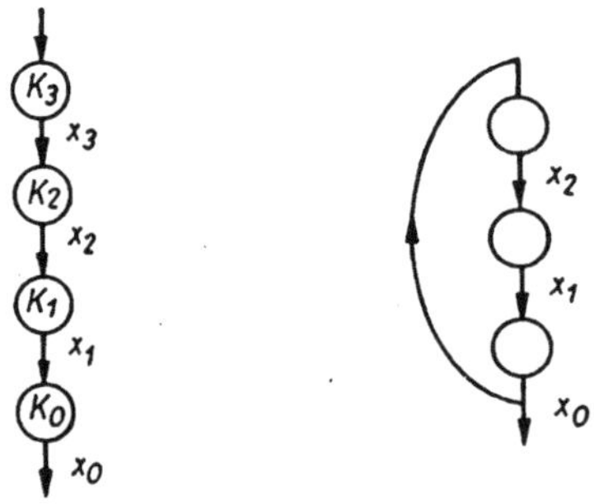


**Fig. 21** Exponential chain and hypercycle of order $r = 3$

The notion of the hypercycle in our sense is completely analogous to the hypercycle-notion introduced by Eigen and Schuster [9]. We consider chains and hypercycles built with rate-coupled systems as significant substructures (clusters) of complex systems. Complex ecological systems, for example, the food system of the virgin forest or of the antarctic sea, possess as basic structures chains and cycles with a relatively small number of inner feedbacks or connections between them. We have the imagination to demand such a system concept for the structure of all sufficiently complex systems.

Because, from the functional point of view, we assume that nature prefers to work with applying homogenous elements and rate-coupling so that our structural picture can be projected from ecological observation. A structure model for a complex system should be composed of chains and cycles, and rate-coupling between consecutive elements should be the most important determining factor for functional interaction. In this sense, we generally consider the dynamic behaviour in this world, at least in regard of continuous growth processes, always from an ecological point of view. The term *ecological* in this context is considered to be under the restriction that not the full richness of structures and connections in ecological systems are meant. However, we also use this term in a more general sense, as in *ecological systems*, because we generalise such processes as competition for a common resource, the organisation of an ecosystem into well-defined species, requisites and resources and the change of species by irreversible growth processes to much more general situations, such as we meet in ecosystems. Specifically we apply these considerations to all multi-particle/multi-species systems in an irreversible energy flow. Examples of this are scatter processes in the atmosphere, complex energy cascades, and many other homogeneous physical and nonphysical processes.

Let us take for example the two-dimensional Lotka-Volterra system in its simplest form, with two species $x$ and $y$ and the simplest coupling $x \cdot y$ between them

$$\mathrm{d}x/\mathrm{d}t = Ax - By \cdot x\,, \qquad A, B, C, D \geqq 0\,,$$
$$\mathrm{d}y/\mathrm{d}t = -Cy + Dy \cdot x\,, \qquad x, y \geqq 0\,.$$

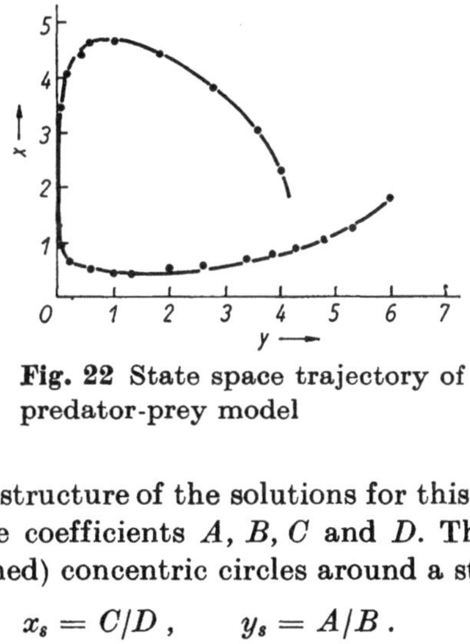


**Fig. 22** State space trajectory of second-order Lotka-Volterra system, the simplest predator-prey model

The structure of the solutions for this equation is completely independent of the concrete coefficients $A$, $B$, $C$ and $D$. The solutions in the plane $x$, $y$ are given by (deformed) concentric circles around a stationary point

$$x_s = C/D\,, \qquad y_s = A/B\,.$$

Ecologically interpreted $x(t)$ and $y(t)$ may be species densities in a given region. The prey $x$ (for example, rabbits) grows without taking into account the predators (for example, foxes) exponentially with a growth rate $A$. The necessary and unrestricted available resource for this growth is implicitly contained in the positive growth rate, which is comprised of a certain natural death rate being overcompensated for by a larger birth rate. The interaction is proportional to the probability or frequency of meeting between both types of species: $Bx \cdot y$, because we can consider $x$ and $y$ as the corresponding concentrations or densities of the species in the given region and accept the individuals meeting independently as an approximation.

The harvested prey $Bx \cdot y$ will be accounted for by a certain factor $D/B$ (with $\Phi = Bx \cdot y$ being a so-called *trophic function*), and will be converted into a corresponding growth increment for the predators. The predators also have a certain natural death rate, $C$. We are willing to use this model as a reasonable system concept in every instance where we deal with two state-variables and a density-dependent competition between them, including applications which are beyond the context of ecology.

## *1.8. Can Hypercycles Really Exist?*

The existence of hypercycles is a very difficult question. The polemic discussions of this question have not stopped since it was first raised by Eigen and Schuster in 1979 [9]. In this section, we will only gather some arguments for and some arguments against the real existence of hypercycles. We will consider this question quantitatively later in this book.

From the point of view that chains are coupled settlements along downflow-trails of energy, it is not understandable how a feedback should be established from the bottom level, where we find highly degenerated energy, back to the highest level of a chain, where we find the most valuable energy form. We meet chain structures, however, also in chemical cascade reactions, for example, enzyme chains. Here we do not find bottom or top elements, but rather a first element and a last element, a buffering feedback mechanism between two consecutive elements. It could be possible here that the last element of such a chain suppresses or stimulates the first element of the chain, the initiator of the whole production along the chain.

In the brains of primates, the existence of closed circles built of neurons has been proven anatomically. In such a circle, we find nervous activations also based on chemical reactions. If any form of rate-coupling plays a role there then we would have to consider these structures as hypercycles.

Some models have been made for autonomous structure selection out of a medium with a lot of elements and a mechanism for random connections. There are models of this type where chains and cycles preferably all of a certain length can be found. However, all these models have the disadvantage that the result we get from them is forced by a certain statistical assumption and we do not know how the relevance of the introduced assumptions could be proved.

We think that the most convincing approach to this question is stability, meaning dynamic arguments for the existence of these structures. This provides us with the following way to study and decide this question. We must analyse the dynamic behaviour of chains and cycles for different assumptions for the sets of coefficients $K_i$. Then we must introduce parameter disturbances in a realistic way, for example, by applying independent random and amplitude-bounded disturbances for the parameters. If we then meet a behavioural type of a chain or a cycle which is stable against these disturbances and which does not have the property of cutting off a chain or a cycle because the corresponding control signal becomes extinct, then we may conclude that the corresponding structure can exist.

A structure which cannot preserve the previous pattern of behaviour under small disturbances cannot exist under real conditions. It is our opinion that existence is bound to the robustness of the behaviour under a certain class of disturbances.

## *1.9. Continuous State-Transitions and Cooperative Effects*

Onedimensional continuous growth processes usually appear in the form of *s*-formed continuous and monotonous state-transitions as can be seen schematically in Fig. 23 for $x' = F(t)$.

If $F(t)$ is differentiable the transition process can be written in the form of an ordinary differential equation

$$dx/dt = f(t)$$

or, more generally, as

$$dx/dt = f(t, x) .$$

The right-hand side of this differential equation we call the "driving force". Fig. 24 shows a typical time function of such a driving force.

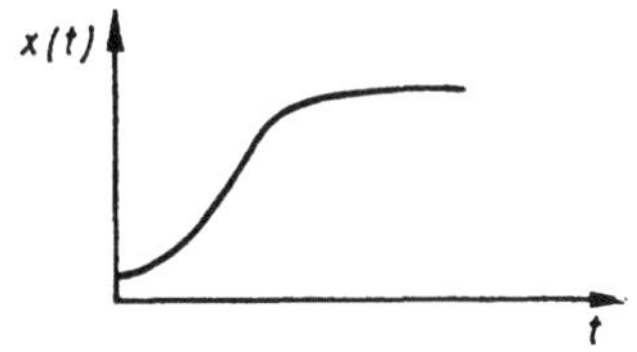


**Fig. 23** Transition process of a growth indicator

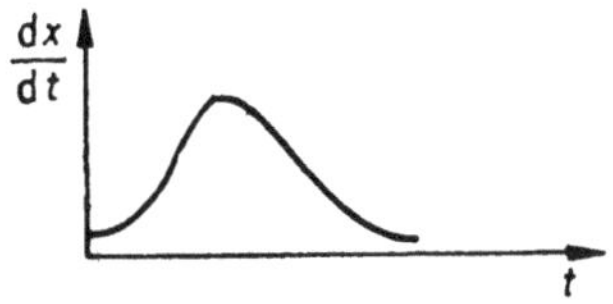


**Fig. 24** The function of the driving force of a growth indicator $x$

Because of the monotonous connection between $t$ and $x$ the inverse function, $F^{-1}$, for $F(t)$ exists, and therefore, in principle, it is always possible to write the growth equation in the autonomous form

$$dx/dt = g(x) .$$

There is an obvious analogy between the pulse-formed curve of the driving force $f(t)$ or $g(x)$ and the probability density function of a random variable. In many cases for *s*-formed transition functions, unimodal functions $f(t)$ or $g(x)$ for the corresponding driving forces are of interest to us. Therefore, it is usually assumed that if we empirically meet multimodal driving forces we are dealing with different driving force-pulses shifted against each other over time. A distinct separation between unimodal and multimodal curves of a growth process is not possible perhaps because of the existence of random disturbances.

Cooperative effects are frequently determined by a corresponding structure-building in the background of the measurements of the growth indicators. From the point of view of the observable phenomenon, cooperative effects can be seen implicitly in simultaneously acting driving forces which can amplify or damp each other or which are coupled and follow consecutively one after the other with a certain overlapping. Very often, we have to model cooperative effects in the phenomena in a reasonable way. The following proposal to do that arises from the theory of fuzzy sets. For the building of a complex driving force model, we usually need operations of the type *disjunction* (disj.) and *conjunction* (conj.).

*Disjunction* is the combination of individual driving forces adjoined to one another for generating a common resulting force. This operation is useful for the description of consecutively acting forces (with or without overlapping).

*Conjunction* is a synchronous interference of different driving forces with the forces restricting or enhancing each other. This operation can be very useful for the description of cooperative effects because of a simultaneous influence from different acting driving forces. General driving force models can be described as a *disjunction of conjunctions* (in analogy to the so-called *disjunctive normal form* from the theory of automata), because we meet under real conditions simultaneous influences at every point in time, as well as consecutively acting forces at every point in time. Usually, it is demanded that the disjunction and conjunction operations fulfill axiomatic conditions which will reflect well the contentional orientations formulated above. For example, very often the following axioms are used.

*Axioms of Disjunction* $\quad \text{disj.}(f_1, f_2) = f_1 \vee f_2$

$$\text{D1}: f_1 \vee f_2 = f_2 \vee f_1\,,$$
$$\text{D2}: f_1 \vee (f_2 \vee f_3) = (f_1 \vee f_2) \vee f_3\,,$$
$$\text{D3}: f \vee 0 = f\,.$$

*Axioms of Conjunction* $\quad \text{conj.}(f_1, f_2) = f_1 \,\&\, f_2$

$$\text{C1}: f_1 \,\&\, f_2 = f_2 \,\&\, f_1\,,$$
$$\text{C2}: f_1 \,\&\, (f_2 \,\&\, f_3) = (f_1 \,\&\, f_2) \,\&\, f_3\,,$$
$$\text{C3}: f \,\&\, 0 = 0\,.$$

Usually, a coupling between the two operations is demanded with the following so-called distributive law

$$f_1 \,\&\, (f_2 \vee f_3) = (f_1 \,\&\, f_2) \vee (f_1 \,\&\, f_3)\,.$$

Important special cases for disjunction and conjunction in the past were

$$f_1 \vee f_2 = \max(f_1, f_2)\,, \qquad f_1 \,\&\, f_2 = \min(f_1, f_2)\,.$$

This originates on one hand from the multivariable logics of Lukasiewicz, but Justus von Liebig [56] had earlier introduced these operations as the so-called limitative laws for production functions in agriculture. Recently, these operations won great importance as basic operations between so-called membership functions in the theory of fuzzy sets and systems (Zadeh [44]).

There is an interesting connection between the operations max and min, for it holds that

$$(f_1^a + f_2^a)^{1/a} \to \begin{cases} \min(f_1, f_2) & \text{for} \quad a \to -\infty\,, \\ \max(f_1, f_2) & \text{for} \quad a \to \infty\,. \end{cases}$$

This is at the same time a bridge to another group of disjunction/conjunction operations, namely

$$f_1 \vee f_2 = f_1 + f_2\,, \qquad f_1 \,\&\, f_2 = f_1 \cdot f_2\,.$$

Also these operations for the aggregation of different influence factors with production functions played an important role for the modelling of production in agriculture.

With a serious polemics in the last century, Mitscherlich [55], [57], [58] attacked the limitation theory of v. Liebig. He tried to show that max/min operations are not relevant in agriculture for production models, resulting in Mitscherlich proving his point of view, so that today the usual arithmetics are used mainly for the construction of production functions in agriculture and elsewhere. In macroeconomies in connection with the so-called Cobb-Douglas production functions, the power-product driving force concept used is also based on the arithmetic disjunction and conjunction.

The annual production $P$ is determined by the working force $A$ and the investments $I$ with the help of the following growth equation

$$dP/dt = KA^a I^i .$$

Here $a$ and $i$ are non-negative exponents, the so-called *elasticities*. $A$ and $I$ have the character of state-variables, meaning they are the overall working force, and the realised and productive investments. Related to the arithmetic concept for the generation of production functions is the following generalisation already proposed by Kolmogoroff [70]

$$f_1 \vee f_2 = \Phi^{-1}(\Phi(f_1) + \Phi(f_2)) ,$$

$$f_1 \,\&\, f_2 = \Phi^{-1}(\Phi(f_1) \cdot \Phi(f_2))$$

with an arbitrary fixed monotonous function $\Phi$.

These two operations generate something like a generalised arithmetics and can be used instead of $+$ and $\cdot$ for the analysis. The separation between $\vee$ (vel) and & (et) is relative as can be seen from the example

$$\Phi(u) = \ln u ,$$

because in this case $f_1 \vee f_2 = f_1 \cdot f_2$ holds.

## 1.10. The EVOLON: A Model Concept for an Evolutionary Step

We call every process or mechanism of growth in which the growth indicator is expressed by an $s$-formed monotonous increase an EVOLON; an $s$-formed monotonous decrease we call an ANTIEVOLON. Obviously, this notion is a generalisation of the INSTANTON vs. ANTIINSTANTON concept of physics for an ideal step-formed transition in the sense of a Heaviside step function. Fig. 25 shows the Heaviside step function.

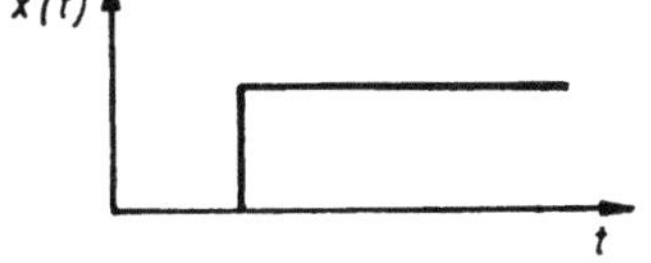


**Fig. 25** Heaviside's unit step function

An EVOLON is a highly aggregated form of the transition of a growth indicator from one stationary state to the next as a consequence of a quantitative accumulation and, perhaps as a preparation for a consecutive qualitative change.

Frequently, unimodality is the result of smoothing through aggregation. With a rough resolution, such a transition process may seem to be unimodular, but if we apply a better resolution to it, it may turn out that one evolutionary step is in fact a sequence of these steps, a so-called *evolutionary staircase*. The roughly resolved transition, then, blunts the shape of the staircase as an envelope of its contents. Taking this relation into account, it makes sense to use a unimodal system concept for modelling an EVOLON, and, in cases where we meet multimodal driving forces, to try to identify different influencing variables which produce the considered driving force as a consequence of their co-action.

Usually, the future of a system approaching a steady state at the end of an evolutionary step is not uniquely determined; frequently, because of random influences or system instabilities, there may be a splitting into several possible futures. The resulting structure corresponds to a tree structure. In other words, bifurcations might occur. Fig. 26 schematically shows such a situation.

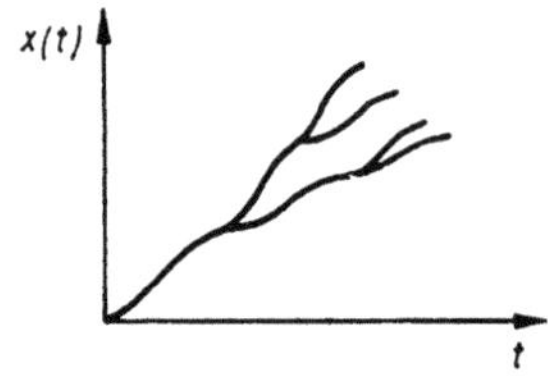


**Fig. 26** The branches of an event tree

Sometimes the evolution of a system, even on the background of a homogeneous process mechanism, can be connected with necessary switches between several different growth indicators. It may be that a given growth indicator is relevant only in a certain phase of the growth process. After that phase, the corresponding variable can be submerged, giving way to another growth indicator which is better suited to reflect in aggregated form the ongoing evolution process. For example, in the evolution of primates, at least for a certain time, the brain volume was an essential growth indicator. But in later evolutionary phases, the brain volume is no longer a relevant indicator for higher evolution. In this case, we now need an appropriate structural parameter of the brain as a growth indicator.

Another example is that, for a long time, an essential parameter for the efficiency of computers was the tact frequency. However, with the increase of parallel processing and with the growing importance of the network of inner data flows, the tact frequency loses its significance for the evaluation of the power of a computer.

Very often the sources for possible discontinuous branching in an evolutionary tree are bifurcations which depend on the thresholds of certain system parameters and lead to a splitting of the future in some evolutionary pipelines. Semantically behind these phenomena might be hidden changes in the systems behaviour originating from mutations by randomly passing critical thresholds of some system parameters. We expect such mutations not only on the level of microbiological structures, but also we like to assume such discontinuous changes of behaviour in more global structural regions. Such changes can become fixed and autonomous, and, they may be inherited by later generations. We will not discuss here whether such global structural changes can alter the genetic code; this question is obviously connected with the highly polemic problem of possible inheritance of acquired properties, a question already raised by Darwin. He had no doubt of the reality of such an effect [69].

In many cases, the reason for mutations are random fluctuations, which, although they might be very small, are able to switch the system's behaviour in the neighbourhood of critical thresholds. Because the consequences of such mutations near a saturation state can pierce deep into the interior of a system and cause there structural switches and degeneration of structure, these random fluctuations implicitly lead to pressures which force the system to seek a new niche in which it could continue its existence with the help of newly discovered resources.

## *1.11. The Hyperlogistic Differential Equation: A Good Model for the EVOLON*

We have to distinguish between the different successive phenomenological evolutionary phases contained in the EVOLON, for example, the *extensive phase* when the system flourishes and the *intensive phase* when the system enters saturation. This presents us with a rough picture of an evolutionary step connected with an EVOLON. From these background mechanisms, we must model the phenomenon, differentiating between the *s*-formed transition between an early stage and that of a later stage of the system, as expressed by the values of a growth indicator. For this purpose, we propose that suitable system's concepts with free parameters be specified having at hand concrete time-series for the corresponding growth indicator. The minimum demand for such system concepts is that they lead to a good fit with the observed data.

A system concept which reflects the mechanisms in the background relevantly with concrete analytic relationships, and with parameters which possess a semantic interpretation in terms of the systems behavioural features is by far the best system concept to have. We are strongly convinced by analytical studies which we will present in this book, and by much experience with many concrete time-series from very different areas that the system concept based on the *hyperlogistic differential equation* is of great importance, not only from the point of view of a good fit, but also from that of its relevance for aggregated growth in soft systems.

The hyperlogistic differential equation is given by the expression

$$\mathrm{d}x/\mathrm{d}t = K(x-a)^k \left(B - (x-A)^w\right)^l .$$

Obviously, $a$ and $A$ are shift parameters, $B^{1/w}$ is the saturation value or steady state. The exponents $k$ and $l$ are very important parameters, $k$ characterizes the autonomous growth in the extensive phase and $l$ the autonomous (but forced by restrictions) saturation in the intensive phase. According to our experience, these exponents play the role of complexity measures of the corresponding subprocesses.

With the analysis of many time-series, we observed that many natural growth processes are highly cooperative. This is expressed in the values of $k$ and $l$ by the result:

> The more $k$ resp. $l$ deviates from 1, the higher is the cooperativity within the considered system.

These phenomenological connections are proposed as a theoretical concept by Mende in [49], [78].

Limiting cases of the hyperlogistic growth model in the sense of limiting some parameter configurations are

$$\mathrm{d}x/\mathrm{d}t = K(x - a)\,\mathrm{e}^{-\varkappa(x-A)^w},$$

$$\mathrm{d}x/\mathrm{d}t = K(x - a)^k \left(\ln\,(B - A)/(x - A)\right)^l .$$

Obviously, we can get from this hyperlogistic model with its limiting cases a lot of very well-known and frequently used growth functions. Some of them are given below.

| | |
|---|---|
| $\mathrm{d}x/\mathrm{d}t = Kx^k$ | power growth, |
| $\mathrm{d}x/\mathrm{d}t = K(B - x)^l$ | power saturation, |
| $\mathrm{d}x/\mathrm{d}t = K(B - x^w)$ | allometric saturation, |
| $\mathrm{d}x/\mathrm{d}t = Kx(B - x)$ | logistic saturation, |
| $\mathrm{d}x/\mathrm{d}t = Ax^m - ax^n$ | Bertalanffy model, |
| $\mathrm{d}x/\mathrm{d}t = Kx^w(B - x^w)$ | Weibull distribution, |
| $\mathrm{d}x/\mathrm{d}t = Kx\ln\,(B/x)$ | Gompertz growth model, |
| $\mathrm{d}x/\mathrm{d}t = Kx(\ln\,(B/x))^{1/2}$ | Ecological growth function, |
| $\mathrm{d}x/\mathrm{d}t = Kx(\ln\,(B/x))^l$ | generalised ecological growth function, |
| $\mathrm{d}x/\mathrm{d}t = Kx^k(B - x)^l$ | incomplete $\beta$-distribution. |

$$k, l > -1$$

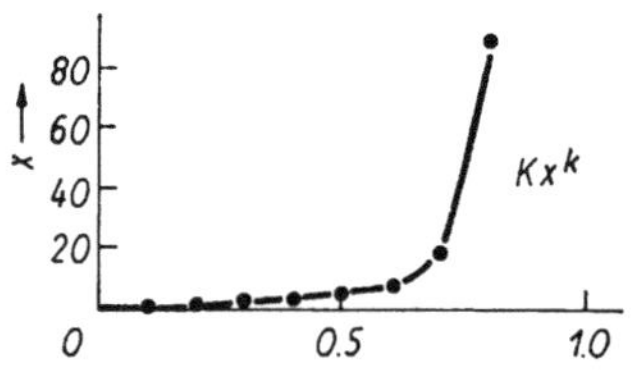

Fig. 27 Power growth

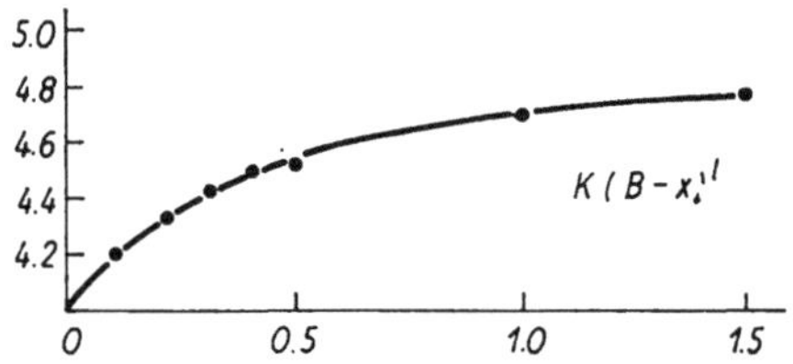

Fig. 28 Power saturation

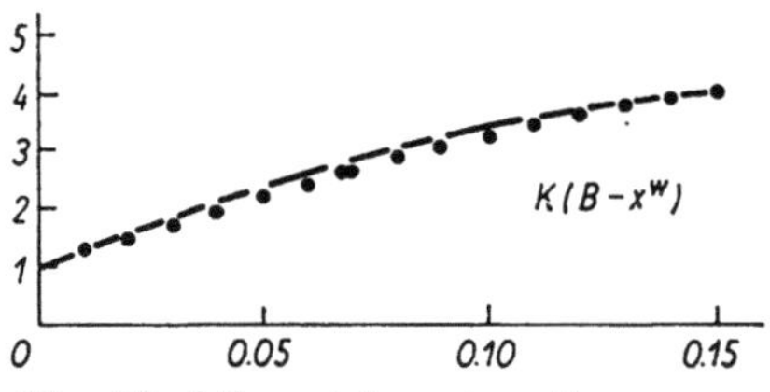

Fig. 29 Allometric saturation

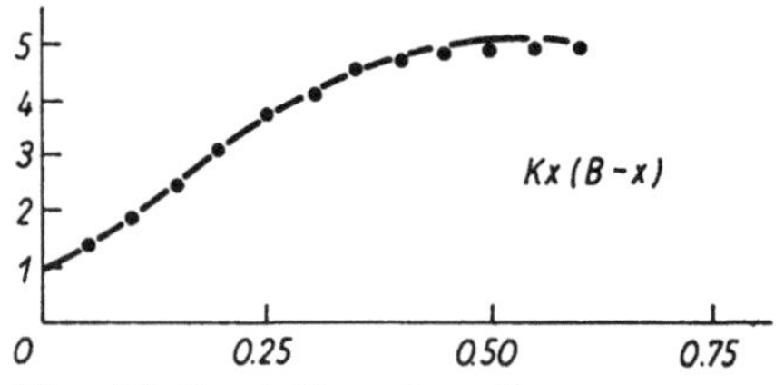

Fig. 30 Logistic saturation

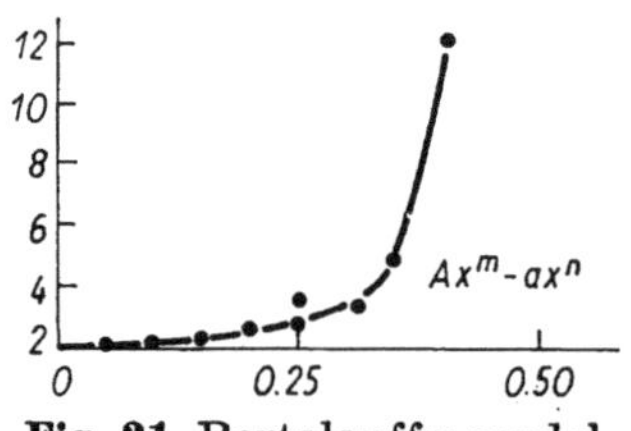

Fig. 31 Bertalanffy model

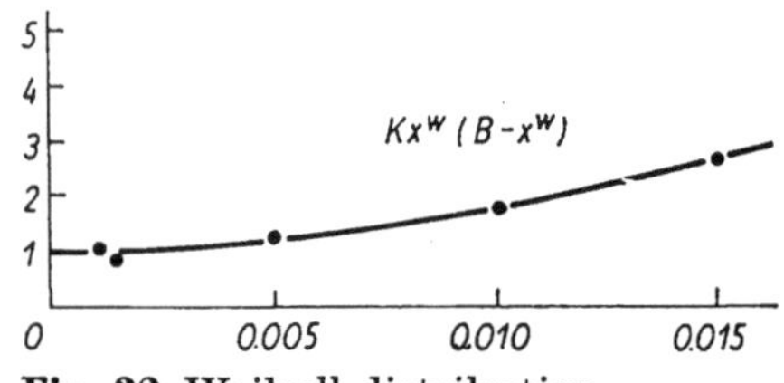

Fig. 32 Weibull distribution

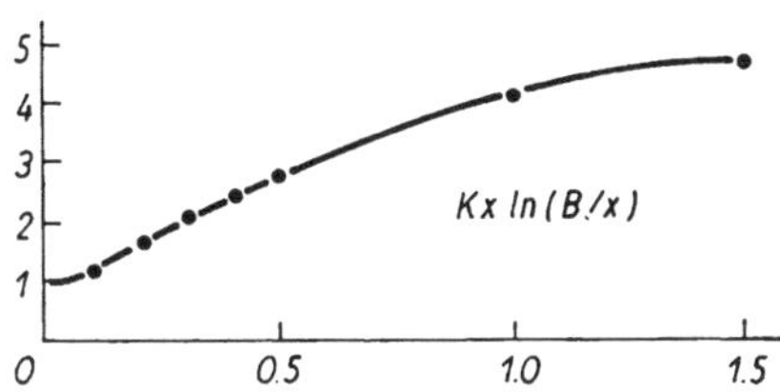


**Fig. 33** Gompertz growth model

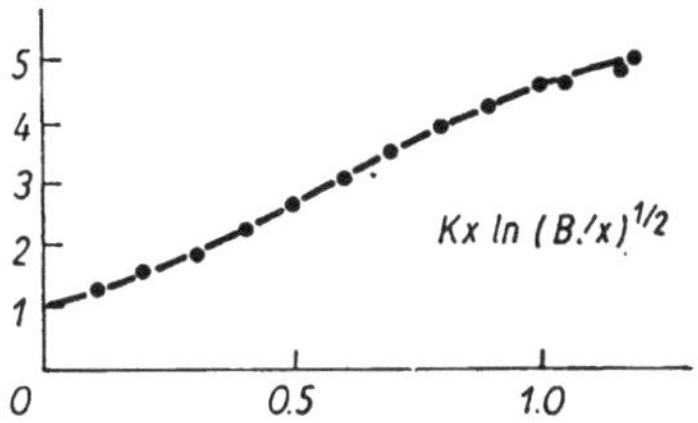


**Fig. 34** Ecological growth function

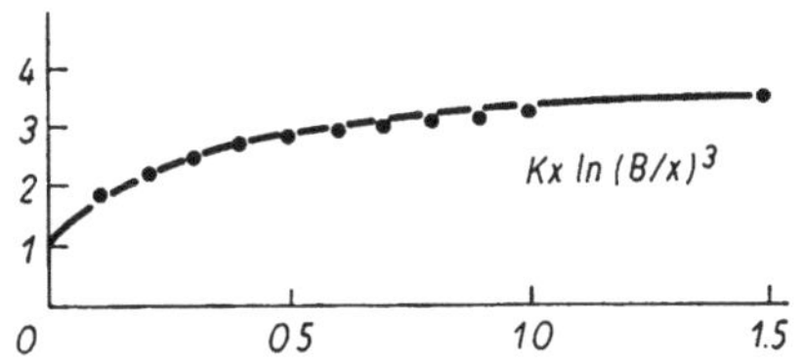


**Fig. 35** Generalised ecological growth function

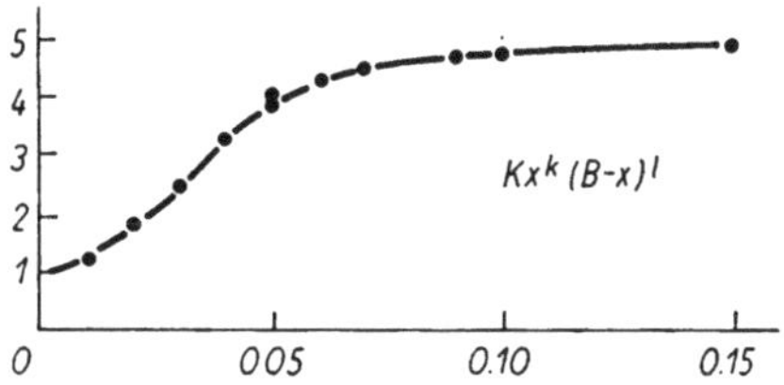


**Fig. 36** Incomplete $\beta$-distribution

For practical purposes, the question of interest is how we can maximally simplify the hyperlogistic differential equation without changing its form. In the usual understanding of the equation we assume

$$k, l, w \quad \text{and} \quad B > 0 .$$

Very often, though we must also take into account only one shift parameter, meaning that we can assume

$$A = a .$$

We can then introduce a new variable by

$$u = (x - a)/B^{1/w} \qquad \text{and get} \qquad du/dt = K'u^k(1 - u^w)^l .$$

Of the three exponents $k$, $l$ and $w$, only two have an independent significance.

We can introduce a second substitution

$$z = u^w$$

and get the simplified hyperlogistic differential equation

$$dz/dt = wK'z^{(k-1+w)/w}(1 - z)^l .$$

## *1.12. Case-Studies Demonstrating the Relevance of the Hyperlogistic Model for an EVOLON*[1])

1. *An evolutionary analysis of world energy consumption and world population* [80]

In this section we will show that the hyperlogistic evolutionary model (HLEM) successfully describes the development of world population and global primary energy consumption over the past century; we also use the model to provide projections of

[1]) We heartily thank Dr. Kriegel, ZKI, for his constructive cooperation in elaborating the collaborative paper for IIASA 80.

world population and primary energy consumption up to the year 2100. Our approach is based on the assumption that the investigated system develops as a whole and that the external perturbations are not very strong. Here, the interrelationships between the subsystems are such that they fulfill a certain uniform law of evolution for the entire system. Generally, this approach drastically reduces the complexity of the problem.

As the system concept for the growth model, we apply the hyperlogistic differential equation

$$dx/dt = Kx^k(B - x)^l \tag{A}$$

in which we normalize the values of $x$ with a certain factor to get reasonable values for the parameter $K$ a priori.

Fig. 37 and 38 present observed data on world primary energy consumption and world population growth and give also the corresponding model curves. Except for

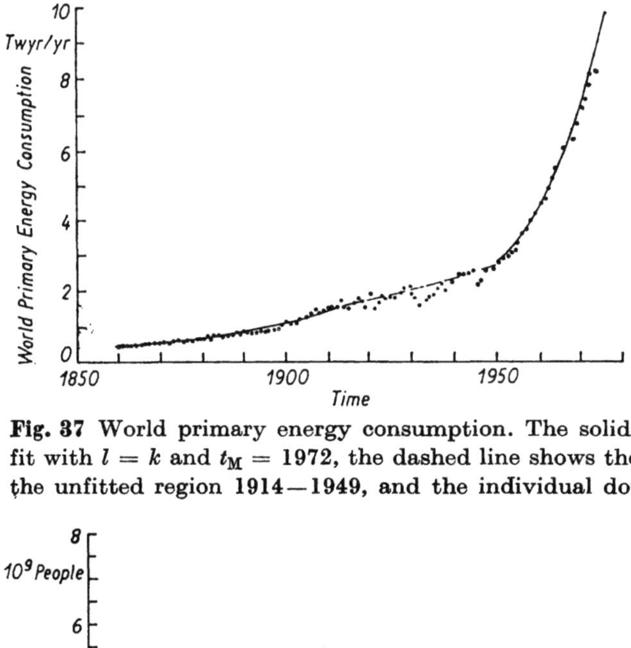


**Fig. 37** World primary energy consumption. The solid line represents the HLEM fit with $l = k$ and $t_M = 1972$, the dashed line shows the behaviour of the model in the unfitted region 1914—1949, and the individual dots represent observed data.

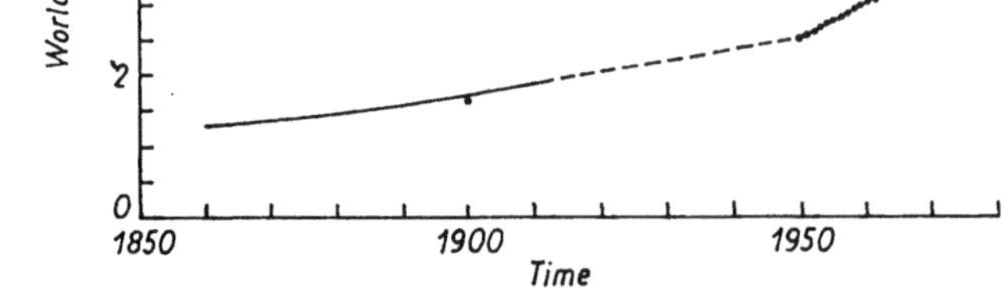


**Fig. 38** World population. The solid line represents the HLEM fit with $l = k$ and $t_M = 1975$, the dashed line shows the behaviour of the model in the unfitted region 1914—1949, and the individual dots represent observed data.

the period between the years 1914 and 1949 in Fig. 38, the data show a smooth, regular growth pattern with no sign of saturation. The irregular pattern between 1914 and 1949 is probably due to the effects of World War I, the Depression, World War II, and post-war depression. However, we assume that these events only slow down the evolutionary process without changing its structural parameters, e.g. the exponents.

Thus equation (A) with $K = K(t)$ can be used to describe the pattern of evolution over this period, and we assume that all parameters remain constant except that parameter $K$ is reduced by a constant factor $0 \leqq \varepsilon \leqq 1$ between 1914 and 1949. This is equivalent to a definite delay in the evolutionary process. There is no information on the saturation function in either of these data, so our parameter identification problem appears to be underdetermined. However, we can get some idea about the saturation process from other sources. For example, we know that the maximum rate of population increase has now been reached [115], and that the maximum rate of increase in energy consumption has probably been surpassed. Knowing the time $t_M$ when the maximum growth rate $dx/dt : x|_{t=t_M}$ occurs, we may express the saturation level $B$ in terms of parameters $k$ and $l$, and the value $x_M$ of the state variable $x$ at time $t_M$:

$$B = x_M \frac{k + l - 1}{k - 1}. \tag{B}$$

We further assume that the strength of interactions in driving and saturation processes are equal, i.e., $k = l$. Thus, we must still evaluate

— the time scale factor $K$,
— the delay factor $\varepsilon$,
— the measure of interaction $k$,
— the value of $x_M$ corresponding to the maximum growth rate.

In order to identify the values of these parameters, the hyperlogistic differential equation must be first integrated. In general, an analytic integration of this equation is impossible if the exponents are arbitrary real values. We therefore used a fourth-order Runge-Kutta-Merson method [94], to perform a numerical integration. We excluded the data between 1914 and 1949 and adjusted the integrals by minimization of the $\chi^2$-distance with the help of the program system MINUIT [95].

Since we had no information on the uncertainty of the data, we minimized the $\chi^2$-function with an assumed error of 10% and then renormalized the $\chi^2_{\min}$-value in such a way that an error $\sigma$ in the data would correspond to an error of one standard deviation in the parameters. Using this approach we obtained deviations of $\sigma = 3\%$ and $\sigma = 1\%$ for energy resp. population data.

In order to estimate how the errors $\delta_{\text{HLEM}}$ of our model curve depend on the errors $\delta\alpha$ of the fitted parameters $\alpha$, we used a numerical approximation of the Gaussian error propagation law [109]

$$\delta_{\text{HLEM}}(t) = \{\sum_{\alpha_i} \left(x(t, \alpha_i + \delta\alpha_i) - x(t, \alpha_i)\right)^2\}^{1/2}$$

where $x(t, \alpha_i + \delta\alpha_j)$ denotes the model output obtained at time $t$ with parameter values $(\alpha_i + \delta\alpha_j)$ and $\alpha_i$ $(i \neq j)$.

Table 1 presents the parameter values obtained by fitting the model to world primary energy consumption data, under three different assumptions as to when the

**Table 1.** Values of parameters $A$, $\varepsilon$, $k$, and $x_M$ obtained by fitting world primary energy consumption data under different assumptions for $t_M$. The calculated values of the saturation level $B$ are also given.

| Parameter | $t_M = 1970$ | $t_M = 1972$ | $t_M = 1974$ |
|---|---|---|---|
| $A$ ($\times$ 10² TWyr/yr) | $1.704 \begin{smallmatrix} + 0.007 \\ - 0.2 \end{smallmatrix}$ | $1.501 \begin{smallmatrix} + 0.006 \\ - 0.2 \end{smallmatrix}$ | $1.312 \begin{smallmatrix} + 0.006 \\ - 0.2 \end{smallmatrix}$ |
| $\varepsilon$ | $0.38 \begin{smallmatrix} + 0.10 \\ - 0.05 \end{smallmatrix}$ | $0.39 \begin{smallmatrix} + 0.02 \\ - 0.05 \end{smallmatrix}$ | $0.40 \begin{smallmatrix} + 0.02 \\ - 0.05 \end{smallmatrix}$ |
| $k$ | $1.55 \begin{smallmatrix} + 0.01 \\ - 0.04 \end{smallmatrix}$ | $1.53 \begin{smallmatrix} + 0.01 \\ - 0.04 \end{smallmatrix}$ | $1.51 \begin{smallmatrix} + 0.01 \\ - 0.04 \end{smallmatrix}$ |
| $x_M$ (TWyr/yr) | $7.11 \begin{smallmatrix} + 0.06 \\ - 0.2 \end{smallmatrix}$ | $7.89 \begin{smallmatrix} + 0.08 \\ - 0.3 \end{smallmatrix}$ | $8.79 \begin{smallmatrix} + 0.11 \\ - 0.4 \end{smallmatrix}$ |
| $B$ (TWyr/yr) | $27 \pm 1$ | $31 \pm 1$ | $35 \pm 2$ |
| $\chi^2_{10\%}$/NDF | 7.3/76 | 7.7/76 | 8.1/76 |

maximum growth rate occurred ($t_M = 1970, 1972, 1974$). The fits show that the period before 1914 is governed by a hyperlogistic evolution law with essentially the same parameters as the post-war period. Within the error bounds, parameters $\varepsilon$ and $k$ are the same for different values of $t_M$. The value of the interaction factor $k \approx 1.5$ shows that the HLEM curves should be quite different from logistic curves, and should have in the driving phase a hyperbolic from. As an additional check, a hyperbola was fitted to the data under the assumptions made above

$$x(t) = K(t_p - t)^b .$$

This resulted in a value of $b$ about 2.3, corresponding to $k \approx 1.4$. Comparing the significance value $\sigma = 4\%$ ($\chi^2_{10\%}$/NDF $= 13/76$) of the hyperbola fit with corresponding values for the HLEM fits, it is clear that the assumption of a hyperbolic saturation process improves the fit.

From the delay factor $\varepsilon$, we can conclude that the evolution of the energy system was retarded by about $21 \pm 1$ years due to the two world wars and the several economic depressions. Again this was checked by fitting a hyperbola, which suggested a delay of $19 \pm 1$ years.

Unlike $k$ and $\varepsilon$, which remain constant as $t_M$ increases, the parameter $x_M$ clearly increases with $t_M$, since we start our integration procedure at time $t_M$ with value $x_M$. Equation (B) shows that for fixed values of $k$ the saturation level $B$ depends only on $x_M$ and therefore increases with increasing $t_M$ (see the results in Table 1). Therefore, the choice of the $t_M$ value has a strong influence on $x_M$ and $B$ values. The goodness of fit is not affected by the choice of $t_M$, however, due to the decoupling of driving and saturation processes and the lack of information on the latter. A similar result may be obtained by varying $l$ according $l = \eta k$ where $0.5 \leqq \eta \leqq 2$. In this case only the saturation level $B$ will be influenced, all other parameters will remain unchanged.

Thus, with the present data, it is impossible to check hypotheses concerning the value of $t_M$ and the strength of interactions in the saturation process. As an example, Fig. 37 shows the integral obtained with $t_M = 1972$. The solid line represents the model output in the fitted region, while the dashed line representing values between 1914 and 1949, was obtained by assuming a reduced speed of evolution. The HLEM curve reproduces the shape of the data in the fitted region quite accurately, although in the unfitted interval between 1914 and 1949 there are large deviations from the observed data. It is interesting to note that, with the exception of a few points, our model gives values for primary energy consumption which are higher than the actual data. This implies that the real deceleration is stronger than $\varepsilon$. However, after an event such as a war, there seems to be an upsurge which temporarily increases the velocity of the increase of the values to greater than normal. The depression of the data relative to the delayed model curve is plausible and can be explained by the fact that a certain part of the existing system being put out of action during crises: and hence, it does not contribute to the output of the system. When the crisis is over these reserves are reactivated and used to bring the evolutionary rate back to normal relatively quickly. Nevertheless, this cannot bring back what has already been lost. The most recent data show a level of energy consumption which is considerably lower than that suggested by our model, reflecting the socalled energy crisis of the 1970's. According to our model, this behaviour should not represent the start of an entirely new trend, but rather the effects of a temporary depression; there are already hints of a recovery. We therefore argue that the current energy crisis is only an interruption in the normal long-term behaviour of the system as projected by our model.

**Table 2.** Values of parameters $A$, $\varepsilon$, $k$, and $x_M$ obtained by fitting world population data. The calculated value of the saturation level $B$ is also given.

| Parameter | $t_M = 1975$ |
|---|---|
| $A$ (people) | $4 \times 10^9$ |
| $\varepsilon$ | $0.5 \begin{smallmatrix} -0.04 \\ +0.03 \end{smallmatrix}$ |
| $k$ | $3.1 \pm 0.3$ |
| $x_M$ (people) | $(4.0 \pm 0.1) \times 10^9$ |
| $B$ (people) | $1 \times 10^{10}$ |
| $\chi^2_{10\%}$NDF | 0.7/29 |

The results obtained by fitting the model to world population data are given in Table 2 and Fig. 38. The relative lack of historical data means that the parameter errors are larger than in the case discussed earlier, but, even within these larger parameter errors, the fit shows a large deviation from a logistic growth ($k \sim 3$). The value of the delay factor $\varepsilon \sim 0.5$ corresponds to a loss of about 17 years of evolution, which is comparable with the delay obtained for world primary energy consumption. However, a comparison of the values of $k$ in Tables 1 and 2 shows that the population system is more complex, with more internal interactions, than the energy system. Since smaller systems are, in general, less complex than larger ones [116], and the energy system can be considered as a component of the world population system, this result is not surprising. The HLEM curve shown in Fig. 38 again increases smoothly with increasing time and provides a good description of the data.

*2. Projection of world energy consumption and world population up to the year 2100.*

There are several ways of making long-term projections of the evolution of large-scale systems. For example, predictions can be made by estimating the increase in per capita consumption together with population growth in an appropriate scenario, or by estimating local energy densities and making certain assumptions about urbanization [117]. Another quite different approachi s to assume that observed dynamics of growth propagate according to a particular evolution model.

The hyperlogistic evolution model provides, despite its simplicity, a good description of historical data. We will now use our model to make predictions using the following assumptions about growth dynamics:

- The long-term growth dynamic is a process of transition from one stationary level to another.
- In the absence of external disturbances, the transition follows a hyperlogistic evolution law.
- Driving and saturation processes have the same degree of complexity and level of interval interactions ($k = l$).
- An estimate of the time $t_M$ at which the maximum growth rate occurs is obtained by analyzing historical data.

We also assumed that the present trend of evolution will continue undisturbed in the future.

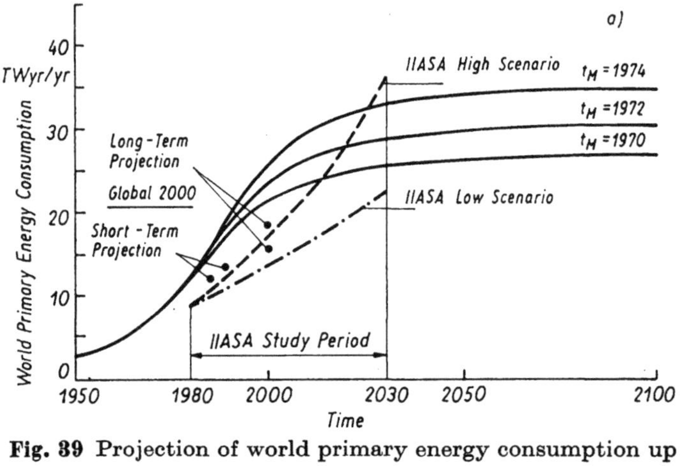


**Fig. 39** Projection of world primary energy consumption up to the year 2100 under the assumption $l = k$ and with $t_M = 1970$, 1972, and 1974; the IIASA high and low scenarios and the Global 2000 projections are also shown.

Fig. 39 presents a projection of world primary energy consumption up to the year 2100 for three different values of $t_M$; the corresponding error ranges are given in Fig. 40. The values obtained by an IIASA study [92] (IIASA: International Institute of Applied Systems Analysis, Laxenburg, Austria) and those from the Global 2000 Report [96] are also included for comparison.

After a slightly braked rise between 1970 and 2000, the saturation process dominates, resulting in a very weak increase towards the saturation level from about 2020 onwards. The height of the saturation level rises strongly as the value of $t_M$ increases. The point of inflection of the curves, i.e., the point at which driving and saturation processes are of equal strength, lies between 1984 and 1988, depending up on the $t_M$ value. Around 1990, the HLEM curves separate (the error bars no longer overlap)

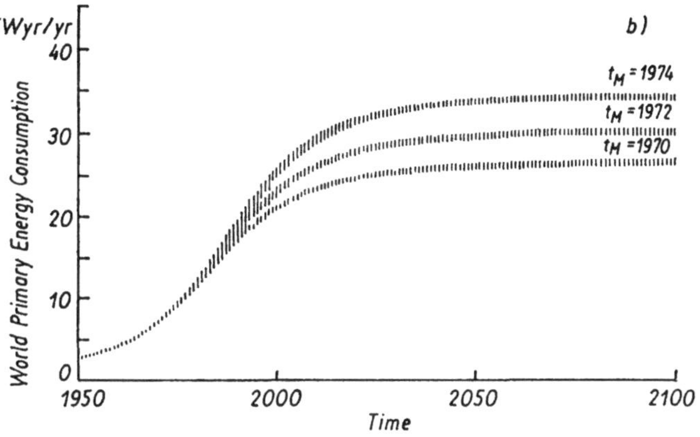


**Fig. 40** Error ranges corresponding to the HLEM projections given in fig. 39

and one can try to decide which of our assumptions for the $t_M$ value is the best, and, therefore, which value of energy consumption may be expected to exist in the steady state. A comparison with the IIASA energy scenarios [92] shows that for the year 2000, the energy consumption predicted by our model is more than 20% higher than that suggested by the IIASA high energy consumption scenario, while all our projections for 2030 lie between the high and low energy consumption IIASA scenarios. However, the growth rates predicted by HLEM behave more steadily than the IIASA predictions. This is shown in Table 3, which gives the growth rates corresponding to the various projections. As we go further into the future, the growth rates calculated by our model steadily decrease, reaching a value of about 0.05%/year for the period 2060–2100. Between 1975 and 2000 the IIASA scenarios suggest lower growth rates than does HLEM, although for the next period the IIASA predictions are more than twice as high as our values.

Fig. 41 shows the projection of world population up to the year 2100; the corresponding error bars are given in Fig. 42. It is clearly be seen that there are only small differences between the HLEM projections and the IIASA estimates [92], [93]. Until the year 2000, our predictions are slightly higher than IIASA's; then this situation reverses. Having projected world primary energy consumption and world population,

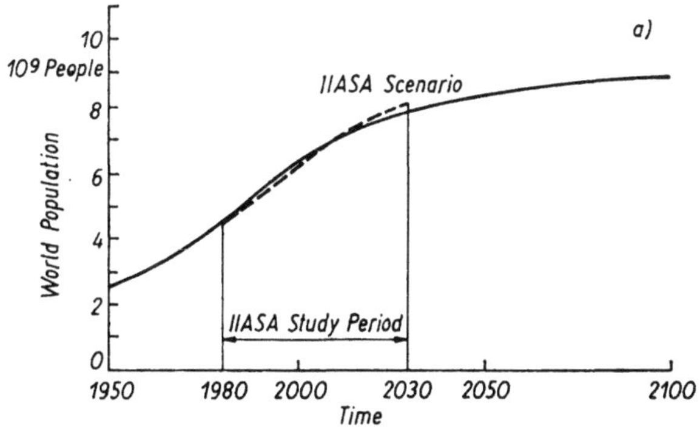


**Fig. 41** Projection of world population up to the year 2100 under the assumptions $l = k$ and $t_M = 1975$. The dashed line represents IIASA projections.

**Table 3.** Annual growth rates (as percentages) of world primary energy consumption calculated from historical data and projected using HLEM. The corresponding values for two IIASA projections are given for comparison

| Basis of projection | Historical 1950–1973 | Projection | | | | |
|---|---|---|---|---|---|---|
| | | 1950–1975 | 1975–2000 | 2000–2030 | 2030–2060 | 2060–2100 |
| HLEM | | | | | | |
| $t_M = 1970$ | | 4.98 ± 0.09 | 3.38 ± 0.10 | 0.63 ± 0.06 | 0.12 ± 0.04 | 0.04 ± 0.04 |
| $t_M = 1972$ | | 4.99 ± 0.10 | 3.75 ± 0.12 | 0.73 ± 0.08 | 0.13 ± 0.04 | 0.04 ± 0.03 |
| $t_M = 1974$ | 4.80 | 5.00 ± 0.10 | 4.11 ± 0.15 | 0.84 ± 0.10 | 0.14 ± 0.04 | 0.04 ± 0.03 |
| IIASA | | | | | | |
| High scenario | | — | 2.92 | 2.53 | — | — |
| Low scenario | | — | 2.04 | 1.68 | — | — |

**Table 4.** Annual growth rates (as percentages) of world per capita primary energy consumption calculated from historical data and projected using HLEM. The corresponding values for two IIASA projections are given for comparison

| Basis of projection | Historical 1950–1973 | Projection | | | | |
|---|---|---|---|---|---|---|
| | | 1950–1975 | 1975–2000 | 2000–2030 | 2030–2060 | 2060–2100 |
| HLEM | | | | | | |
| $t_M = 1970$ | | 3.06 ± 0.09 | 1.58 ± 0.010 | −0.09 ± 0.06 | −0.14 ± 0.04 | −0.08 ± 0.04 |
| $t_M = 1972$ | | 3.07 ± 0.10 | 1.94 ± 0.12 | 0.004 ± 0.08 | −0.13 ± 0.04 | −0.08 ± 0.03 |
| $t_M = 1974$ | 2.80 | 3.07 ± 0.10 | 2.30 ± 0.15 | 0.11 ± 0.10 | −0.12 ± 0.04 | −0.08 ± 0.03 |
| IIASA | | | | | | |
| High scenario | | — | 1.15 | 1.63 | — | — |
| Low scenario | | — | 0.29 | 0.76 | — | — |

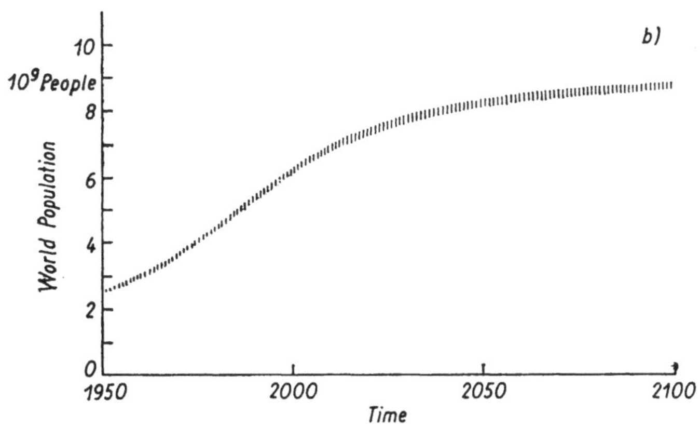


**Fig. 42** Error ranges corresponding to the projections given in fig. 41

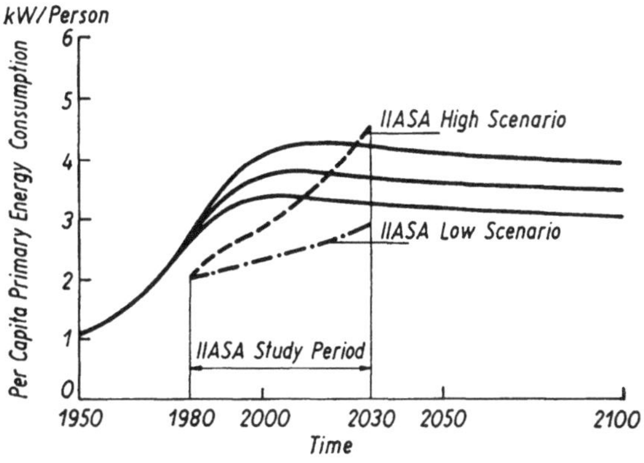


**Fig. 43** Projection of per capita primary energy consumption up to the year 2100. The IIASA predictions are shown for comparison.

it is now easy to calculate future per capita primary energy consumption. The results are presented in Fig. 43. After rising to a maximum of between 3.3 and 4.2 kW per capita around the year 2010, the curves decrease smoothly to the plateau values. From the results given in Tables 3 and 4 we calculate that these values lie between 2.7 and 3.5 kW per capita. This points to the need to increase the efficiency of energy use if we are to guarantee increasing standards of living. Since HLEM and the IIASA study predicted nearly the same population values, the discrepancies between the two sets of projections must be due to different predictions of primary energy consumption. This is also reflected in the growth rates given in Table 5: after the year 2000 the HLEM growth rates decrease very smoothly, with values of between −0.1% per year and 0.2%/year for the period 2000−2030, compared with a range of 0.8%/year to 1.6%/year for the IIASA scenarios.

In summary, HLEM predicts that world primary energy consumption and world population will undergo a very smooth transition to saturation during the next 50 years. This differs in several respects from the IIASA projections. The actual level at which saturation occurs depends strongly on the value of $t_M$ (the time corresponding to the maximum relative growth rate). Our model thus implies that the future development of the human race will depend upon whether or not the maximum growth rates have already been reached.

### *3. A possible description and interpretation of population growth data in German towns*[1])

The population growth of a model town over a 100-year period has been investigated. This *model town* was constructed by summing up and averaging the population of ten big (population of about 800,000 in 1970) German towns. The growth results from this model have been compared with the logistic model and with the EVOLON model [98]. In the process of parameter identification, we removed the period from 1910 to 1955. This period has been strongly influenced by the two world wars.

The EVOLON model describes the data substantially better than does the logistic model. Special parameters are related with the cooperativity of the urban growth process. This shows that the growth process at points far from saturation is a hyperbolic curve, and differs clearly from exponential growth.

The possibility of using the EVOLON model for prognostic estimations of population growth in towns has been investigated. Results of prognostic experiments performed with the data only from 1871 to 1910 based on the EVOLON model are considerably closer to reality than those based on the logistic model, which yields completely absurd results.

In biology, ecology, physics etc. many processes show indicators with *s*-shaped transition behaviour. This is the same phenomenon which we meet for the population of towns. Limited resources in this case are:

— supply capacity of the surroundings for flows of village population, energy, raw materials, products, etc. into the town
— transport capacity for food, garbages, materials and energy
— carrying capacity of the sometimes limited area of a town.

Therefore, we expect *s*-shaped behaviour of the population growth for the towns over time.

Data for the initial growth phase as well as of the saturation phase are of special interest for testing the applicability of the EVOLON model. As for German towns, there are annual data available from 1871, gathered in *Statistisches Jahrbuch deutscher Städte* [97]. This 100-year period reflects both the high initial growth phase and a part of the saturation phase. However, due to World Wars I and II, we must expect considerable deviation from the undisturbed growth process, which forced us to exclude a period of 46 years of the time-series for fitting the model to the data.

For the purpose of comparing these models, we used two data sets. The data set of a *model town* by summing up and averaging the annual population of ten big German towns. These towns are: Hamburg, Munich, Cologne, Frankfurt am Main, Hannover, Stuttgart, Bremen, Düsseldorf, Nuremberg, and Dortmund. Thus, we consider the growth of the model-town from about 100,000 inhabitants in 1870 to about 800,000 in 1970. The data for the model-town are listed in Table 5.

---

[1]) We thank Dr. Albrecht, ZKI, for providing the results of the collaborative paper of IIASA [98] for this monograph.

**Table 5.** Population numbers for the model town [$10^4$]

| year: $t$ | $t+1$ | $t+2$ | $t+3$ | $t+4$ | $t+5$ | $t+6$ | $t+7$ | $t+8$ | $t+9$ | $t+10$ |
|---|---|---|---|---|---|---|---|---|---|---|
| 1870 | 11.351 | 11.737 | 12.143 | 12.559 | 13.032 | 13.448 | 13.977 | 14.346 | 14.775 | 15.166 |
| 1880 | 15.578 | 15.959 | 16.331 | 16.727 | 17.419 | 17.695 | 18.239 | 19.767 | 20.539 | 21.414 |
| 1890 | 22.327 | 23.227 | 23.530 | 24.175 | 25.059 | 25.875 | 26.723 | 27.652 | 28.952 | 30.657 |
| 1900 | 31.400 | 32.230 | — | — | 35.429 | 36.444 | — | — | 39.144 | 41.441 |
| 1910 | — | 43.847 | 44.986 | 46.058 | — | — | — | — | 47.372 | 48.959 |
| 1920 | 49.910 | 50.885 | 50.612 | 51.226 | 51.224 | 52.185 | 52.823 | 55.580 | — | 57.142 |
| 1930 | 57.217 | — | 57.923 | 58.261 | 58.316 | 58.590 | 59.084 | 67.495 | 67.923 | 67.906 |
| 1940 | — | — | — | — | — | — | 56.878 | 59.184 | — | 63.197 |
| 1950 | 66.323 | 68.108 | 70.135 | 72.118 | 73.812 | 75.671 | 76.117 | 76.741 | 78.572 | 79.702 |
| 1960 | — | — | — | 81.904 | — | 81.889 | 81.258 | — | — | 81.216 |
| 1970 | 81.555 | — | 81.385 | 83.128 | 82.025 | — | — | — | — | — |

**Table 6.** Population numbers for Düsseldorf [$10^4$]

| year: $t$ | $t+1$ | $t+2$ | $t+3$ | $t+4$ | $t+5$ | $t+6$ | $t+7$ | $t+8$ | $t+9$ | $t+10$ |
|---|---|---|---|---|---|---|---|---|---|---|
| 1870 | 6.86 | 7.07 | 7.34 | 7.63 | 7.92 | 8.21 | 8.49 | 8.78 | 9.08 | 9.39 |
| 1880 | 9.73 | 10.10 | 10.49 | 10.89 | 11.51 | 11.78 | 12.27 | 12.94 | 13.60 | 14.22 |
| 1890 | 14.57 | 14.49 | 15.97 | 16.68 | 17.20 | 18.07 | 19.00 | 19.86 | 20.42 | 21.40 |
| 1900 | 22.11 | 22.69 | — | — | 25.33 | 25.75 | — | — | 29.00 | 35.32 |
| 1910 | — | 38.59 | 39.74 | 41.30 | — | — | — | — | 41.39 | 41.86 |
| 1920 | 42.66 | 43.18 | 41.99 | 42.72 | 43.11 | 43.71 | 44.13 | 44.38 | — | 47.63 |
| 1930 | 47.28 | — | 49.46 | 50.69 | 50.93 | 51.50 | 52.26 | 53.58 | 54.80 | 54.28 |
| 1940 | — | — | — | — | — | — | 43.97 | 46.04 | — | 50.06 |
| 1950 | 54.15 | 56.62 | 59.52 | 62.15 | 64.42 | 66.40 | 67.72 | 76.99 | 69.27 | 69.75 |
| 1960 | — | — | — | 70.28 | — | 69.34 | 68.25 | — | — | 66.36 |
| 1970 | 65.43 | — | 62.85 | 67.54 | 66.43 | 61.53 | — | — | — | — |

If the data should reflect a general trend, it is essential to apply aggregation for about ten towns. The data for single towns show, in principle, the same effect. But it is problematic to fit and interpret model parameters for single towns because of special influences (for instance, administrative reforms). We shall demonstrate this during our discussion of Düsseldorf. These difficulties are even greater for other towns. The data for Düsseldorf are contained in Table 6.

As an EVOLON model, we propose the hyperlogistic evolution model

$$dx/dt = V(t)\, x^k (B - x)^l$$

with velocity parameter

$$V(t) = K\varepsilon(t) \quad \text{with } \varepsilon(t) \text{ as velocity reduction factor.}$$

In the region of nonautonomous disturbances during wars and crises (from 1910 to 1955) we set $\varepsilon$ = constant, $0 < \varepsilon \leqq 1$. Taking into account that the two world wars and the crisis in between heavily disturbed the process of urbanisation, we exclude this period from our fitting.

The EVOLON concept assumes that the basic growth process is active also during the excluded period, but it operates there with a reduced velocity. Therefore, the indicator of the evolutionary process (population) is disturbed by the impact of the wars and the crisis. During this time interval, our indicator is not able to reflect the real evolution because many people from the towns were evacuated, but they remained in close contact with their town through certain life activities, and they tended to return. This reason forbids one using the indicator for a detailed investigation of the period from 1910 to 1955. Consequently, we could not use the data from this period for the fitting process or for predictions. Later on, we will try to investigate this period using data from other countries which were not so strongly influenced by the world wars.

The parameters $K$, $B$ and the initial value $x_0$ are necessary for scaling the $x$- and $t$-axes, respectively. Only the remaining parameters $k$ and $l$ are independent of scaling and reflect intrinsic structural properties of the process.

There are indications that the EVOLON step in the evolution of the towns investigated here is not the first one. However, the saturation level of the preceding EVOLON step is very small in comparison with our data and, therefore, we can only use an approximation of one step. We should mention that the obtained value for $k$ is mainly determined by the data of the initial growth process, whereas $l$ and $B$ are mainly determined by the data in the period of saturation. The identification of parameters based on data with limited accuracy is an *inverse problem*, which is often indeterminate, or the parameters may have all possible values within a semi-infinite interval. In our case, the data offered sufficient information about the initial growth phase to identify the corresponding parameter $k$ within a narrow error range. The parameters $l$ and $B$ related to the saturation phase cannot be determined with such accuracy because of the lack of sufficient information about the saturation phase in our available data.

For parameter identification we minimized the function

$$\chi^2(K, B, k, l, x_0) = \sum_{i=1}^{n} (x_i - x_i^d)/\sigma_i^2$$

where

$i$ is the running index for the corresponding year,
$x_i$ is the population number obtained by the integration of the hyperlogistic differential equation,
$x_i^d$ is the real population number from the data for the year indicated by $i$,
$\sigma_i^2$ is the variance of $x_i^d$ (see [98]),
$K$, $B$, $k$, $l$ and $x_0$ are free parameters in this minization process.

We minimized this objective function with the help of the function minimizing program system MINUIT [95]. The minimizing procedure is the same as used in [80].

We got $x_i$ by Runge-Kutta-Merson integrating of the driving force in the following way

$$x_i = \int_{t_0}^{t_i} \mathrm{d}x/\mathrm{d}t \, \mathrm{d}t + x_0$$

where

$x_0$ is the initial value for the integration,
$t_0$ is the time corresponding to the starting year with initial value $x_0$,
$t_i$ is the time corresponding to the year index $i$,
$t_0$ is used as fixed parameter which might be chosen freely,
$x_0$ is the model value corresponding to $t_0$, and is a free parameter for the fitting process.

We tried to choose the value of $\sigma_i$ in such a way that $\chi^2$ is defined in analogy to statistics. For this purpose, we assume a constant relative error $R$ of the real data and get thereby $\sigma_i$ as follows:

$$\sigma_i = Rx_i \,.$$

We estimated $R$ by assuming $R = 2\%$ for all population data, independent of the different years. For a given $R$, the fit quality depends on the obtained minimal $\chi^2$ value, $\chi_0^2$, and the number of degrees of freedom, NDF. Using another method of estimation for the relative error, $R_\chi$, we obtain its value by comparison of data with the corresponding model values. For this purpose, we assume a standardized goodness of fit by using a normalized $\sigma_\chi^2$:

$$\chi^2 = \chi^2 \cdot \mathrm{NDF}/\chi_0^2$$

where $\chi^2$ and $\chi_0^2$ correspond to $R = 0.02$. The relative data error in this case is:

$$R_\chi = R\sqrt{\chi_0^2/\tilde{\chi}_0^2} = R\sqrt{\chi_0^2/\mathrm{NDF}} \,.$$

In this way, $\sigma_R$ denotes the relative data error on the assumption that the model truely describes the idealized underlying process from which the data deviate because of error influences. In accordance with $\tilde{\chi}^2$ and $\chi^2$, we obtain two different values of errors (confidence intervals) $\sigma_R$ and $\sigma$, for the identified parameters. $\sigma_R$ denotes the standard deviation on the assumption that $R(= 0.02)$ is the real relative data error. We got this error as follows

$$\sigma_R^+ = p_i^+ - p_i^0 \quad \text{and} \quad \sigma_R^- = p_i^- - p_i^0$$

where $p_0 = (p_1^0, \dots, p_r^0)$ is the parameter vector corresponding to $\chi_0^2$, and $p_i^+$, $p_i^-$ are values of the $i$-th parameter leading to $\chi^2 = \chi_0^2 + 1$ under the condition that the other parameters $p_j$, $j \neq i$, minimize $\chi^2$ in the $(r - 1)$-dimensional parameter subspace. $\sigma_\chi$, we obtain in the same way using $\chi^2$ now, rather than $\tilde{\chi}^2$. $\sigma_\chi$ denotes the standard deviation on the assumption that $R$ is the real relative data error.

### *Results and Interpretation*

The results of the fits are summarized in Tables 7 and 8, and compared with the population data in Figs. 44—50.[1]) It is obvious from Figs. 44—47 and Table 7 that population growth can be adequately described by the EVOLON model in the fitted period. The relative error of the data in regard of the model is $Rx_i \approx 1.7\%$.

The impact of the world wars on the data is also visible from the figures. We recognize that the real data in the nonfitted period from 1910 to 1955, except for the data shortly after administrative reforms in 1910 and 1938, are below the EVOLON model curves already reduced in velocity. This indicates that, due to the wars, there may be essentially two components which reduce the actual population:

[1]) All BASIC programs in this monograph are checked and can be run on the Home Computer SINCLAIR ZX 81. Of course they can be run with small modifications also on higher developped products of the SINCLAIR family as for example ZX SPECTRUM with or without a ZX MICRODRIVE. (SJNCLAJR Vertrieb, Österreich: Peter Paweska, Gmünden)

**Table 7.** a) Model town: $1871 \leqq t < 1910$, $1955 < t \leqq 1970$; $t_s = 1899$; $R = 0.02$

| Parameter | $k \neq l$ | $k = l$ | $\sigma_R$ | $\sigma_\varkappa$ | $k = l = 1$ |
|---|---|---|---|---|---|
| $k$ | 1.59 | 1.60 | +0.09<br>−0.08 | +0.07<br>−0.07 | 1. |
| $l$ | 1.58 | 1.60 | | | 1. |
| $B\,[10^5]$ | 9.40 | 9.41 | +0.29<br>−0.29 | +0.24<br>−0.24 | 10.00 |
| $X_0\,[10^5]$ | 2.88 | 2.88 | +0.01<br>−0.01 | +0.01<br>−0.01 | 2.87 |
| $C\,[10^{-3}]$ | 6.37 | 5.99 | +5.04<br>−2.79 | +3.96<br>−2.42 | 423.10 |
| $\varepsilon$ | 0.73 | 0.72 | +0.11<br>−0.08 | +0.09<br>−0.07 | 0.80 |
| $\chi^2$/NDF;<br>$R_\chi$ | 26.5/38;<br>0.017 | 26.5/39;<br>0.017 | | | 76.7/40;<br>0.027 |

b) Düsseldorf: $1871 \leqq t < 1910$, $1955 < t \leqq 1970$; $t_s = 1900$; $R = 0.02$

| Parameter | $k \neq l$ | $k = l$ | $\sigma_R$ | $\sigma_\chi$ | $k = l = 1$ |
|---|---|---|---|---|---|
| $k$ | 1.41 | 1.37 | +0.06<br>−0.05 | +0.05<br>−0.02 | 1. |
| $l$ | 1.67 | 1.37 | | | 1. |
| $B\,[10^5]$ | 8.31 | 7.89 | +0.56<br>−0.32 | +0.41<br>−0.30 | 10.18 |
| $X_0\,[10^5]$ | 2.10 | 2.10 | +0.01<br>−0.01 | +0.01<br>−0.01 | 2.08 |
| $C\,[10^{-3}]$ | 11.42 | 50.90 | +2.40<br>−1.14 | +1.73<br>−1.26 | 445.39 |
| $\varepsilon$ | 0.93 | 0.87 | +0.13<br>−0.15 | +0.13<br>−0.13 | 0.72 |
| $\chi^2$/NDF;<br>$R_\chi$ | 19.0/35;<br>0.015 | 19.8/36;<br>0.015 | | | 55.9/37;<br>0.026 |

1. The actual population carrying potential (i.e. infrastructure and other population supporting structures) evolves more slowly (taken into account by $\varepsilon$).
2. The existing population carrying potential is not totally exhausted because parts of the population left the towns.

Figs. 45 and 47 show that the logistic model (indicated by $k = l = 1$) does not describe the data time series adequately. The $\chi^2$/NDF value is about three times larger than for the EVOLON model (Table 7).

The fit of the EVOLON model to the population data of the model town gives $k \approx l$. Therefore, we will get nearly the same result if we require $k = l$ during the fit process (Table 7a, Fig. 45). That means that the cooperativity of the growth process

**Table 8.** a) Model town: $k = l$; $t_s = 1899$; $R = 0.02$; $\varepsilon = 1$

| Parameter | EVOLON model | | | $\sigma_R$ | $\sigma_\chi$ | logistic model | | |
|---|---|---|---|---|---|---|---|---|
| | $t_c = 1903$ | $t_c = 1907$ | $t_c = 1910$ | | | $t_c = 1903$ | $t_c = 1907$ | $t_c = 1910$ |
| $k = l$ | 1.80 | 2.04 | 2.20 | +0.60<br>−0.33 | +0.45<br>−0.27 | 1. | 1. | 1. |
| $B\,[10^5]$ | 8.75 | 7.20 | 6.73 | +0.27<br>−0.25 | +0.22<br>−0.20 | 661.27 | 101.63 | 102.39 |
| $X_0\,[10^5]$ | 2.91 | 2.91 | 2.91 | +0.02<br>−0.02 | +0.02<br>−0.02 | 2.88 | 2.88 | 2.87 |
| $C\,[10^{-3}]$ | 1.61 | 0.50 | 0.20 | +1.02<br>−0.16 | +0.69<br>−0.14 | 5.10 | 33.82 | 33.42 |
| $\chi^2$/NDF;<br>$R_\chi$ | 19.2/28;<br>0.016 | 19.4/30;<br>0.016 | 20.0/31;<br>0.016 | | | 30.8/29;<br>0.021 | 32.4/31;<br>0.020 | 33.7/32;<br>0.020 |

b) Düsseldorf: $k = l$; $t_s = 1900$; $R = 0.02$; $\varepsilon = 1$

| Parameter | EVOLON model | | | $\sigma_R$ | $\sigma_\chi$ | logistic model | | |
|---|---|---|---|---|---|---|---|---|
| | $t_c = 1903$ | $t_c = 1907$ | $t_c = 1910$ | | | $t_c = 1903$ | $t_c = 1907$ | $t_c = 1910$ |
| $k = l$ | 1.89 | 2.03 | 1.92 | +0.20<br>−0.21 | +0.12<br>−0.12 | 1. | 1. | 1. |
| $B\,[10^5]$ | 4.75 | 4.53 | 4.81 | +0.47<br>−0.32 | +0.25<br>−0.21 | 606.70 | 795.25 | 257.69 |
| $X_0\,[10^5]$ | 2.12 | 2.12 | 2.12 | +0.01<br>−0.01 | +0.01<br>−0.01 | 2.10 | 2.09 | 2.08 |
| $C\,[10^{-3}]$ | 5.38 | 2.61 | 4.11 | +7.17<br>−0.48 | +3.46<br>−0.38 | 6.52 | 4.94 | 15.28 |
| $\chi^2$/NDF;<br>$R_\chi$ | 9.82/28;<br>0.012 | 10.47/30;<br>0.012 | 11.2/31;<br>0.012 | | | 25.7/29;<br>0.019 | 27.7/31;<br>0.019 | 28.6/32;<br>0.019 |

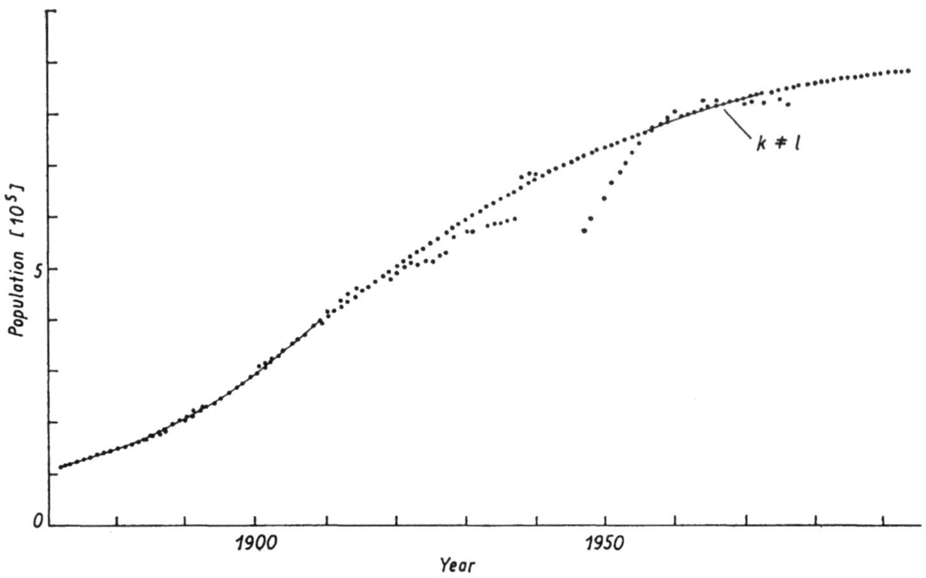

Fig. 44 Model parameter estimations

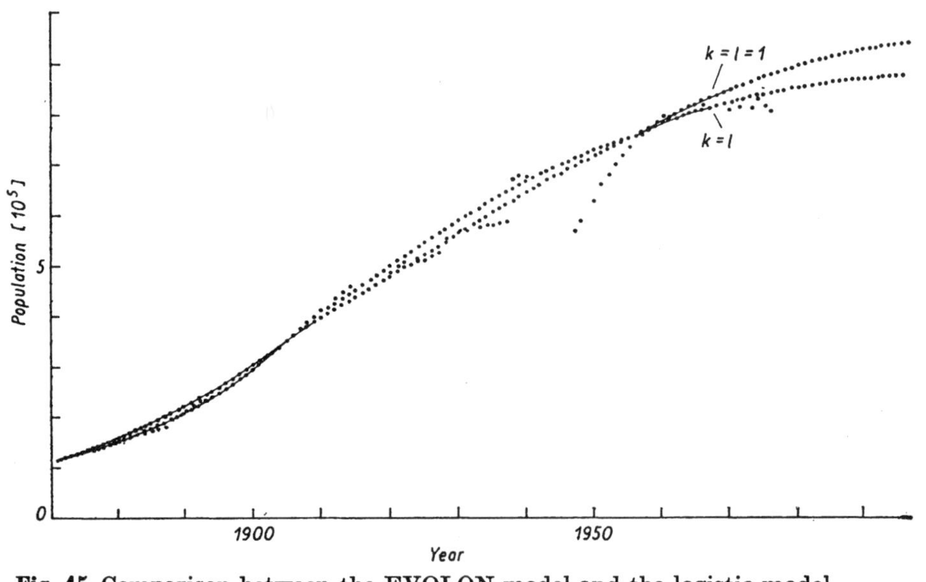

Fig. 45 Comparison between the EVOLON model and the logistic model

is approximately the same as it is for the saturation process. This phenomenon could be interpreted as the cooperativity which generated the growth process being responsible also for the control of the saturation process because of the same structures acting within the system.

The value $k \approx l \approx 1.6$ differs considerably from one (see $\sigma_\chi$ in Table 7a) and indicates relatively complex mechanisms for growth and saturation. This again supports our hypothesis that the initial growth of evolving systems is hyperbolic. The hyper-

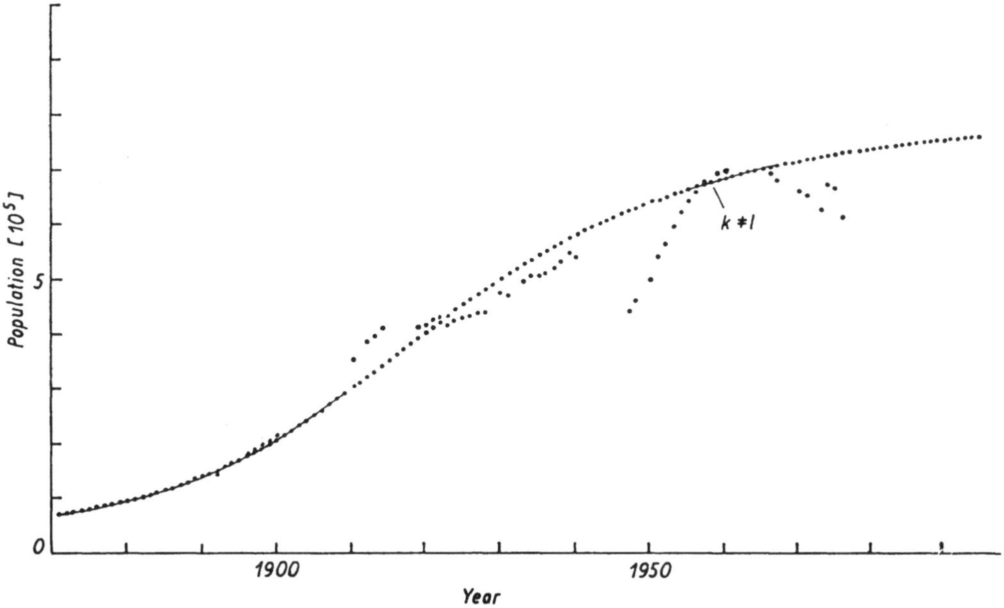


**Fig. 46** The data fitting by the EVOLON model for Düsseldorf

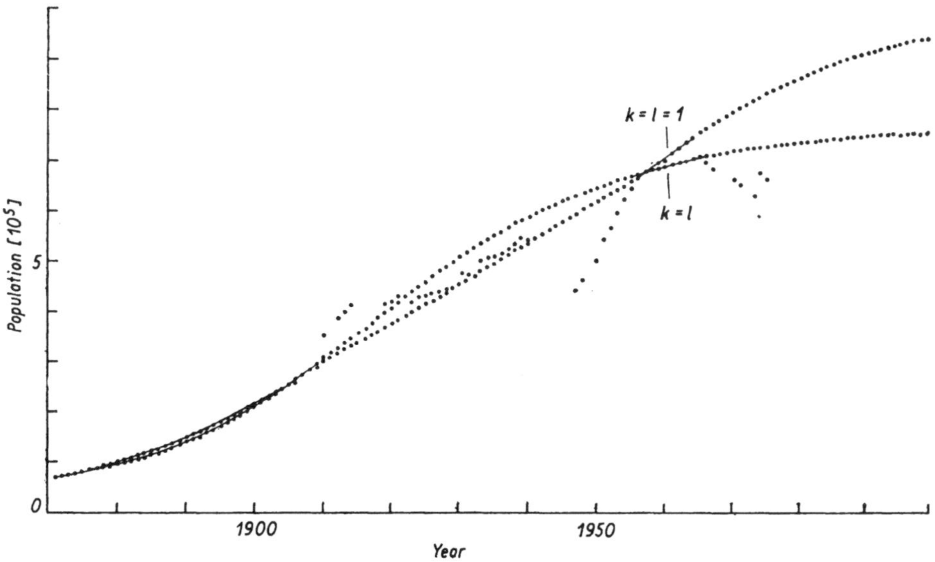


**Fig. 47** The data fitting by the logistic model for Düsseldorf

logistic growth reflects the cooperativity of the elements of the system, and corresponds to the relative growth rates of urban growth processes being positive density correlated. In terms of the exponential tower concept [50], this means that the growth process is based on a hierarchical chain of exponentially growing systems.

We can calculate the maximum growth rate and its data in the framework of the EVOLON model. For $k = l$, we can derive the relation

$$B = \frac{2k - 1}{k - 1} x_{\mathrm{m}}$$

where $x_{\mathrm{m}}$ is the $x$-value for which the growth rate $\mathrm{d}x/\mathrm{d}t : x$ has its maximum. From the parameters indicated by $k = l$ in Table 7a, it follows that $x_{\mathrm{m}} = 25.7 \cdot 10^5$. This value was reached in 1896. Since then, the relative growth rate of the model town has been decreasing.

The value of $\varepsilon = 0.73$ (Table 7a) may be interpreted as follows: If we compare the real disturbed growth process of the ten towns used to construct our model town with an undisturbed normal growth process, we may conclude that the velocity depression during 46 years is equivalent to a time lag $\Delta t = (1 - \varepsilon)\, 46 = 12$ years. This means the influence of the two world wars and the crisis in between on the population growth of the ten towns may be expressed as a time loss of about 12 years, in comparison with the undisturbed normal growth process.

The corresponding results for a single town (Düsseldorf) are contained in Table 7b and Figs. 46 and 47. As mentioned above, an interpretation of these results is problematical. The population of Düsseldorf has been decreasing since 1964. This is a reflection of a new process of population loss (coalescence) which is not formalized by our model in its present form. For this reason, we cut off the fit period at 1964 and got a $\chi^2$-value of only 2/3 of the value we would have obtained had we extended the fit period up to 1970. In the case of Düsseldorf, the coalescence begins relatively early, but we have to expect the same phenomenon influences also in the population data of the model town. We cut off the fit region in the case of the model town in 1970 (Table 7a).

Another problem is an administrative reform occurring in 1910 which caused an increase of about 20% of the population of Düsseldorf. This led to a deviation of the population data from the model values around 1910 (Figs. 46 and 47). Immediately before 1910, the population growth was inhibited by infrastructural limitations and corresponding administrative regulations. Immediately after 1910, we have a population growth spurt because the administrative reforms were already oriented towards future needs. The large $\varepsilon$ values 0.93 and 0.87 in Table 7b are caused by the inhibited growth before 1910. For instance, if we included the years 1911 and 1912 in the fit region, we would get $\varepsilon = 0.52$. In the case of the aggregated model town, the $\varepsilon$ values are less sensitive to such variations of the fit region.

### *About the Possibility of Predicting the Population of Towns: Prognostic Experiments*

In order to test the ability of the EVOLON model to obtain reliable prognoses, we examined to what extent an urban planning department applying our model would have been able to predict the evolution of the town's population as early as at the beginning of the century. For that purpose, we used for the fit only data from before a certain cut-year $t_c$ out of the first decade. To check the influence of the selected cut-year on the predictions, we used three different cut-years $t_c = 1903$, 1907, and 1910.

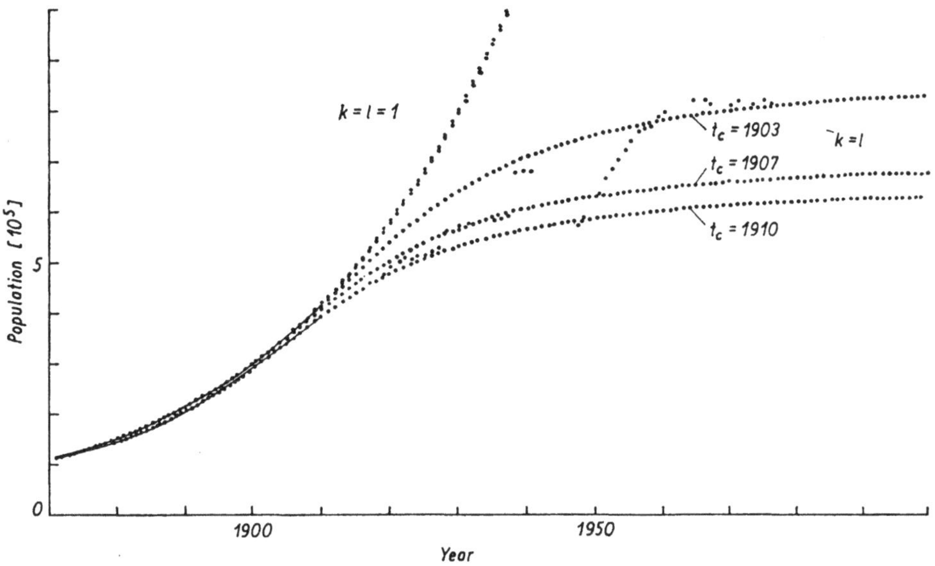


**Fig. 48** Projection of the population for the model town

The results are compared with corresponding logistic fits (Tables 8 a and b, Figs. 48 and 50).

The prognostic experiments for the model town, based on the EVOLON model, lead, depending on $t_c$, to 93%, 77% and 72% (Table 8a and Fig. 48) of the saturation value, $B = 9.41 \times 10^5$, obtained for the whole data set (in the period $1871 \leqq t \leqq 1910$ and $1955 \leqq t \leqq 1970$, Table 7a). These predictions yield values close to the real data. The curve for $t_c = 1903$ coincides especially well with the data in the 1960's. This is a good result for a prediction extended over sixty years.

In contrast to this, the logistic model leads to $\geqq 1100\%$ of the mentioned $B$-value obtained by the EVOLON model for the whole data set (Table 7a). For that reason, the logistic model is, in our estimation, absolutely unsuitable for long-range forecasting.

To get an impression for the accuracy of our prognostic experiments, we plotted the $\chi^2$-, $B$- and $K$-dependence on the parameter $k = l$. Fig. 49 shows that the value $k = l = 2.2$, obtained for $t_c = 1910$, is two confidence intervals larger than the value $k = l = 1.6$, obtained for the overall fit region $1871 \leqq t \leqq 1910$ and $1955 \leqq t \leqq 1970$ (marked by *all*), whereas the value $k = l = 1.8$, obtained for $t_c = 1903$, completely agrees with the value $k = l = 1.6$ in the limits of only one confidence interval. In our prognostic experiments, the confidence intervals, according to one standard deviation, are large. Nevertheless, the $k = l$ values are obviously larger than one. We can thus conclude with high reliability, on the basis of data only for $t < t_c$, that the cooperativity of the growth process of a town's population is hyperbolic, and it cannot be described adequately by exponential growth. The $\chi^2$-, $B$- and $K$-values for the logistic model with $k = l = 1$ are also indicated in Fig. 49.

The $\chi^2$-dependence on $k = l$ is weak for the prognostic experiments (marked by $t_c = 1910$ and $t_c = 1903$), in comparison with the $\chi^2$-dependence for the overall fit region. This means the parameter $k = l$ cannot be determined with high accuracy in

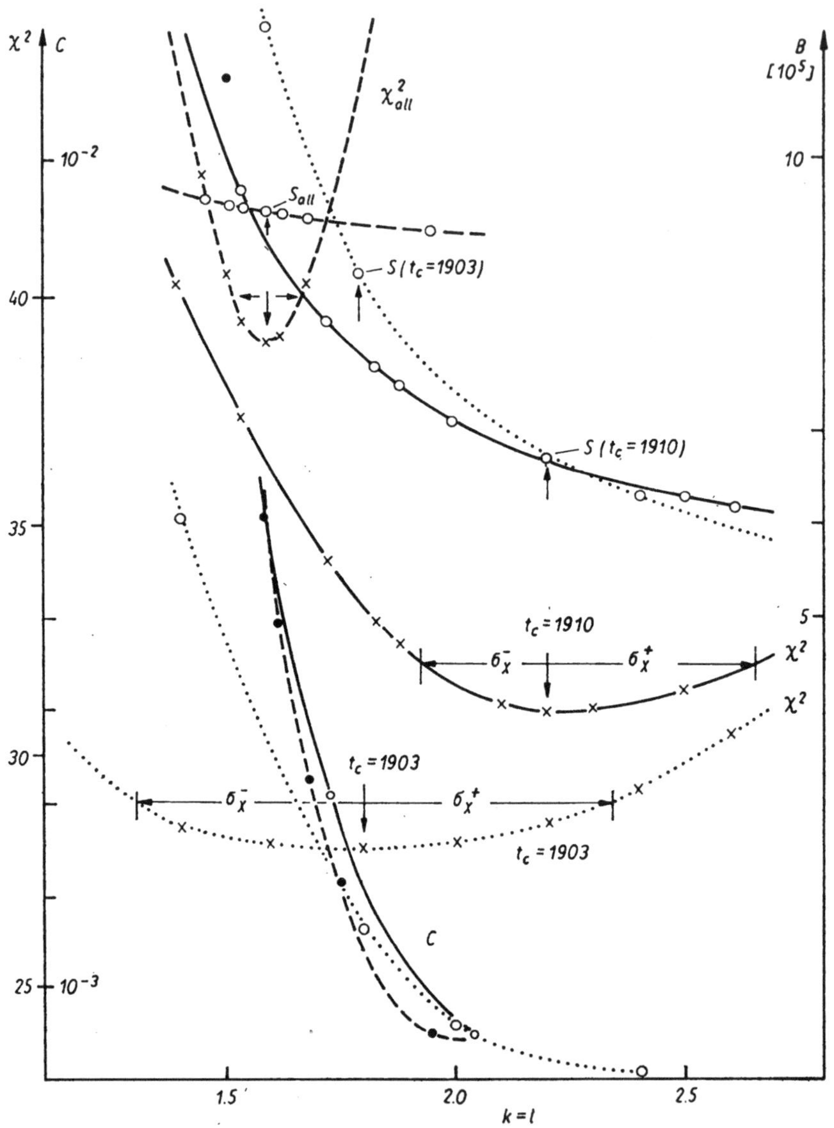


**Fig. 49** The $\chi^2$-dependence on $k = l$ for the model town

our prognostic experiments. Furthermore, we recognize from Fig. 49 that strong changes of the velocity parameter $K$ and the saturation parameter $B$ are coupled with variations of the parameter $k = l$. Because of the flat $\chi^2$-dependence for the prognostic experiments, a small systematic deviation of the data can easily shift the value of $k = l$. This seems to be the case for $t_c = 1910$. We notice that the parameters obtained in prognostic experiments for the model town converge to those values obtained for the overall fit region (Table 7a) if $t_c$ goes from 1910 to 1903 (Table 8a and Fig. 49). This indicates once more that the data are disturbed near 1910 by an inhibitation of growth before the administrative reform (see also [98]).

In the prognostic experiments based on the Düsseldorf data, there is no such trend. We do not get a sensitive variation for the model parameters in dependence on $t_c$ in this case (Table 8b, Fig. 50). This could be explained eventually in such a way that the inhibitation of the growth of Düsseldorf begins yet earlier than that of the model town. That may also be the reason for the necessity for an especially large increase of Düsseldorf's population during the administrative reform of 1910. We did not check this supposition.

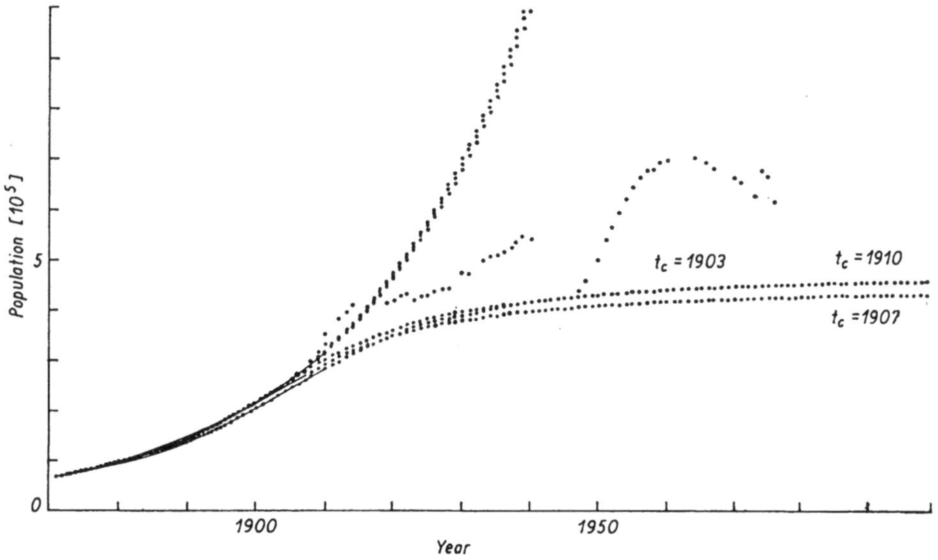


**Fig. 50** The projection of population for Düsseldorf

Sixty years later, the prediction for the population of towns deviates from the real data of the model town by 2—25% and, in the case of Düsseldorf, by 40%. The good form of the process with only a few parameters suggests that the evolutionary model reflects (in a simplified form) essential characteristics of the real growth process. This opinion is supported by the results of applications of the EVOLON model to other complex cooperative growth processes [118].

## *1.13. Other Approaches for the Systematisation of Growth Processes*

### *Pearson's Systematisation of Growth Processes*

Pearson tried to obtain most of the more frequently used probability distributions from a unique model [53], [54]. The fact that the most frequently used probability density functions are solutions of a certain ordinary differential equation of the first order (hereafter called *Pearson's differential equation*) served as his instrument for their unification

$$d\varphi/du = \varphi(u + a)/(b_0 + b_1u + b_2u^2) .$$

According to the decision tree represented in Fig. 51 we can distinguish some important special cases belonging all to well-known probability density functions.

*Case 1*: $b_2 \neq 0$, $D = -4b_2b_0 + b_1^2$.

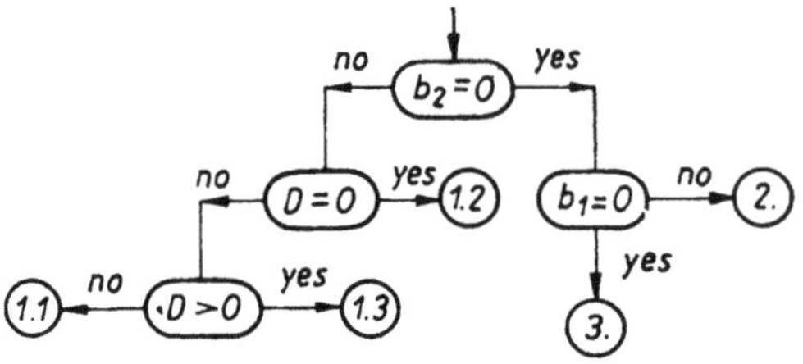


**Fig. 51** The decision tree for Pearson's differential equation

The quadratic polynomial in the denominator has the following roots

$$u_{1/2} = -\frac{b_1}{2b_2} \pm \left(\frac{b_1^2}{4b_2^2} - \frac{b_0}{b_2}\right)^{1/2} = \frac{1}{2b_2}(-b_1 \pm \sqrt{D}).$$

1.1. $D < 0$, two conjugated complex roots.

We introduce the following notations

$$u_{1/2} = \gamma \pm \delta \quad \text{with} \quad \gamma = -b_1/2b_2\,, \qquad \delta = \frac{\sqrt{D}}{2b_2}, \qquad k = \frac{1}{2b_2},$$

$$\varphi(u) = C\left((u-\gamma)^2 + \delta^2\right)^k \exp\left(2k\frac{a+\gamma}{\delta}\arctan\frac{(u-\gamma)}{\delta}\right).$$

Special cases:

$$b_2 \neq 0\,, \qquad b_1 = 0\,, \qquad b_0 \neq 0 \qquad \text{2 imaginary roots,}$$

$$\varphi(u) = C(u^2+\delta^2)^k \exp\left(2k\frac{a}{\delta}\arctan u\right);$$

$$b_2 = -1/(n+1)\,, \qquad b_1 = 0\,, \qquad b_0 = n/(n+1)\,, \qquad a = 0\,, \qquad n \geqq 1\,,$$

$$\varphi_n(u) = \frac{\Gamma\left((n+1)/2\right)\,(1+u^2/n)^{-(n+1)/2}}{\Gamma(1/2)\,\Gamma(n/2)\,n}$$

density of Student's $t$-distribution.

1.2. $D = 0$, two coincident real roots.

$$u_{1/2} = \gamma\,,$$

$$\varphi(u) = C(u-\gamma)^k \exp\left(-k\frac{a+\gamma}{u-\gamma}\right).$$

1.3. $D > 0$, two real roots.

$$u_{1/2} = \gamma_1 \quad \text{and} \quad \gamma_2 \quad \text{with} \quad \gamma_1 < \gamma_2\,,$$

$$\gamma_1 = k(-b_1 + \sqrt{D})\,, \qquad \gamma_2 = k(-b_1 - \sqrt{D})$$

(for $k = 1/(2b_2) > 0$, otherwise both roots exchange their roles),

$$\varphi(u) = C(u-\gamma_1)^{k(\gamma_1+a)/(\gamma_1-\gamma_2)}\,(\gamma_2-u)^{k(\gamma_2+a)/(\gamma_2-\gamma_1)}\,.$$

Special cases

$$b_2 \neq 0\,, \qquad b_1 \neq 0\,, \qquad b_0 = 0\,, \quad \text{therefore} \quad \gamma_1 = 0 \quad \text{for} \quad k > 0\,,$$

$$\varphi(u) = Cu^{-ka/\gamma_2}(\gamma_2 - u)^{k(\gamma_2+a)/\gamma_2};$$

$$b_2 \neq 0\,, \quad b_1 = 0\,, \quad b_0 \neq 0\,, \quad \gamma_1 = -\delta\,, \quad \gamma_2 = \delta\,, \quad \delta > 0\,,$$

$$\varphi(u) = C(u + \delta)^{k(\delta - a)/(2\delta)} (\delta - u)^{k(\delta + a)/(2\delta)}\,;$$

$$b_2 = -b_1\,, \quad b_0 = 0\,, \quad \gamma_1 = 0\,, \quad \gamma_2 = 1\,,$$

$$\varphi(u) = \frac{\Gamma(p+q)}{\Gamma(p)\,\Gamma(q)}\, u^{p-1} (1 - u)^{q-1} \quad \text{density of the } \beta\text{-distribution}$$

with the parameters $p = 1 - ka$, $q = 1 + k(1 + a)$.

*Case 2*: $b_2 = 0$, $b_1 \neq 0$, one simple root.

Here we put

$$\gamma = -b_0/b_1 \quad \text{and} \quad k = 1/b_1\,,$$

$$\varphi(u) = D(u - \gamma)^{k(a+\gamma)}\, e^{ku}\,.$$

Special case

$$b_2 = 0\,, \quad b_1 = 1\,, \quad b_0 = 0\,,$$

$$\varphi(u) = \frac{q^p}{\Gamma(p)}\, u^{p-1}\, e^{-qu} \quad \text{density of the } \Gamma\text{-distribution}$$

with the parameters

$$p = 1 + ka\,, \qquad q = -k\,.$$

*Case 3*: $b_2 = 0$, $b_1 = 0$, $b_0 \neq 0$,

$$\varphi(u) = C \exp\left(k((u + a)^2 - a^2)\right)\,.$$

This is for $k < 0$, the density of the Gaussian normal distribution.

There are two different ways to use the Pearson concept for the construction of growth functions which could serve as EVOLON-models.

### *Pearson I*

In this case, the growth indicator $x$ is identified with the random variable $u$ in the Pearson concept, and the probability density function $\varphi$ is interpreted as the driving force function of a growth model. Therefore, we get the following system concept for a growth model

$$\mathrm{d}x/\mathrm{d}t = \varphi(x)\,,$$

$$\mathrm{d}\varphi(x)/\mathrm{d}x = \varphi(x + a)/(b_0 + b_1 x + b_2 x^2)\,.$$

For an EVOLON in the proper sense, meaning with a finite saturation value $B$, it is necessary that $\varphi(x)$ vanishes outside of a certain finite interval. This condition is for the Pearson I concept, only satisfied by the case

$$\varphi(x) = C(x - \gamma_1)^{k(\gamma_1 + a)/(\gamma_1 - \gamma_2)} (\gamma_2 - x)^{k(\gamma_2 + a)/(\gamma_2 - \gamma_1)}\,.$$

This is obviously the power-product of the driving force model according to the hyperlogistic differential equation. Nevertheless, the case of Pearson I deserves special interest, because it leads, in connection with our structure design principle presented in Chapter 3 to interesting structural consequences.

### *Pearson II*

This is the original Pearson concept, which he proposed in analogy to the probability distribution functions as a system concept for *s*-shaped growth transitions. In this case we must substitute into the Pearson differential equation, instead of the random variable $u$, the time variable $t$. We then get the following system concept for growth functions

$$\mathrm{d}x/\mathrm{d}t = \varphi(t) \,,$$
$$\mathrm{d}\varphi(t)/\mathrm{d}t = \varphi(t + a)/(b_0 + b_1 t + b_2 t^2) \,.$$

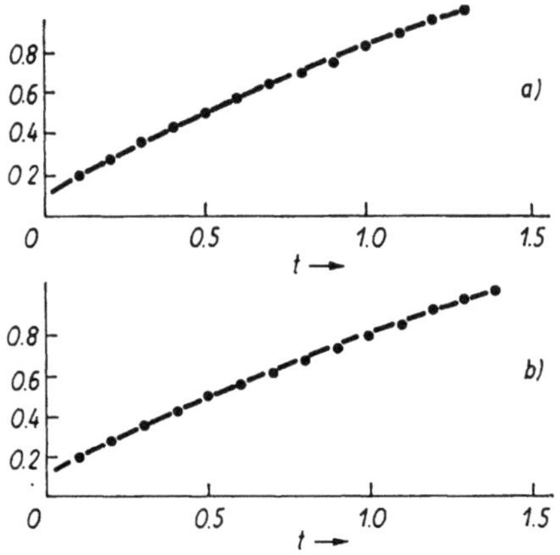


**Fig. 52** Two special cases of the solution of Pearson's differential equation according to the decision tree
a) $a = -1$, $b_0 = 1$, $b_1 = 1$, $b_2 = 1$,
b) $a = -1$, $b_0 = 1$, $b_1 = 0$, $b_2 = 1$.

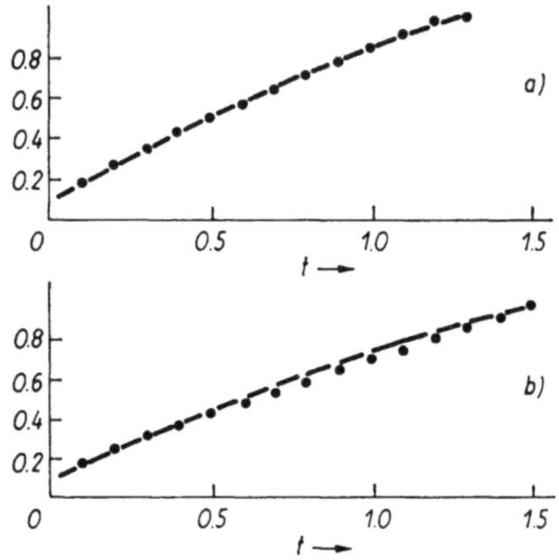


**Fig. 53** Two special cases of the solution of Pearsons's differential equation according to the decision tree
a) $a = -1$, $b_0 = 1$, $b_1 = 2$, $b_2 = 1$,
b) $a = -1$, $b_0 = 0.5$, $b_1 = 2$, $b_2 = 0.5$.

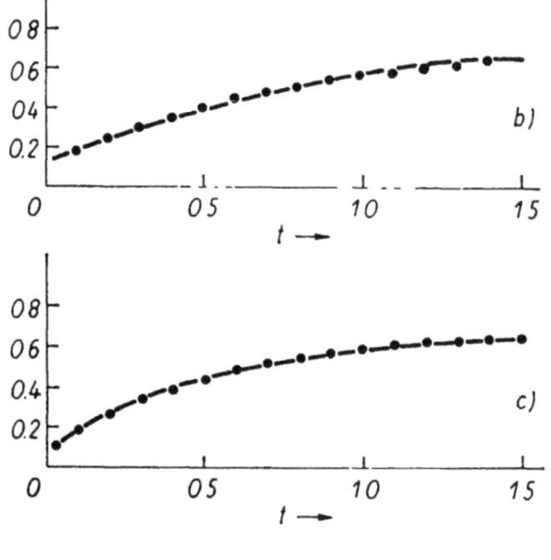


**Fig. 54** Three special cases of the solution of Pearson's differential equation according to the decision tree
a) $a = 1$, $b_0 = -1$, $b_1 = -1$, $b_2 = 0$,
b) $a = 0.1$, $b_0 = 0$, $b_1 = -1$, $b_2 = 0$,
c) $a = 1$, $b_0 = -1$, $b_1 = 0$, $b_2 = 0$.

# 2. Dynamic Behaviour of Chains

## 2.1. *Chain Models*

We meet chain structures frequently in nature. Every element of a chain has a unique successor and a unique predecessor. Energy transfer and food supply structures are often organised in chains in nature. In chemistry we meet an interaction of chains in reaction cascades. In the course of structure development in the ontogeny of any organism we find chain constructions because the growth of a predecessor usually is the supposition that a successor can start its development. In this chapter, we present, based on earlier studies in unified theory of linear systems [24], some properties of chain structures, stressing in these considerations rate-coupling in chains.

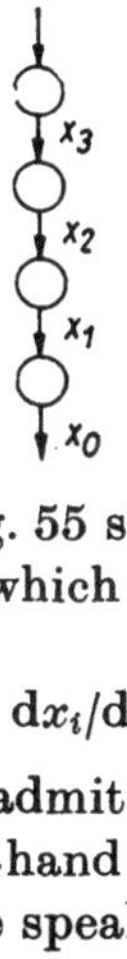


**Fig. 55** Structure scheme of a general chain

Fig. 55 shows the structural scheme of a chain constructed from deterministic modules which are connected with each other according to the following coupling equation

$$dx_i/dt = f_i(K_i, x_i, x_{i+1}) .$$

We admit in general that the chain is built of different modules, meaning that the right-hand sides $f_i$ can correspond to different functional concepts.

We speak of a homogenous chain if we meet the same module on all places of the chain, perhaps with only different values of the parameter $K_i$, meaning that the following equation holds

$$f_i = f \quad \text{for all } i .$$

In this case, the portion of the chain for which the following equation holds

$$dx/dt = f(K, x; x'),$$

is called the *base module* of the chain.

## 2.2. *Taylor Series Expansion as a Chain Concept with a Linear Base Module*

Following the traditions of cybernetics, we will use the signal notion frequently in the following.

A *signal* is, in its form, always a time function $x(t)$ of a one-dimensional or vectorial (multivariable) variable; in its semantics, it is always a physical variable with a well defined dimension. The value of the signal $x$ at time point $t$, $x(t)$, is well determined if we have chosen a certain physical unit for the corresponding physical variable. We get a pure time function always by normalisation. In this book, we will use frequently the following normalisation

$$x(t) = x(0)\,\xi\,.$$

After this normalisation procedure, we get, in the corresponding equations (for the description of the input/output behaviour), correspondences between variables without dimensions, meaning between mathematical functions. Of course, the parameters in the equations must then take over the change of dimensions.

We now consider a chain for the generation of a signal $x_0 = x(t)$.

We are looking for a corresponding signal generator. The chain shall consist of a linear base module at all places with the form

$$\mathrm{d}x/\mathrm{d}t = K + x'\,.$$

The whole chain is then described by the following recursion

$$\mathrm{d}x_i/\mathrm{d}t = K_i + x_{i+1}\,, \qquad i = 0, 1, \ldots\,.$$

For separating uniquely the influences between $K_i$ and $x_{i+1}$ from each other we allow only for normalised initial conditions in the following form

$$x_i(0) = 0 \quad \text{for} \quad i = 0, 1, \ldots\,.$$

For a given signal $x_0 = x(t)$ with continuous derivatives of sufficiently high order formally a unique chain for the generation of $x(t)$ can be constructed recursively in the following way

$$\mathrm{d}x_i/\mathrm{d}t = K_i + x_{i+1}\,.$$

Because of the assumed normalisation condition we have

$$K_i = \mathrm{d}x_i/\mathrm{d}t|_{t=0}; \qquad x_{i+1} = \mathrm{d}x_i/\mathrm{d}t - K_i \quad \text{for} \quad i = 0, 1, \ldots$$

and $\mathrm{d}x_{i+1}/\mathrm{d}t = \mathrm{d}^2x_i/\mathrm{d}t^2$ yields $K_i = x_0^{(i+1)}|_{t=0}$.

Therefore, the coefficients $K_i$ of the chain can be calculated from $x_0(t)$ in the same way as the coefficients of the Taylor expansion of $x_0(t)$. This means that the construction of such a chain is nothing but the well-known Taylor expansion of $x_0(t)$. This can be shown with the following considerations.

If we cut the structure expansion on the level $N$, we get for the top signal $x_{N+1} = y$ passing through the chain bottom-up. Obviously we have

$$x_{N+1} = F^{(N+1)}x_0 - K_N \quad \text{with} \quad F = \mathrm{d}/\mathrm{d}t\,.$$

Now we will study separately the influence of the rest $x_{N+1}$ and all coefficients $K_i$ on the signal $x_i(t)$ being expanded into the chain. Therefore, we are now passing through the chain top-down. To avoid the influence of $K_N$, we go one step beyond the highest level and get then

$$Fx_{N+1} = F^{(N+2)}x_0\,.$$

By back-integration top-down, we get, using the inverse operator, $F^{-1} = I = \int_0^t$

the following sequence of expressions.

$$x_{N+1} = F^{-1}F^{(N+2)}x_0 \qquad x_{N+1} = Fx_N - K_N$$

$$x_N = F^{-2}F^{(N+2)}x_0 + K_N t \qquad x_N = Fx_{N-1} - K_{N-1}$$

$$x_{N-1} = F^{-3}F^{(N+2)}x_0 + K_N t^2/2 + K_{N-1} \qquad x_{N-1} = Fx_{N-2} - K_{N-2}$$

$$\ldots \qquad x_1 = Fx_0 - K_0$$

$$x_0 = F^{-(N+2)}F^{(N+2)}x_0 + \sum_{i=1}^{N+1} K_{i-1}t^i/i\,!$$

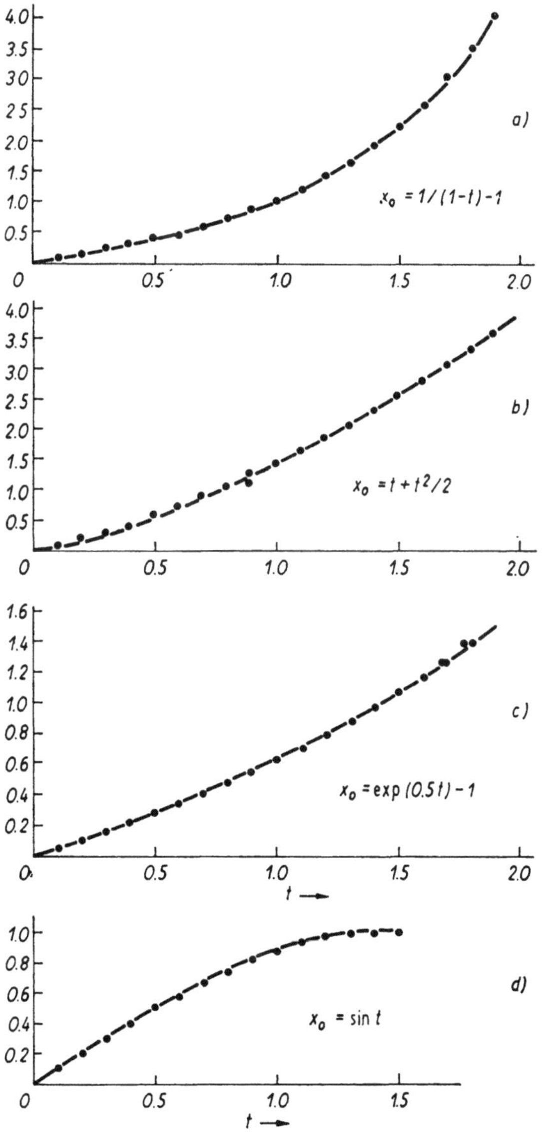


**Fig. 56** Some function expansions with the help of a Taylor chain

Obviously, this is the Taylor expansion of a function $x_0(t)$ which vanishes at $t = 0$, thus fulfilling the normalised initial condition and for which the expansion is extended up to the power $t^{N+1}$. The corresponding rest formula is then

$$R_{N+1} = F^{-(N+2)}F^{(N+2)}x_0 .$$

This is the well-known integral representation of the rest of a Taylor expansion [83].

It is remarkable that this expansion of a given function $x_0(t)$ into a linear chain leads us to a more general point of view for signal representations, of which the above-mentioned series expansion is only a special case. To be sure, we must get rid of the thought that we could in general expect a linear superposition of the rest and of the series. Further on, we even cannot assume that the signal parts of the chain determined by the different parameters $K_i$ superimpose linearly to generate the overall signal. The composition of these signal parts will then be much more difficult. In a certain sense, we shall see that generalised superposition laws play a role already studied from different authors [70], [74].

It is essential for such a chain expansion that the influence of the rest signal $x_{N+1}$ on the bottom-signal $x_0$ decreases with growing truncation index $N + 1$. In the case of the Taylor expansion it is sufficient that the absolute value of all $F^i x_0$ is uniformly bounded.

## 2.3. *Exponential Chains with the Exponential Integrator as a Base Module*

In nature, rate-coupling is a broadly distributed phenomenon. This leads to the idea of constructing chains with a base module of the following form

$$\mathrm{d}x/\mathrm{d}t = K \times x' .$$

Basic operator of this module is the logarithmic derivative

$$F = \mathrm{d}\ln/\mathrm{d}t .$$

A chain built up with such a base module is then given by

$$Fx_i = K_i x_{i+1} \quad \text{or} \quad \mathrm{d}x_i/\mathrm{d}t = K_i x_i x_{i+1} , \qquad i = 0, 1, \ldots .$$

Such a chain we call an exponential chain or exponential tower. Again we have the task of generating a given signal $x(t) = x_0$ using such a chain with suitable chosen parameters $K_i$ and corresponding initial values $x_i(0)$. To get a unique separation between the influences caused by the parameters $K_i$ on one hand and by the initial values $x_i(0)$ on the other hand, we again pass to normalized initial values $x_i(0)$. In this case, we demand the following initial value normalisation:

$$x_i(0) = 1 \quad \text{for} \quad i = 0, 1, \ldots .$$

Then, quite similar to the Taylor series case, the parameters $K_i$ can be calculated in the following way:

$$K_i = Fx_i|_{t=0} = F^{(i+1)}x_0|_{t=0}$$

using here only $F = \mathrm{d}\ln/\mathrm{d}t$ instead of formerly $F = \mathrm{d}/\mathrm{d}t$. The state-variables of the consecutive steps of the exponential chain are determined by

$$x_{i+1} = Fx_i/K_i \ .$$

If we run through the chain bottom-up, we get the following sequence $x_1 = Fx_0/K_0$, $x_2 = Fx_1/K_1 = F^2x_0/K_1, \ldots$,

$$x_{N+1} = F^{(N+1)}x_0/K_N \ .$$

This again is something like a rest of the signal $x_0(t)$, to be expanded on the top level.

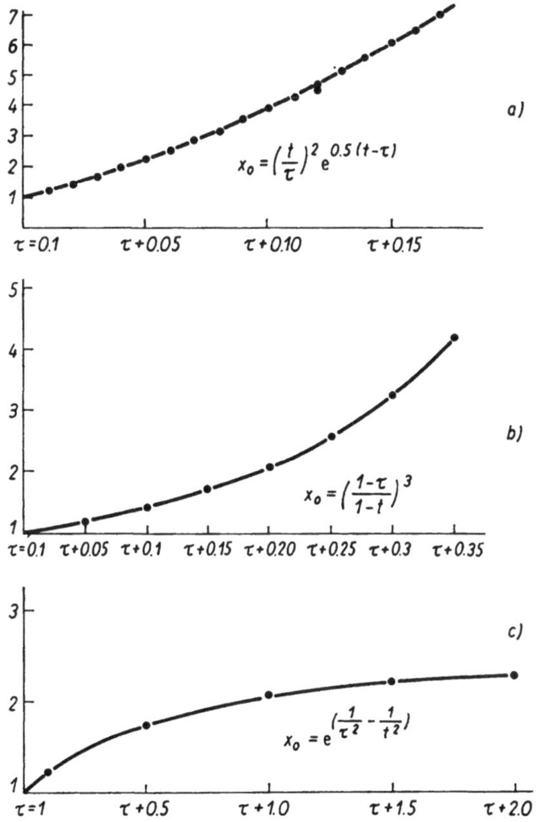


**Fig. 57** Some function expansions with the help of an exponential chain

If this rest deviates from 1 by only a little, its importance is very small. Therefore, a realistic rest concept for the expansion of $x_0(t)$ measured at the bottom level results from the following considerations: The complete bilance at the base of the chain after climbing to the level $N$ and returning top-down is obviously determined by

$$x_0(t) = F^{-1}(K_0F^{-1}(K_1 \ldots F^{-1}(F^{N+1}x_0)) \ldots) \ .$$

We get a feasible approximation of the bottom-signal $x_0(t)$ if we break the chain expansion on the $N$-level putting there

$$x_{N+1} \equiv 1 \ .$$

This leads to the approximation

$$x_{0a}(t) = F^{-1}(K_0F^{-1}(K_1 \dots F^{-1}(K_N.1)) \dots) .$$

This gives us a rest concept for discussion of the form

$$R_{N+1} = x_0(t) - x_{0a}(t) .$$

In the case of the Taylor series expansion, this is identical with the above-mentioned integral representation of the rest at the bottom level. In the case of the exponential chain such a simplification of the expression for the rest does not exist because the linear superposition law no longer holds.

We will discuss convergence considerations later in the book.

## 2.4. *Hypercycles for Taylor Chains and Exponential Chains*

For a Taylor chain the behaviour of a hypercycle of order $r$ is determined in the following way

$$\mathrm{d}x_i/\mathrm{d}t = K_i + x_{i+1} , \qquad i = 0, 1, \dots , r - 1,$$

$$x_r \equiv x_0 \quad \text{and} \quad x_i(0) = 0 .$$

To avoid degenerations, the sequence of coefficients

$$K_0, K_1, \dots , K_{r-1}$$

is assumed not to contain any subperiod of the coefficients.

Let us now, for the Taylor case analytically characterize the hypercycle motions.

By backward integration, top-down over one period, and taking into account $x_r \equiv x_0$ and the normalised initial conditions $x_i(0) = 0$, we get the following formula

$$x_0 = I^r x_0 + I\left(K_{r-1}\frac{t^{r-1}}{(r-1)!} + K_{r-2}\frac{t^{r-2}}{(r-2)!} + \dots + K_0\right).$$

From this, we get by applying the integral operator series [24], the general solution

$$x_0 = \frac{I}{(1 - I^r)}\left(K_{r-1}\frac{t^{r-1}}{(r-1)!} + K_{r-2}\frac{t^{r-2}}{(r-2)!} + \dots + K_0\right).$$

A hypercycle especially of order $r = 1$ has the solution

$$x_0 = K_0(\mathrm{e}^t - 1) .$$

Let us consider the degenerated case, where we follow a chain with equal coefficients $K_i = K$ up to the level $r$ and then close by a feedback in the sense of a hypercycle of order $r$ (remember the subperiod 1 !). The corresponding solution is then

$$x_0 = \frac{I}{(1 - I^r)} K(I^{r-1} + I^{r-2} + \dots + 1)\, 1(t) = K\frac{I}{1 - I} 1(t) .$$

Here, $1(t)$ is the Heaviside step function with

$$1(t) = \begin{cases} 1 & \text{for} \quad t \geqq 0 , \\ 0 & \text{for} \quad t < 0 . \end{cases}$$

Because of the assumed degeneration, this solution is identical with the solution of a hypercycle of order $r = 1$.

Of interest is also the behaviour of a homogeneous chain $K_i = K$ ending after $r$ levels with $x_r \equiv 0$. For the corresponding non-homogenous chain, we get the solution

$$x_0 = I\left(K_{r-1}\frac{t^{r-1}}{(r-1)!} + K_{r-2}\frac{t^{r-2}}{(r-2)!} + \dots + K_0\right)$$

and, from this expression, for the special case $K_i = K$ respectively

$$x_0 = K\,(I^r + I^{r-1} + \dots + I^2 + I)\,1(t) = K\left(\frac{(I^{r+1}-1)}{(I-1)} - 1\right)1(t)\,.$$

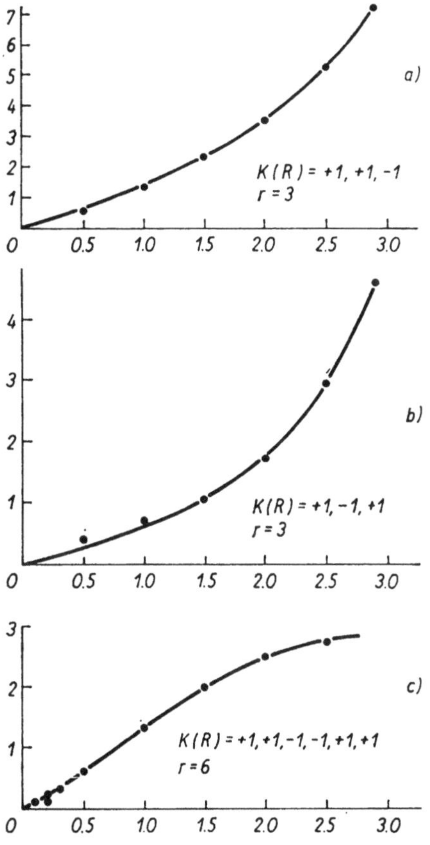


**Fig. 58** Some examples of Taylor chain hypercycles

The integral operator series in this expression apparently for $r \to \infty$ converges against the hypercycle solution with order $r = 1$

$$x_0 = K\frac{I}{1-I}\,1(t)\,.$$

From this discussion we derive the important connection that the hypercycle solution for $r = 1$ coincides with the bottom solution $x_0$ of a homogenous chain of infinite length. Therefore, in practice, long homogeneous chains may mimic the behaviour of a hypercycle $r = 1$ without the corresponding structure necessarily really having a hypercycle structure, meaning a feedback after the first step.

In this context, it is also of interest to study the behaviour of periodic chain structures with the basic period $r$ in the coefficients

$$K_0, K_1, \dots, K_{r-1} .$$

If we repeat this basic period $m$ times and then stop the expansion, putting the top signal equal to zero, we get, by backward integration and going top-down, the following result

$$\begin{aligned}
x_{0,m} &= P_r(t) = I\left(K_{r-1}\frac{t^{r-1}}{(r-1)!} + K_{r-2}\frac{t^{r-2}}{(r-2)!} + \dots + K_0\right),\\
x_{0,m-1} &= I^r P_r(t) + P_r(t) ,\\
x_{0,m-2} &= I^{2r} P_r(t) + I^r P_r(t) + P_r(t) ,\\
&\dots\dots\dots\dots\dots\dots\dots\dots\dots\dots\\
x_0 &= x_{0,0} = (I^{mr} + I^{(m-1)r} + \dots + I^r + 1)\, P_r(t) .
\end{aligned}$$

Because of the convergence of the integral operator series [24],

$$1/(1 - I^r)$$

this expression necessarily converges against the hypercycle solution of the hypercycle with the order $r$ and the coefficients

$$K_0, K_1, \dots, K_{r-1} .$$

In the case of an exponential chain the behaviour of the corresponding hypercycle of order $r$ is obviously characterized by the conditions

$$\mathrm{d}x_i/\mathrm{d}t = K_i x_i x_{i+1} \quad \text{for} \quad i = 0, 1, \dots, r-1$$

and

$$x_r \equiv x_0 \quad \text{and the normalized initial conditions} \quad x_i(0) = 1 .$$

To avoid a degeneration of the hypercycle into a hypercycle of lower order we have to demand that the sequence of coefficients

$$K_0, K_1, \dots, K_{r-1}$$

does not contain a subperiod.

In comparison with the linear Taylor chain, an easily surveyed analytical solution by an explicit formula is, in the case of the exponential chain impossible.

By backward integration top-down with taking into account the condition $x_r \equiv x_0$ and the initial conditions $x_i(0) = 1$ we get recursively

$$\begin{aligned}
x_{r-1} &= F^{-1}(K_{r-1}x_0) ,\\
x_{r-2} &= F^{-1}\big(K_{r-2}F^{-1}(K_{r-1}x_0)\big) ,\\
&\dots\dots\dots\dots\dots\dots\dots\dots\\
x_0 &= F^{-1}\big(K_0F^{-1}(K_1F^{-1}(\dots F^{-1}(K_{r-1}x_0)\dots))\big) .
\end{aligned}$$

From this implicit expression we cannot get an analytically closed formula for $x_0$, but can determine $x_0(t)$ with only the help of computational simulation for a given set of coefficients, $K_i$. Only for the hypercycles of the lowest orders $r = 1, 2$ an analytical closed solution for $x_0$ can be given. Thus, for example, we get for the hypercycle of order $r = 1$ the condition

$$dx_0/dt = K_0 x_0^2$$

with the explicit solution

$$x_0(t) = 1/(1 - K_0 t) .$$

Therefore the hypercycle of order $r = 1$ shows hyperbolic growth, a behaviour what we could observe in most time series of growth indicators from different fields.

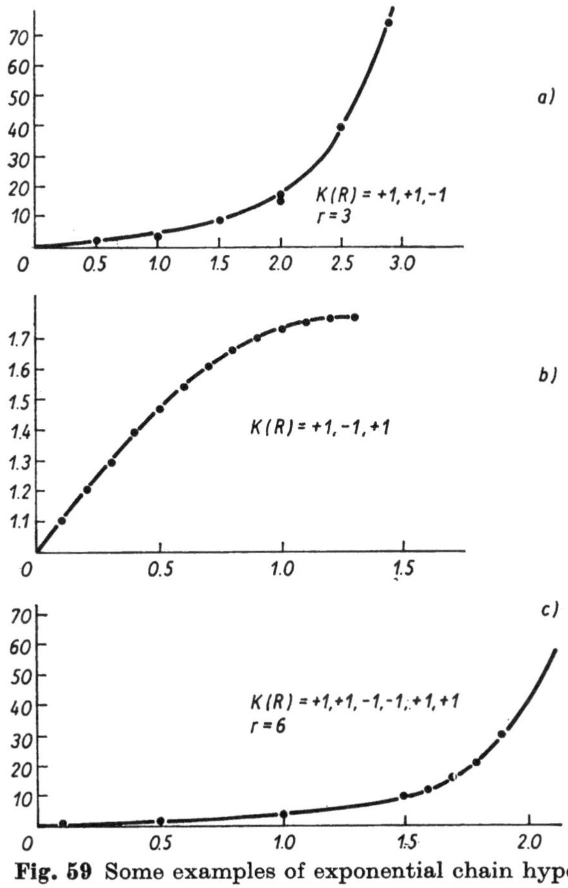


**Fig. 59** Some examples of exponential chain hypercycles

The hypercycle of order $r = 1$ describes the special hyperbolic growth function with the hyperbola exponent 1. However, we should mention, that these practical observations do not prove that we are, in reality, confronted with the structure of a hypercycle because we know that a homogenous chain with $K_i = K$ and sufficiently large length exposes a behaviour which is a very good approximation of the hypercycle behaviour.

For the hypercycle of order $r = 2$, we get the following equations

$$dx_0/dt = K_0 x_0 x_1 ,$$

$$dx_1/dt = K_1 x_1 x_0 .$$

They resemble a degenerated Lotka-Volterra predator/prey model, which is determined by the following equations

$$dx_0/dt = a_0x_0 - g_0x_0x_1\,, \qquad x_0 \text{ concentration of the prey,}$$

$$dx_1/dt = g_1x_0x_1 - a_1x_1\,, \qquad x_1 \text{ concentration of predators.}$$

Behind the Lotka-Volterra predator/prey-model is that the prey increase according to a fertility rate, $a_0$, and will be fed on by the predators according to the interaction coefficient, $g_0$, the predators are fed by the prey according to the interaction coefficient, $g_1$, and possess a death rate, $a_1$. For the hypercycle of order $r = 2$ we observe the degeneration that the prey is no longer increasing, but will only be eaten (if $K_0 < 0$). The predators eat as long as the prey exist (if $K_1 > 0$), but they will not die. It is obvious that under those conditions the prey will extinct, leaving at the end a constant predator concentration.

The case of the hypercycle of order $r = 2$ allows for an explicit analytical solution. We recognize at one the existence of a first integral. Because of

$$\frac{dx_0/dt}{K_0} = \frac{dx_1/dt}{K_1}$$

and the initial value condition $x_i(0) = 1$, we have

$$(x_0 - 1)/K_0 = (x_1 - 1)/K_1\,.$$

Therefore, the motion of the system is bound, as a stright line, with the equation given above.

After substitution of the first integral into the first of the two equations, we get

$$dx_0/dt = x_0\big(K_1x_0 + (K_0 - K_1)\big)\,.$$

This equation can be easily integrated after expanding into partial fractions, with the result

$$x_0(t) = (K_1 - K_0)/(K_1 - K_0\,e^{(K_1-K_0)t})\,,$$

$$x_1(t) = (K_0 - K_1)/(K_0 - K_1\,e^{(K_0-K_1)t})\,.$$

Fig. 60 shows the types of growth functions for $x_0(t)$ depending upon the corresponding parameter constellation ($K_0$, $K_1$). We find here the special cases of the logistic growth and also simple hyperbolic growth, the latter being in a state of degradation.

For the hypercycle of order $r = 3$, we get the following equations

$$dx_0/dt = K_0x_0x_1, \qquad dx_1/dt = K_1x_1x_2, \qquad dx_2/dt = K_2x_2x_0\,.$$

Already for this case, no explicit analytical solution exists.

Of high interest also is the behaviour of a chain which ends after $r$ steps by $x_r \equiv 1$. In this case, we get for $x_0(t)$ the following expression, which can be evaluated with the help of the computer by a sequence of quadratures

$$x_0(t) = F^{-1}\big(K_0F^{-1}(K_1F^{-1}(\ldots\, F^{-1}(K_{r-1}1(t))\,\ldots))\big)\,.$$

Let us consider the homogeneous case where we have $K_i = K$ for all $i$. This exponential tower converges for $r \to \infty$ uniformly in any closed interval $[0, t]$ with $t < 1/K$ for $K > 0$, and everywhere for $K < 0$ against the hypercycle solution for $r = 1$, meaning that it converges against the following hyperbola

$$x_0(t) = 1/(1 - Kt)\,.$$

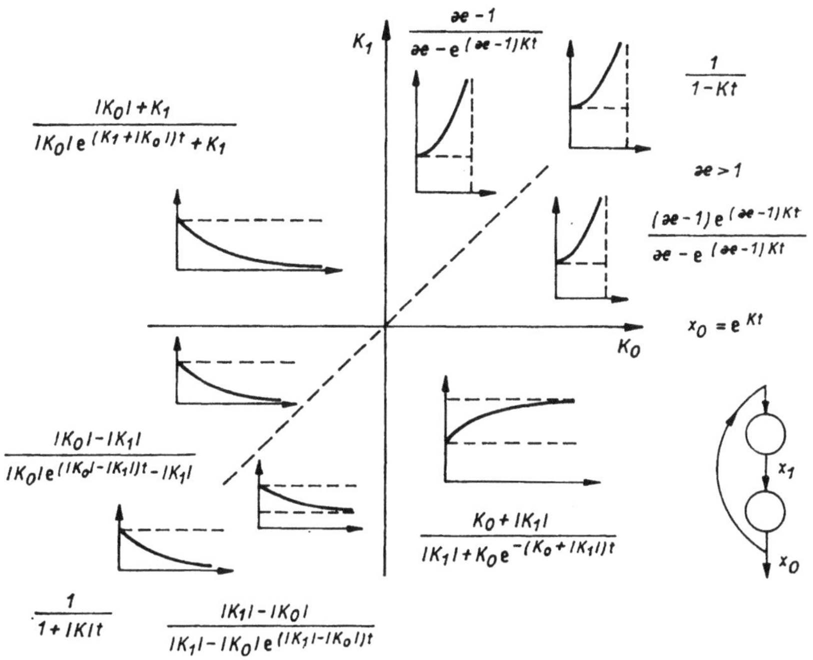


**Fig. 60** Representation of the behavioural types of a hypercycle of order $r = 2$ in the parameter plane

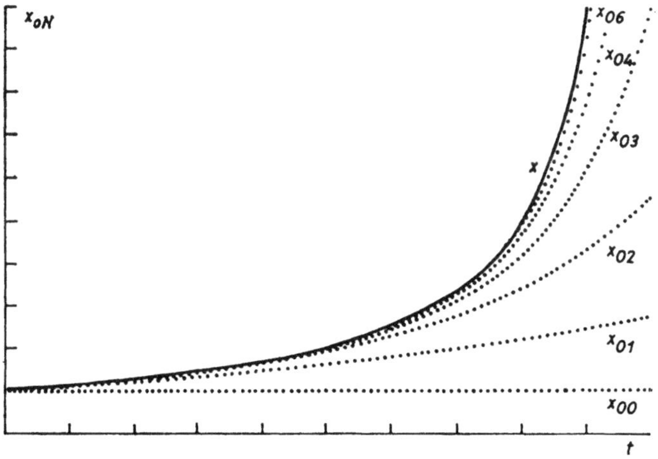


**Fig. 61** Convergence of an exponential tower against the hyperbola $x = 1/(1 - t)$

Fig. 61 shows some computer simulations for this case, in order to give a rough impression of the quality of this convergence.

We now give more information about the convergence in this case. Further details can be found in Section 8.8.

Perhaps it may be that $x_{0N}(t)$ is the bottom signal of an exponential tower which has been cut at the level $N$ by putting $x_{N+1} \equiv 1$. From

$$x_{0N} = F^{-1}(Kx_{0\,N-1})$$

and the inequality

$$\exp\left(\sum_{i=0}^{N} K^{i+1}t^{i+1}/(i+1)\right) \geqq \sum_{i=0}^{N} K^{i+1}t^{i+1} + 1$$

we get the following inclusion property, from which the convergence and an estimation of the rest of the expansion follows

$$\sum_{i=0}^{N} K^i t^i \leqq x_{0N}(t) < 1/(1 - Kt) .$$

The expansion into an exponential tower represents quite a new expansion of signals (with the structure coefficients $K_i$) for the case where for $N \to \infty$ (infinitely increasing length of the chain) the bottom signal $x_{0N}(t)$ converges against a signal $x(t)$. This expansion is especially suitable for strongly monotonous functions which possess a logarithmic derivative of any order. Functions which can be generated by an exponential tower in this manner we will call *exponentially analytical functions.*

It is remarkable that usually a very small number of levels of the approximating exponential tower is already sufficient to make it difficult to distinguish the behaviour of the hypercycle of order $r = 1$ from the corresponding finite exponential tower. The hypercycle solution with $r = 1$ is obviously the eigen-solution of the operator $F = \mathrm{d}\ln/\mathrm{d}t$ with the eigen-value $K_0$, for it holds

$$Fx_0 = K_0 x_0 .$$

Therefore, we got the following property, which we will call the *ecological law of large numbers*: The hypercycle solution for $r = 1$ coincides with the solution $x_0$ at the bottom of an infinitely long chain.

In practice, chains of very restricted length can mimic the hypercycle behaviour; it is not necessary that we have a hypercycle structure in reality.

In this context, it is interesting to study the behaviour of periodic chain structures with the basic period $r$ and the coefficients

$$K_0, K_1, \dots, K_{r-1} .$$

In analogy to the Taylor chains, we now denote with $P_r$ the following operator

$$P_r = P_r(K_0, K_1, \dots, K_{r-1}) = F^{-1}\big(K_0 F^{-1}(K_1 F^{-1}(\dots F^{-1}(K_{r-1}(\ )) \dots))\big)$$

and with $P_r(t)$ the result of the application of $P_r$ on the Heaviside's step function $1(t)$.

$$P_r(t) = P_r 1(t) .$$

If we repeat the basic chain with the period $K_0, K_1, \dots, K_{r-1}$ $m$ times and then terminate this tower with the top signal $1(t)$, we get the following sequence of expressions

$$\begin{aligned} x_{0,m} &= P_r 1(t) = P_r(t) , \\ x_{0,m-1} &= P_r\big(P_r(t)\big) = P_r^2 1(t) , \\ &\dots\dots\dots\dots \\ x_{0,0} &= x_0 = P_r^{m+1} 1(t) . \end{aligned}$$

In the case that convergence holds for $m \to \infty$, this construction obviously leads to a solution of the hypercycle equation of order $r$ meaning that it leads to a solution of the nonlinear iteration equation

$$x_0(t) = P_r\big(x_0(t)\big) .$$

However, it is not clear under which conditions we get this convergence. Further information about this problem is given later.

If we have convergence and the velocity of that convergence is sufficiently large, then, from practical observations, we cannot easily differentiate between a real hypercycle solution and a finite repetition of such a periodical chain structure. Therefore, from observations alone, it should be very difficult to prove that the corresponding signal $x_0(t)$ is really generated by a hypercycle structure.

## 2.5. *Imbedding Chain Structures into a Unified Theory of Linear Systems*

Early in the 1970's we developed a unified theory of linear systems and used it for the computation of control systems [25]. Here, we will present some basic considerations of this theory. This is, on one hand, a generalisation of the Taylor series chains described above, and, on the other hand, this theory gives a certain background for the approach to nonlinear chain structures as we analysed above for the case of the exponential chains.

A basic notion for the unified theory of linear systems is the notion of a *Taylor operator*:

$F$ is a linear Taylor operator, if the homogenous equation

$$Fx = 0$$

has nontrivial solutions $x_0 \not\equiv 0$, and if there exists, corresponding to the given linear operator $F$, another linear operator $G$, with the help of which the different solutions of the homogenous equation can be distinguished from each other uniquely, meaning that the following linear system of equations has a unique solution

$$Fx = 0\,, \qquad Gx = z$$

for a given (admissable) value $z$ which is taken from a suitable linear space (usually something like a state-space). $F$ has the order $n$, if it is possible to compose the uniqueness operator $G$ for $F$ from exactly $n$ linearly independent linear functionals.

For example, $F = p = \mathrm{d}/\mathrm{d}t$ is such a linear Taylor operator of order 1. Here we can choose $G$ as that linear functional which takes out of $x(t)$ the value $x(0)$ (fixation of the initial value).

In multivariable linear systems, we meet the operator

$$F = \begin{pmatrix} p & 0 & \dots & 0 \\ 0 & p & \dots & 0 \\ \cdot & \cdot & \cdot & \cdot \\ 0 & 0 & \dots & p \end{pmatrix}$$

which is a diagonal matrix with the diagonal element $F = p = \mathrm{d}/\mathrm{d}t$. As uniqueness operator in this case can serve any matrix of $n$ linear functionals which assign to the state-vector $x(t)$ component-wise the corresponding initial values $x(0)$. For the aims of this book we will have to do preferably with Taylor operators of order $n = 1$.

For a given value of $z$ we can construct for the Taylor operator $F$ a pseudoinverse operator $F_z^{-1}$ by solving the equation

$$Fx = y$$

with the condition $Gx = z$ (assuming of course that the space of feasible condition values $Z$ can be chosen independently of $y$)

$$x = F_z^{-1},$$

then obeying the following conditions

$$FF_z^{-1}y = y \quad \text{and} \quad GF_z^{-1}y = z\,.$$

For the case of linear systems, it is often useful to work with the condition $z = 0$. In this case, we omit the index $z$ and write simply

$$F_0^{-1} = F^{-1}\,.$$

Important for the unified theory of linear systems is the notion of an *interpolation sequence*. We begin with a nontrivial solution of the homogeneous equation

$$Fx_0 = 0 \quad \text{and} \quad x_0 \neq 0$$

and construct recursively a sequence of signals $x_i$ with the help of the following equation

$$Fx_{i+1} = x_i \quad \text{for} \quad i = 0, 1, \ldots$$

always using the condition $Gx_i = 0$ for $i = 1, 2, \ldots$. Then we get obviously

$$x_{i+1} = F^{-1}x_i = F^{-(i+1)}x_0\,.$$

This construction can apparently be continued as long as the recursion equation

$$Fx_{i+1} = x_i$$

has a solution.

Obviously, for the elements of the interpolation sequence thus constructed the following conditions hold

$$F^l x_i = \begin{cases} x_{i-l} & \text{for} \quad l \leqq i\,, \\ 0 & \text{for} \quad l > i\,, \end{cases}$$

$$GF^l x_i = \begin{cases} Gx_0 & \text{for} \quad l = i\,. \\ 0 & \text{for} \quad l \neq i\,. \end{cases}$$

It can be shown easily [24], that all signals $x_i$ constructed in this way are linearly independent of each other. Therefore, it seems to be useful, to decompose given signal time-functions $x(t)$ into a superposition of these signals $x_i$, taking them as coordinate functions. Therefore, series expansions of the following form are considered

$$x(t) = \sum a_i x_i(t)\,.$$

It is obvious that all these series are generalisations of the Taylor series for their coefficients can be computed by similar expressions

$$a_i = GF^i x(t)/Gx_0\,.$$

If we stop the expansion of such a series at the level $N$, we can write the corresponding rest of such a series with the formula

$$R_N = F^{-(N+1)}F^{(N+1)}x(t)\,.$$

From this expression, the similarity to the Taylor chains discussed above can be seen.

For the linear case, we can show a close connection to the Taylor chains with the help of the unified theory of linear systems. For this purpose, we must invert the direction in the enumeration of the indices. Chain structures relying consequently on the unified theory of linear systems shall be called linear chain structures.

Now, we repeat the considerations which led us to the Taylor chains under the conditions of the unified theory of linear systems. For constructing linear chain structures, we now choose as a basemodule a linear Taylor operator $F$. We consider first chains of finite length which are cut at the level $N$ with the signal $x_N$. At this level, we feed in a signal $x_{N+1}$, being a nontrivial solution of the homogenous equation

$$Fx_{N+1} = 0 .$$

Every level of the corresponding chain shall hold an equation of the form

$$Fx_i = K_i x_{i+1} \quad \text{for} \quad i = 0, 1, \dots, N$$

(assuming, of course, that, in dependence on the concrete chosen base operator $F$, the corresponding chain can be continued up to the level $N$). The chain construction will be performed under the normalisation conditions

$$Gx_i = 0 \quad \text{for} \quad i = 0, 1, \dots, N .$$

We are then obviously dealing with an interpolation sequence $x_i$ of the Taylor operator $F$ enumerated with opposite running direction of the indices.

For such a linear chain we will now repeat all considerations which we formerly presented for the Taylor chains. The behaviour of a hypercycle of order $r$ for a linear chain is characterized by the following set of equations:

$$Fx_i = K_j x_{i+1} \quad \text{for} \quad i = 0, 1, 2, \dots, r-1,$$

$$x_r \equiv x_0 \quad \text{and} \quad Gx_0 \neq 0$$

(If we allowed $Gx_0 = 0$ this would lead to the trivial solution $x_0 \equiv 0$ for the whole hypercycle). We now apply backward integration top-down and admit for in general different pseudoinverse operators, and we get the following solution

$$x_{r-1} = K_{r-1} F_1^{-1} x_0 ,$$

$$x_{r-2} = K_{r-2} K_{r-1} F_2^{-1} F_1^{-1} x_0 ,$$

$$\dots\dots\dots\dots\dots\dots$$

$$x_0 = K_0 K_1 \dots K_{r-1} F_r^{-1} F_{r-1}^{-1} \dots F_1^{-1} x_0 .$$

The lower index $j$ for $F_j^{-1}$ designates the possible different choice of the uniqueness operations, meaning, that in this case, we can violate the conditions $Gx_i = 0$. Especially we could admit

$$F_j^{-1} x = L_{r-j} + F^{-1} x .$$

In such a case, $F_j^{-1}$ would be a pseudoinverse (but nonlinear) operator for $F$, if for every constant $K$ holds $FK = 0$. Thus, we could introduce on every level of the chain an additive inhomogenity by the conditions (which allows to work with different initial values for all state-variables of the chain).

This brings us into the neighbourhood of the formerly considered Taylor chains where we had such an additive inhomogenity on every level. For the case of a homogeneous chain, meaning that we apply only one uniqueness operator, we would get

$$x_0 = KF^{-r}x_0 \quad \text{with} \quad K = K_0K_1 \dots K_{r-1}\,.$$

Because of our normalisation, $x_0$ is at the same time a solution of

$$F^r x_0 = Kx_0\,.$$

This is obviously an eigen-solution equation for the operator $F^r$ with the eigen-value K.

To characterize the full ambiguity of possible solutions, we consider the case of inhomogenous uniqueness operators defined by

$$F_j^{-1} = L_{r-j} + F^{-1} \quad \text{with} \quad GF^{-1}x = 0\,.$$

In this general case, we can determine the solution with the help of the following assumed superposition [24]:

$$x_0 = F^{-r}u + \sum a_i\xi_i\,.$$

Here $\xi_i$ is an interpolation sequence for $F$ corresponding to the changed direction of indexing.

$$F\xi_r = 0 \quad \text{with} \quad G\xi_r \neq 0$$

and

$$\xi_i = F^{-1}\xi_{i+1} \quad \text{and} \quad G\xi_i = 0 \quad \text{for} \quad i = r-1, \dots, 0\,.$$

If we introduce this superposition into the hypercycle equation

$$F^r x_0 = Kx_0$$

we get (slightly more general case than in the case of the Taylor chain)

$$u = K\big(F^{-r}u + P_r(t)\big) \quad \text{with} \quad P_r(t) = \sum a_i\xi_i$$

with the formal solution

$$u = \frac{1}{(1 - KF^{-r})}KP_r(t)\,.$$

Frequently, the operator $1/(1 - KF^{-r})$ can be evaluated with the help of the corresponding Neumann series

$$\frac{1}{(1 - KF^{-r})} = \sum_{i=0}^{\infty} K^iF^{-ir}\,.$$

For $x_0$, we then get the following solution

$$x_0 = \frac{1}{(1 - KF^{-r})}P_r(t)\,.$$

It is no problem to transform the $a_j$ into the parameters $L_j$ and vice versa. Especially the hypercycle of order $r = 1$ has the solution

$$x_0 = \frac{1}{(1 - KF^{-1})}a_0\xi_1\,.$$

Let us now consider the behaviour of an inhomogenous chain (in respect for the $L_j$) which is cut off after $r$ steps, meaning that we assume

$$F^r x_0 = 0\,.$$

According to the considerations above, the corresponding general solution of this equation is

$$x_0 = \sum_{i=0}^{r-1} a_i \xi_i .$$

For a homogenous chain in respect of the $L_j$ (all $L_j = L$ and therefore all $a_j = a$) we get

$$x_0 = a(F^{-(r-1)} + F^{-(r-2)} + \dots + F^{-1} + 1)\,\xi_r$$

with $F\xi_r = 0$ but $\xi_r \not\equiv 0$.

In the case of our meeting convergence, this solution with increasing $r$ obviously approaches the hypercycle solution with $r = 1$. In this connection, the behaviour of periodic chain structures in which the initial value distributions, meaning the parameters $L_j$, iterate periodically is also of interest. This tower will be cut after a finite number of repetitions of the basic period and fed at the top level with the zero signal. For this case we must discuss the corresponding input/output relation of the form

$$F^r x = Ky .$$

Again, we decompose the general solution in the following form

$$x = F^{-r}u + P_r(t) \quad \text{with} \quad P_r(t) = \sum_{i=0}^{r-1} a_i \xi_i$$

which leads to the solution

$$x = KF^{-r}y + P_r(t) .$$

Now we descend through the $m$ repetitions of the $r$-periodical tower top-down, starting with the top input signal $y = 0$ and get the following sequence of expressions

$$x_{0,m} = P_r(t) ,$$

$$x_{0,m-1} = KF^{-r}P_r + P_r ,$$

$$x_{0,m-2} = (KF^{-r})^2 P_r + (KF^{-r}) P_r + P_r ,$$

. . . . . . . . . . . . . . . . . . . . . . . . . . . . . . .

$$x_0 = x_{0,0} = \left((KF^{-r})^{(m-1)} + (KF^{-r})^{(m-2)} + \dots + 1\right) P_r .$$

In the case of convergence for $m \to \infty$, we obviously get the solution

$$x_0 = \frac{1}{(1 - KF^{-r})} P_r$$

meaning the solution of a hypercycle of order $r$. Therefore, in this case also, a periodical chain with a finite but sufficiently large number of repetitions shows a behaviour which we cannot differentiate from the behaviour of a hypercycle of order $r$.

## 2.6. *Chains with a General Continuous Base Module and the Principle of Generalised Superposition*

Connecting the unified theory of linear systems with our chain approach, we arrived at a generalisation of the Taylor chains for which the essential hypercycle properties are preserved. Now we proceed to a generalised nonlinear case. This will be accompanied by a generalisation of the exponential chains. With the introduction of an operation of linear superposition already considered by Kolmogorov in connection with mean value operators [70], we get a similarity to linear system theory.

We introduce a base module by

$$\mathrm{d}\Phi(x)/\mathrm{d}t = Ky\,.$$

Here $\Phi(x)$ is an arbitrary differentiable and strictly monotonously increasing function. Our aim is to construct chain structures with such a base module. In general, we admit that, in a chain, modules with different $\Phi$ functions occur. Usually we shall deal with the case where we use only one $\Phi$ module and inhomogenities are met because of different initial values of the state-variables in the chain.

If we ascend such a chain bottom-up, we apply the given module equation in such a way that we generate from the output signal $x$ of the module (its statevariable) the next higher control signal $y$. For passing the chain from bottom to top, we need the solution of the module equation after the unknown variable $x$. For this solution we get

$$x(t) = \Phi^{-1}\big(\Phi x(0) + \Phi(\xi)\big)$$

with the variable

$$\xi = \Phi^{-1}(KIy)\,, \qquad I = \int_0^t .$$

Obviously $\xi$ is that solution of the module equation which obeys the normalisation condition

$$\Phi\xi(0) = 0\,.$$

$\Phi^{-1}$ is the unique inverse function for the chosen $\Phi$ function. This solution formula has the form of a generalised superposition law in comparison with the linear superposition.

If we have agreed upon a fixed $\Phi$ function, we usually write the superposition in the following form

$$x(t) = x(0) \oplus \xi\,.$$

If a specification of the corresponding $\Phi$ function is necessary, we mark this by a suffix

$$x(t) = x(0) \underset{\Phi}{\oplus} \xi\,.$$

Together with the generalised addition, we can introduce also a generalised multiplication by the definition

$$c = a \otimes b = \Phi^{-1}\big(\Phi(a) \times \Phi(b)\big)\,.$$

Here $\times$ denotes the usual multiplication operation. Both operations — generalised addition and multiplication — are associative, commutative and distributive in the usual sense.

An addition equation

$$a \oplus \xi = b$$

can be solved uniquely after the unknown value $\xi$ in the form

$$\xi = b \ominus a = \Phi^{-1}\big(\Phi(b) - \Phi(a)\big)\,.$$

A multiplication equation

$$a \otimes \xi = b$$

can be solved uniquely after the unknown value $\xi$ in the form

$$\xi = b \odot a = \Phi^{-1}(\Phi(b) : \Phi(a)) .$$

This is possible under the assumption that $\Phi(a) \neq 0$, and, therefore the role of zero is taken over here by the value $0^*$ with $\Phi(0^*) = 0$. The role of the unit element in multiplication is taken over here by a number $e$ with $\Phi(e) = 1$.

Before we consider chains with a $\Phi$ module as basic element, let us consider some examples for generalised additions and multiplications.

a) Let $\Phi(u) = \ln u$.
In this case we have $a \oplus b = a \times b$, but

$$a \otimes b = a^{\ln b} = b^{\ln a} .$$

b) Let $\Phi(u) = \mathrm{e}^u$.
Then we have $a \oplus b = \ln (\mathrm{e}^a + \mathrm{e}^b)$, but

$$a \otimes b = a + b .$$

c) Let $\Phi = \left((\varkappa + u)/(\varkappa - u)\right)^c$ with $c > 0$.
Then we have $a \otimes b = \dfrac{a + b}{1 - ab/\varkappa^2}$.

d) Let $\Phi(u) = c \ln \left((\varkappa + u)/(\varkappa - u)\right)$.
Then we have $a \oplus b = \dfrac{a + b}{1 - ab/\varkappa^2}$.

For a chosen $\Phi$ function, we now construct chains (homogenous in respect to $\Phi$) using the base-module defined above

$$\mathrm{d}\Phi(x_i)/\mathrm{d}t = K_i x_{i+1} \quad \text{for} \quad i = 0, 1, \dots, N$$

with constant input on the highest level

$$x_{N+1} = \text{const.} = x_{N+1}(0) .$$

For a complete and unique parametrisation of the chain, it is suitable to pass to normalised state-variable $\xi_i$ fulfilling the following normalisation condition

$$\Phi(\xi_i(0)) = 0 .$$

To get the description of the chain by the normalised state-variables $\xi_i$, we must substitute into the chain equation the decomposition

$$x_i(t) = \Phi^{-1}\left(\Phi(x_i(0)) + \Phi(\xi_i)\right)$$

and get the following result

$$\mathrm{d}\Phi(\xi_i)/\mathrm{d}t = K_i(L_{i+1} \oplus \xi_{i+1}) \quad \text{with} \quad L_{i+1} = \Phi(x_i(0)) .$$

After this normalisation every module in the chain now possesses two parameters $K_i$ and $L_{i+1}$.

For the corresponding explicit resolution of these equations for the case of running the chain top-down, we get

$$\xi_i = \Phi^{-1}\left(K_i I(L_{i+1} \oplus \xi_{i+1})\right) .$$

To see the relation to the chains discussed previously we now consider the following special cases.

a) Taylor chain
Here we have $\Phi(u) = u$.
In this case, the following holds

$$a \oplus b = a + b \quad \text{and} \quad a \otimes b = a \times b .$$

The normalisation condition is here

$$\xi_i(0) = 0$$

and we get the following chain equations

$$\mathrm{d}\xi_i/\mathrm{d}t = K_i L_{i+1} + K_i \xi_{i+1}$$

which represent a weak generalisation of the Taylor chains completely covered by the chain constructions within the unified theory of linear systems.
b) Exponential chain
Here we have $\Phi(u) = \ln u$.
In this case, we have

$$a \oplus b = a \times b \quad \text{and} \quad a \otimes b = a^{\ln b} = b^{\ln a} .$$

Here, we have the following normalisation condition

$$\ln\big(\xi_i(0)\big) = 0 , \quad \text{that means} \quad \xi_j(0) = 1,$$

and we get the following chain equations

$$\mathrm{d}\xi_i/\mathrm{d}t = K_i L_{i+1} \xi_i \xi_{i+1} .$$

Here the two parameters fuse with each other, forming only one module parameter, and we arrive exactly at the concept of exponential chains.

An interesting generalisation of both concepts a) and b) is given by the so-called power chains.
c) Power chains
Here we introduce $\Phi(u) = u^{\varkappa+1}/(\varkappa + 1)$ with $\varkappa \geqq -1$.
In this case we have

$$a \oplus b = (a^{\varkappa+1} + b^{\varkappa+1})^{1/(\varkappa+1)} .$$

For the normalisation condition we get now

$$\Phi\big(\xi_i(0)\big) = \xi_i(0)^{\varkappa+1}/(\varkappa + 1) = 0 .$$

We must distinguish two cases:
$\varkappa = -1$. Here after applying *L'* Hospital's rule, we get the result

$$\Phi(u) = \ln u \quad \text{and the normalisation condition } \xi_i(0) = 1.$$

$\varkappa > -1$. In this case the chain equations are

$$\xi_i^\varkappa \, \mathrm{d}\xi_i/\mathrm{d}t = K_j(L_{i+1}^{\varkappa+1} + \xi_{i+1}^{\varkappa+1}{}^{1/(\varkappa+1)})$$

and for the normalisation condition we get always $\xi_j(0) = 0$.

This already shows the special role of logarithmic derivative as probably the only operation with which it is possible to reduce multiplicative signal connections into additive signal connections in the case of continuous variables.

We will now repeat the hypercycle considerations also for the case of homogenous chains based on a unique $\Phi$ module. Obviously, the behaviour of a hypercycle of order $r$

from a $\Phi$-chain is given by the following equations

$$\mathrm{d}\Phi(\xi_i)/\mathrm{d}t = K_i(L_{i+1} \oplus \xi_{i+1}) , \quad \text{for} \quad i = 0, 1, \dots , r-1 ,$$
$$\xi_r \equiv \xi_0 \quad \text{and} \quad \Phi(\xi_i(0)) = 0 .$$

To avoid a degeneration into a hypercycle of lower order, we must assume that the coefficient vectors

$$K_0, K_1, \dots , K_{r-1} \quad \text{and} \quad L_1, L_2, \dots , L_r$$

have no synchronous subperiods.

For the determination of the hypercycle motions we again perform a backward integration moving through the chain top-down to get at least an implicit formula for the determination of the corresponding solution $\xi_0$. The result of this is given by the following recursion equations

$$\xi_i(t) = \Phi^{-1}(K_i I(L_{i+1} \oplus \xi_{i+1}))$$

for $i = r-1, r-2, \dots , 0$ and $\xi_i \equiv \xi_0$.

Unfortunately, these formulas can only be evaluated by a computer-based simulation. An explicit analytical solution exists only in the linear case. Nevertheless, it makes sense to perform all considerations as we did formerly for the special cases, omitting for the moment the necessary proofs of convergence. Therefore, let us first consider the case of the hypercycle of order $r = 1$. For this hypercycle the following condition apparently holds

$$\mathrm{d}\Phi(\xi_0)/\mathrm{d}t = K_0\Phi^{-1}(\Phi(L_1) \dotplus \Phi(\xi_0)).$$

We substitute

$$u = \Phi(\xi_0) \quad \text{with} \quad u\,(0^*) = 0$$

and get for the determination of $u$ the following equation

$$\mathrm{d}u/\mathrm{d}t = K_0\Phi^{-1}(\Phi(L_1) + u) .$$

We put

$$\Psi(u) = \int_0^u \mathrm{d}v/\Phi^{-1}(\Phi(L_1) + v) .$$

With this expression, we get for the hypercycle of order $r = 1$ the following explicit solution

$$\xi_0(t) = \Phi^{-1}\Psi^{-1}(K_0 t) .$$

Of great interest is the behaviour of a homogenous chain (not only in respect to the function $\Phi$ but also in respect to the module parameters) which is cut after $r$ steps putting $\Phi(\xi_r) \equiv 0$.

Obviously, this behaviour is determined by the following recursion

$$\xi_i = \Phi^{-1}(KI(L \oplus \xi_{i+1})) \quad \text{for} \quad i = 0, 1, \dots , r-1 \quad \text{and} \quad \Phi(\xi_r) = 0 .$$

Apparently, this is an iteration sequence for the following dynamic nonlinear fixpoint equation

$$\xi_0 = \Phi^{-1}(KI(L \oplus \xi_0))$$

meaning for the hypercycle of order $r = 1$.

In the case of convergence of the recursion for $r \to \infty$ this shows that, with increasing length of such a homogenous chain, it will become more and more difficult to differentiate the chain's behaviour from that of a hypercycle of order $r = 1$. Again, in this context, it is of interest to study the behaviour of periodic chains and especially the behaviour of a finite number of repetitions of the basic period in the chain.

We consider a chain with periodic coefficient sequences

$$K_0, K_1, \dots, K_{r-1} \quad \text{and} \quad L_1, L_2, \dots, L_r .$$

By applying notations similar to those used formerly, we can formalize this problem in the following way:

At first we iterate completely the following sequence

$$\xi_i = \Phi^{-1}\big(K_i I(L_{i+1} \oplus \xi_{i+1})\big) \quad \text{for} \quad i = r-1, r-2, \dots, 0$$

starting with

$$\Phi(\xi_r) = 0 .$$

The result we denote in the following way

$$\xi_0 = P_r(t) = P_r(\xi_r) .$$

Here $P_r$ designates the overall operator of this sequence.

Now we tower $m$ of these sequences with the same sets of parameters one upon the other and introduce again on the top level the control signal with $\Phi(\xi) = 0$. Obviously, the behaviour of this quasi-periodic tower structure is determined by the following recursion

$$\begin{aligned} \xi_{0,m} &= P_r(t) , \\ \xi_{0,m-1} &= P_r(\xi_{0,m}) = P_r\big(P_r(t)\big) , \\ &\dots\dots\dots\dots\dots\dots \\ \xi_{0,0} = \xi_0 &= P_r^m\big(P_r(t)\big) . \end{aligned}$$

For consecutive successive levels, we have

$$\xi_{0,i} = P_r(\xi_{0,i+1}) .$$

The hypercycle solution of order $r$ is characterized by the following nonlinear dynamic fixpoint equation

$$\xi_0 = P_r(\xi_0) .$$

Obviously, the above introduced recursion formula is, for a given value of $m$, an iteration sequence for this fixed-point equation.

If the sequence of $\xi_{0,m}$ converges for $m \to \infty$ we again get the result that, with increasing $m$, it will be more and more difficult to distinguish the behaviour of a period chain from the behaviour of a hypercycle of order $r$. That means that, by observation alone of $\xi_0$, it cannot be decided whether in reality, we are dealing with a hypercycle structure or a quasi-periodic tower with a sufficiently large number $m$ of repetitions of the basic period.

## 2.7. *Difference Chain Concept as a Base for Local Models of Time Functions of Different Orders*

We can adjoin to a given time function $x = f(t)$ an enveloping local assemblage of linear models by local Taylor expansion of $f(t)$ and cutting the expansion with the linear term. The well known result is then

$$x_M = f(t_0) + \dot{f}(t_0)\,(t - t_0)$$

in the reference time point $t_0$.

For this approximation, the following equation describes the family of straight lines

$$f^0 + a(t - t_0)$$

which, for a freely varying parameter $a$, build a field in the plane in the sense that two different points in the plane define exactly one straight line connecting these points.

Now, we repeat this construction after applying a scalarization operation taking ln-value both for the abscissa and the ordinate. That means that we will approximate linearly the dependence of $\ln f(t)$ from $\ln t$ in a certain local neighbourhood of the reference point $t_0$. The corresponding result is given by

$$\ln x_M = \ln f(t_0) + a \ln (t/t_0)\,.$$

The coefficient $a$ can be computed by the following rule

$$a = \frac{\dot{f}(t_0)}{f(t_0)}\,t_0 = f^*(t_0)\,.$$

This expression can be understood as a pseudo-derivation.

Obviously, $a$ is a dimensionless number and, at the same time, a measure for the deviation of the motion $f(t)$ from a uniform motion. Therefore, we now get the following assemblage of local models

$$x_M = f(t_0)\,(t/t_0)^a\,.$$

This is a polar family because all curves of this set go through the points (0, 0) and ($\infty$, $\infty$), respectively (0, $\infty$) and ($\infty$, 0), depending on the sign of $a$. Related to the given time function $f(t)$, this assemblage of curves has the following properties:

$$\dot{x}_m(t_0) = \dot{f}(t_0)\,.$$

That means the local power-function model has a common tangent with the given function $f(t)$ in the reference point $t_0$.

If $f(t)$ is increasing and the tangent in $f(t_0)$ passes the ordinate axis in the positive part, then the local power-function model in $t_0$ has a negative curvature. However, if the ordinate axis is passed by the corresponding tangent in the negative part, the curvature of the local power-function model in $t_0$ will be positive. If $f(t)$ is decreasing in $t_0$ then the local power-function model will be a hyperbola.

The whole assembly of curves

$$f^0(t/t^0)^a$$

also has the field property, with exception of the corresponding polar point. In respect to the curvature, it can be said that the given time function $f(t)$ may be decomposed into parts over intervals respectively, depending on the property, if the signs of the curvatures of $f(t)$ and $x_M(t)$ are equal or opposite. If $x_M$ has a curvature in $t_0$ of the same sign as the curvature of $f(t)$, then $x_M(t)$ is in some aspect preferable in comparison with a linear model; if, on the contrary, the local power model and $f(t)$ have opposite signs of their curvatures, then we have to prefer a linear model in comparison with a local power-function model. This property initiates an approach to use multiobjective decision-making for the choice of the best local models from the power-function concept. Details about the possibilities are presented in Section 2.12.

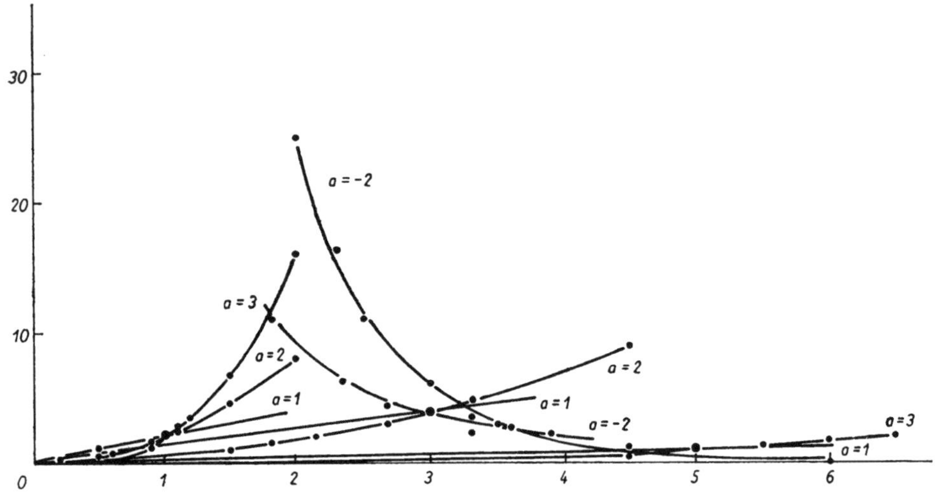


**Fig. 62** Some examples of power functions $f^0(t/t^0)^a$

Now we will imbed the simple result which we derived above into a general concept of local difference chains. For the purpose of approximation of a system's behaviour by local chain models, we change the general chain concept in Section 2.6 based on a scalarizing function $\Phi$ for continuous systems by replacing the derivative $d\Phi(x_i)/dt$ by a suitable difference quotient, meaning that we now construct chains of the following form

$$\frac{\Phi(x_i(t)) - \Phi(x_i(t_0))}{\Psi(t) - \Psi(t_0)} = \Phi^{-1}(\Phi(K_i) + \Phi(x_{i+1}(t))) .$$

Again, we introduce for these chains the normalisation condition

$$\Phi(x_i(t_0)) = 0 ,$$

but now only for $i = 1, 2, \ldots$ and not for $i = 0$, because we want

$$x_0(t_0) = f(t_0)$$

as a consistence condition for the local models. With the use of the normalisation condition, the chain parameters $K_i$ can be uniquely determined by

$$K_i = \frac{d\Phi(x_i(t_0))/dt}{d\Psi(t_0)/dt}$$

using in the limit the rule of de l'Hospital.

**Table 9.** Discrete Taylor Chain $D = 0.1$
$N = 6$ sin $t$
$K = 1, 0, -0.16666, 0, 0.0083333, 0, -0.0001984$

| $x(N)$ | $x(5)$ | $x(4)$ | $x(3)$ | $x(2)$ | $x(1)$ | $x(0)$ | $x$ true |
|---|---|---|---|---|---|---|---|
| $1.9 \cdot 10^{-5}$ | $-1.9 \cdot 10^{-6}$ | $8 \cdot 10^{-4}$ | $8 \cdot 10^{-5}$ | $-1.7 \cdot 10^{-2}$ | −0.0017 | 0.099 | 0.100 |
| $-4 \cdot 10^{-5}$ | $-7.9 \cdot 10^{-6}$ | $1.7 \cdot 10^{-3}$ | $3.3 \cdot 10^{-4}$ | 3.3 | 0.0067 | 0.199 | 0.199 |
| $6 \cdot 10^{-5}$ | $-1.8 \cdot 10^{-5}$ | 2.5 | 7.5 | 5 | 0.015 | 0.296 | 0.296 |
| $8 \cdot 10^{-5}$ | 3.2 | 3.3 | $1.3 \cdot 10^{-3}$ | 6.6 | 0.026 | 0.389 | 0.389 |
| $10^{-4}$ | 5 | 4.1 | 2.1 | 8.2 | 0.041 | 0.479 | 0.479 |
| $1.2 \cdot 10^{-4}$ | 7 | 5 | 3 | 9.8 | 0.059 | 0.564 | 0.565 |
| $1.4 \cdot 10^{-4}$ | 9.7 | 5.8 | 4 | $-10^{-1} \cdot 1.1$ | 0.08 | 0.644 | 0.644 |
| $1.6 \cdot 10^{-4}$ | $1.3 \cdot 10^{-4}$ | 6.6 | 5.3 | 1.3 | 0.10 | 0.717 | 0.717 |
| $1.8 \cdot 10^{-4}$ | 1.6 | 7.4 | 6.6 | 1.4 | 0.13 | 0.783 | 0.783 |
| $2 \cdot 10^{-4}$ | 2 | 8.1 | 8.1 | 1.6 | 0.16 | 0.841 | 0.841 |
| $2.2 \cdot 10^{-4}$ | 2.4 | 8.9 | 9.8 | 1.7 | 0.19 | 0.891 | 0.891 |
| 2.4 | 2.9 | 9.7 | $1.2 \cdot 10^{-2}$ | 1.9 | 0.22 | 0.932 | 0.932 |
| 2.8 | 3.4 | $10^{-2}$ | 1.4 | 2 | 0.26 | 0.964 | 0.964 |
| 2.8 | 3.9 | 1.1 | 1.6 | 2.2 | 0.30 | 0.985 | 0.985 |
| 3 | 4.5 | 1.2 | 1.8 | 2.3 | 0.34 | 0.997 | 0.997 |
| 3.2 | 5.1 | 1.3 | 2 | 2.5 | 0.38 | 0.999 | 1.000 |
| 3.4 | 5.7 | 1.3 | 2.2 | 2.6 | 0.42 | 0.991 | 0.992 |
| 3.6 | 6.4 | 1.4 | 2.5 | 2.6 | 0.46 | 0.973 | 0.974 |
| 3.8 | 7.2 | 1.4 | 2.7 | $-2.7 \cdot 10^{-1}$ | 0.50 | 0.945 | 0.946 |
| $4 \cdot 10^{-4}$ | $-7.9 \cdot 10^{-4}$ | $1.5 \cdot 10^{-2}$ | $3 \cdot 10^{-2}$ | $-2.7 \cdot 10^{-1}$ | −0.55 | 0.908 | 0.909 |

A local model of order $r$ is constructed if we put $\Phi(x_r) \equiv 0$ on the top level of the difference chain related to the reference point $t_0$ by the condition $x_0(t_0) = f(t_0)$. With known optimal coefficients $K_i$ of such a higher order, local models are explicit but long formulas can be given for the corresponding models. The problem is how to get a closed formula for the determination of the coefficients $K_i$ using, for this purpose only, the expression for $f(t)$, or in the case of an estimation, knowing only observations of $f(t)$. Unfortunately, such a closed formula for the general case cannot be presented. We give only some more information for those cases on which our interests are especially concentrated.

1) $\Phi(u) = u$ and $\Psi(u) = u$.
This is the discrete analogy for the Taylor chains constructed above. We get the following difference chain

$$\frac{x_i(t) - x_i(t_0)}{t - t_0} = K_i + x_{i+1}(t)$$

with the normalisation condition $x_i(t_0) = 0$ for $i = 1, 2, \ldots$. For the parameters $K_i$ we get the formula

$$K_i = \dot{x}_i(t_0)\,.$$

Unfortunately, this formula is not explicit because of the chain recursion. We shall try to derive in this case an explicit formula for the determination of $K_i$. For this purpose, we introduce the following functions[1])

$$L_i(t) = \dot{x}_i(t)\,.$$

From the chain recursion, we get for the function $L_i(t)$ the following parameter-free recursion

$$L_{i+1}(t) = \frac{\mathrm{d}}{\mathrm{d}t}\left(\frac{\int\limits_{t_0}^{t} L_i(u)\,\mathrm{d}u}{t - t_0}\right).$$

We need as parameters for the chain the values $L_i(t_0)$ explicitly expressed with the help of the bottom function $x_0(t)$ in the neighbourhood of $t_0$. For the determination of these parameters, we can assume that, in the neighbourhood of $t_0$, $x_0(t)$ is described by a Taylor polynomial of sufficiently high order

$$x_0(t) \sim \sum_{k=0}^{N} x_0^{(k)}(t_0)\,(t - t_0)^k/k!\,.$$

Then we have

$$L_0(t) \sim \sum_{k=0}^{N-1} x_0^{(k+1)}(t_0)\,(t - t_0)^k/k!\,.$$

We consider the linear operator

$$O = \frac{\mathrm{d}}{\mathrm{d}t}\,\frac{\int\limits_{t_0}^{t}}{(t - t_0)}\,.$$

[1]) We thank Dr. Spieß, Forschungsbereich Mathematik und Informatik der Akademie der Wissenschaften der DDR, for performing some computational studies for the difference chain concept.

If we apply O an a certain polynomial, it obviously decreases the polynomial order by one degree. Now we can construct iteratively the $L_i(t)$ and get the following sequence of results

$$L_1(t) \sim \sum_{k=0}^{N-2} \frac{x_0^{(k+2)}(t_0)}{(k+2)} (t - t_0)^k/k!\,,$$

$$L_2(t) \sim \sum_{k=0}^{N-3} \frac{x_0^{(k+3)}(t_0)}{(k+3)\,(k+2)} (t - t_0)^k/k!\,,$$

. . . . . . . . . . . . . . . . . . . . . . . . . . . . . . . .

$$L_i(t) \sim \sum_{k=0}^{N-i-1} \frac{x_0^{(k+i+1)}}{(k+i+1)\,(k+i)\ldots(k+2)} (t - t_0)^k/k!$$

for $i = 1, 2, \ldots, N-1$. Therefore, we get the general result

$$L_i(t_0) = \frac{x_0^{(i+1)}}{(i+1)!}\,.$$

This is remarkable. The parameters of the Taylor difference chain are identical with the Taylor coefficients of the bottom signal $x_0(t)$ at $t_0$.

2) $\Phi(u) = \Psi(u) = \ln u$

This is the difference analogy for the exponential chains discussed above.

Now we get the following logarithmic difference chain

$$\frac{\ln\left(x_i(t)/x_i(t_0)\right)}{\ln\,(t/t_0)} = K_i x_{i+1}(t)$$

with the normalisation condition $x_i(t_0) = 1$ for $i = 1, 2, \ldots$ and the consistence condition $x_0(t_0) = f(t_0)$. This local model at the reference point $\left(t_0, f(t_0)\right)$ of a time function $f(t)$ is called a local power-function model of order $N$ if we construct this difference chain up to the level $N$ putting at the top-level $x_{N+1}(t) \equiv 1$. This is a direct generalisation to higher order of the linear model of the form $f(t_0)\;(t/t_0)^a$ previously considered.

For the determination of the parameters $K_i$ we get the formula

$$K_i = \frac{\dot{x}_i(t_0)}{x_i(t_0)}\,t_0\,.$$

Unfortunately, this formula is not an explicit one because of the chain recursion. We also, in this case, try to derive explicit formulas for the determination of the $K_i$. For this purpose we introduce the following functions

$$L_i(t) = \dot{x}_i(t)/x_i(t)$$

and get from the chain recursion the following recursion for the computation of the functions $L_i(t)$

$$L_{i+1}(t) = \frac{\mathrm{d}}{\mathrm{d}t} \ln\left(\frac{\int_{t_0}^{t} L_i(u)\,\mathrm{d}u}{\ln\,(t/t_0)}\right).$$

The unknown coefficients $K_i$ then can be expressed as follows

$$K_i = L_i(t_0)\,t_0\,.$$

There is no simple way to evaluate explicitly the values $L_i(t_0)$.

## 2.8. Connection to Multiobjective Choice of Optimal Local Models

We refer here only to local power-function models of a time function $f(t)$ at a reference point $t_0$. However, it is possible to extend these links also to vector optimization in the case of local models with higher order.

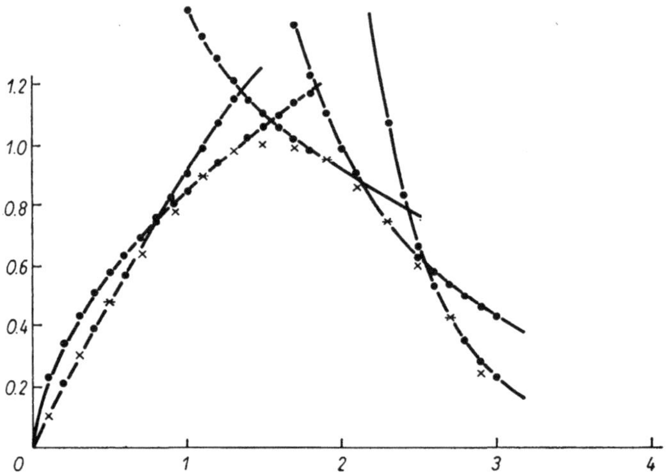


**Fig. 63** Local power models with observation point (0, 0) for sin $t$

With the initial point (0, 0) of the coordinate system as an observation point, we obtain in Section 2.10 the following local power-function model

$$f(t_0)\,(t/t_0)^a$$

with the pseudo-derivation

$$a = \frac{\dot{f}(t_0)}{f(t_0)}\,t_0\,.$$

Depending upon the cut of the tangent to $f(t)$ in $t_0$ with the ordinate axis, the curvature of the local model can have the same or the opposite sign of the curvature of $f(t)$ at the reference point $t_0$.

An interesting link to polyoptimization arises if we use besides the zero-point of the coordinate system some other observation points. If we shift the observation point into $(\tau, B)$ with $B < f(\tau)$, we use as local models the following power-functions

$$\big(f(t_0) - B\big)\,\big((t - \tau)/(t_0 - \tau)\big)^a\,.$$

If we use as an observation point $(\tau, B)$ with $B > f(\tau)$, we additionally invert the direction of the ordinate axis, meaning that we use the following local models

$$\big(B - f(t_0)\big)\,\big((t - \tau)/(t_0 - \tau)\big)^a\,.$$

Also, the corresponding pseudo-derivative $a$ depends upon the chosen observation point, we get

$$a = \frac{\dot{f}(t_0)}{\big(f(t_0) - B\big)}\,(t_0 - \tau)\,.$$

If we simultaneously use several observation points at a given reference point $t_0$, we have the problem of making the optimal choice. This could, but must not, be reduced on a one-dimensional optimization task if, for example, we take the curvature as the objective function. However, if we are interested in planning an optimal set of observation points for the construction of optimum local models for all reference points $t_0$ within a certain interval, we necessarily meet a polyoptimization task with the set of observation points as a restricted resource. Fig. 64 shows schematically as an example how local models change by shifting the observation point.

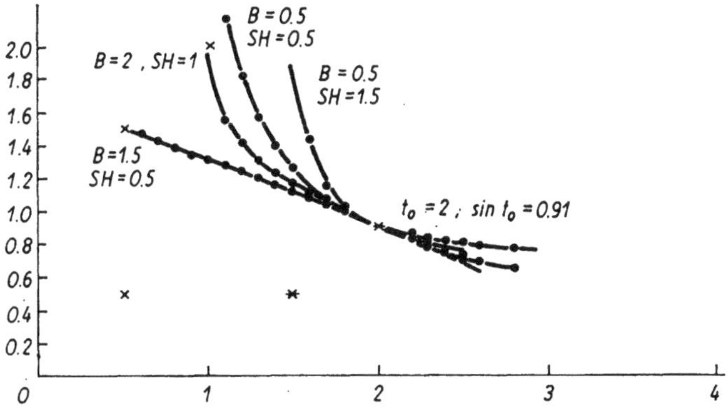


**Fig. 64** Shifted local power models at a given point of the function sin $t$

## 2.9. *Local Models Based upon Difference Chain Concepts for Sets of Ordinary Differential Equations, Especially Local Riccati Models*

We now extend the idea of constructing local power-function models arising from first order approximation with a difference chain on the case of multivariable functions.

Let

$$x = f(z_1, z_2, \dots, z_n) .$$

Applying logarithmic scalarisation and the first order difference chain concept, we get the following model

$$\ln x_M = \ln f(z^0) + \sum a_i \ln (z_i/z_i^0)$$

with the partial pseudo-derivatives

$$a_i = z_i \frac{\partial}{\partial z_i} \ln f(z)|_{z=z^0} .$$

Therefore, we get, in this case, the following assemblage of local power-product models

$$x_m = f_M(z) = f(z_0) \prod (z_i/z_i^0)^{a_i} .$$

Also, in this case, we could shift the observation point to $(\zeta, B)$ and consider the competition of assemblages of the following form

$$\pm (f(z_0) - B) \prod ((z_i - \zeta_i)/(z_i^0 - \zeta_i))^{a_i} .$$

This would be the analogy to the vector optimization task from Section 2.12 for the case of multivariable functions.

The application of this modelling approach is very important to autonomous ordinary differential equations

$$\dot{z} = f(z), \quad \text{or in components,} \quad \dot{z}_i = f_i(z_1, z_2, \dots, z_n)\,.$$

If we replace the right-hand sides of these equations by the corresponding local power-product models, we get, in the neighbourhood of the reference point $t_0$, the following approximation of the dynamic behaviour of the given autonomous system

$$\dot{z}_{iM} = f_i(z^0)\, \Pi\, (z_j/z_j^0)^{a_{ij}}\,.$$

This type of ordinary differential equation, with the driving forces given by exactly one power-product respectively, we call Riccati representation because of the strong connection to usual Riccati differential equations. This similarity will be explained in detail in Chapter 6.[1])

## 2.10. *Robustness of Behavioural Types in Chains Against Amplitude-Bounded Disturbances*

We now consider signal representations by finite exponential chains

$$Fx_i = K_i x_{i+1}\,, \qquad i = 0, 1, \dots, N, \quad \text{and} \quad x_{N+1} = 1$$

with $F = \mathrm{d}\ln/\mathrm{d}t$ and $F^{-1} = \exp\left(\int_0^t\right)$ continuing our analysis of the convergence of these chains.

We are interested in knowing more about the dynamic properties of

$$x_{0N}(t) = F^{-1}\big(K_0 F^{-1}(K_1 \dots F^{-1}(K_N 1(t)) \dots)\big)$$

which is the bottom signal of a tower of finite length $N + 1$. We assume that the parameters $K_i$ are disturbed either by a parameter shift (systematic error) or by random disturbances (independent on the different levels).

Let

$$K_i = k_i + \varepsilon_i\,.$$

For $\varepsilon_i$, we assume amplitude restrictions in the following homogenous form

$$-\varepsilon \leqq \varepsilon_i \leqq \varepsilon\,.$$

We must distinguish three different cases.

I. Strictly positive parameters, $K_i > 0$.

Then the operator

$$F^{-1}(K_i)$$

is strictly positive. We now get the following inequalities. For $w \leqq u \leqq v$ we have

$$F^{-1}\big((k_i - \varepsilon)\, w\big) \leqq F^{-1}(k_i u) \leqq F^{-1}\big((k_i + \varepsilon)\, v\big) \quad \text{for all } i\,.$$

---

[1]) We thank Dr. Jansen for ideas in connection with the convergence of exponential chains, and Dr. Albrecht and Dr. Funke for simulation and stochastic research for parameter-disturbed exponential chains (all ZKI).

We introduce the following notations: $x_{0,N}$, $x_{0,N;-\varepsilon}$, $x_{0,N;\varepsilon}$ are the bottom signals of the exponential chains with the parameters $k_i$, resp. $k_i - \varepsilon$, resp. $k_i + \varepsilon$.

If we apply the above inequality to the sequence $i = N, N - 1, \ldots, 0$ we get

$$x_{0,N;-\varepsilon} \leqq x_{0,N} \leqq x_{0,N;\varepsilon}$$

Because of the monotony property of the generating operators, all these exponential chains produce increasing bottom signals if we adjoin a next level $N + 1$. Therefore, we now have

$$\begin{array}{ccccc} x_{0,N;-\varepsilon} & \leqq & x_{0,N} & \leqq & x_{0,N;\varepsilon}; \\ \wedge & & \wedge & & \wedge \\ x_{0,N+1;-\varepsilon} & \leqq & x_{0,N+1} & \leqq & x_{0,N+1;\varepsilon} \end{array}$$

The upper restriction of parameters is obviously sufficient for the convergence of an exponential chain with non-negative parameters

$$K_i \leqq K \quad \text{for all } i\,.$$

In such a case, the exponential chain with $K_i = K$ apparently is a majorant. Therefore, we get in the limit the following inclusion

$$x_{0,\infty;-\varepsilon} < x_{0,\infty} < x_{0,\infty;+\varepsilon};$$

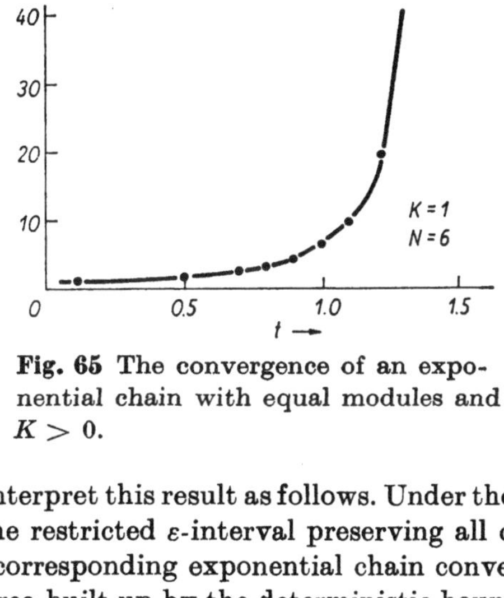


**Fig. 65** The convergence of an exponential chain with equal modules and $K > 0$.

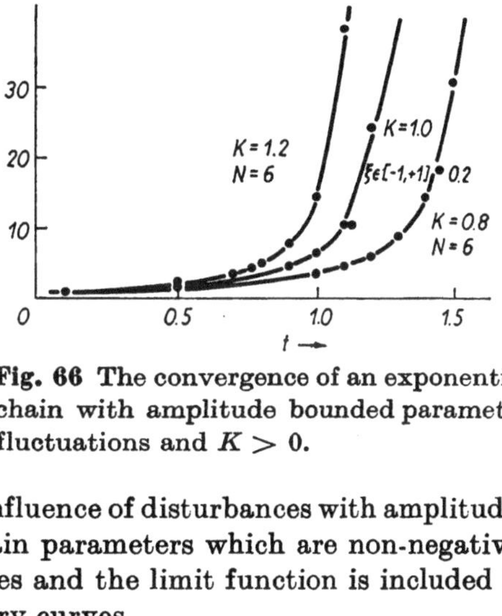


**Fig. 66** The convergence of an exponential chain with amplitude bounded parameter fluctuations and $K > 0$.

we interpret this result as follows. Under the influence of disturbances with amplitudes in the restricted $\varepsilon$-interval preserving all chain parameters which are non-negative, the corresponding exponential chain converges and the limit function is included in an area built up by the deterministic boundary curves

$$x_{0,\infty;-\varepsilon} \quad \text{and} \quad x_{0,\infty;+\varepsilon}\,.$$

In the strong analytical sense, the convergence holds only in a time-interval restricted right-hand sides by the pole of $x_{0,\infty,-\varepsilon}$.

Of great interest is the homogenous disturbed exponential chain with all $k_i = k$. In this case (the law of large numbers in ecology), we get

$$1/\big(1 - (k - \varepsilon)\, t\big) < x_{0,\infty} < 1/\big(1 - (k + \varepsilon)\, t\big)\,.$$

II. Strictly negative parameters, $K_i < 0$.

$F^{-1}(K_i)$ is not a monotonous operator because multiplication with $K_i$ changes the sign of the inequality.

Let

$$w \leqq u \leqq v .$$

Then we have

$$F^{-1}(k_i w) \geqq F^{-1}(k_i u) \geqq F^{-1}(k_i v) .$$

Moreover it holds that

$$F^{-1}\big((k_i + \varepsilon)\, w\big) \geqq F^{-1}(k_i u) \geqq F^{-1}\big((k_i - \varepsilon)\, v\big) .$$

If we, in addition, now apply the operator of the next level $i + 1$, we again get a monotony property

$$F^{-1}\big((k_{i+1} - \varepsilon)\, F^{-1}\big((k_i + \varepsilon)\, w\big)\big) \leqq F^{-1}\big(k_{i+1} F^{-1}(k_i u)\big)$$
$$\leqq F^{-1}\big((k_{i+1} + \varepsilon)\, F^{-1}\big((k_i - \varepsilon)\, v\big)\big) .$$

Thus, a pair of consecutive operators which occur in the chain together again always generate a monotonous operator.

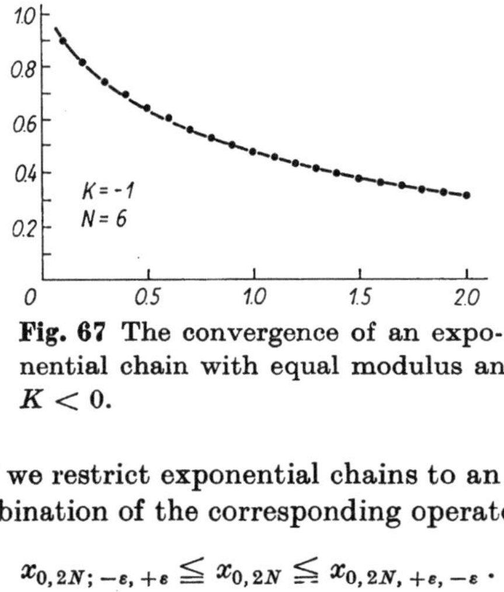


**Fig. 67** The convergence of an exponential chain with equal modulus and $K < 0$.

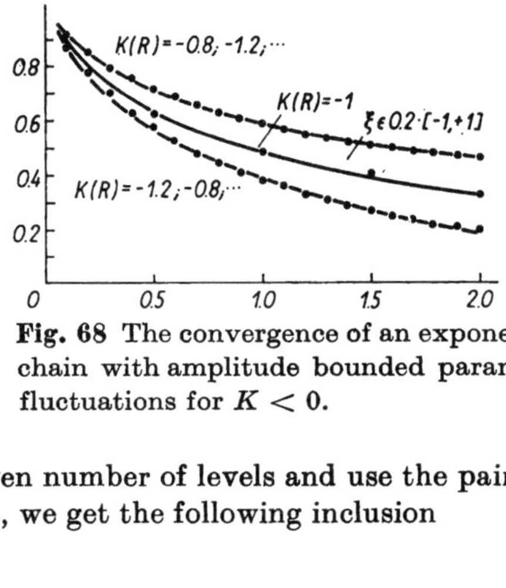


**Fig. 68** The convergence of an exponential chain with amplitude bounded parameter fluctuations for $K < 0$.

If we restrict exponential chains to an even number of levels and use the pairwise combination of the corresponding operators, we get the following inclusion

$$x_{0,2N;\,-\varepsilon,\,+\varepsilon} \leqq x_{0,2N} \leqq x_{0,2N,\,+\varepsilon,\,-\varepsilon} .$$

If we adjoin another pair of levels, we get the result

$$x_{0,2N;\,-\varepsilon,\,+\varepsilon} \leqq x_{0,2N} \leqq x_{0,2N;\,+\varepsilon,\,-\varepsilon} ,$$
$$x_{0,2(N+1);\,-\varepsilon,\,+\varepsilon} \leqq x_{0,2(N+1)} \leqq x_{0,2(N+1);\,+\varepsilon,\,-\varepsilon} .$$

Obviously, the corresponding bottom signals $x_{0,2N}$ form a monotonously decreasing sequence bounded from below by zero. Therefore, this sequence necessarily converges.

For the limit, we get the following inclusion

$$x_{0,\infty;\,-\varepsilon,\,+\varepsilon} < x_{0,\infty} < x_{0,\infty;\,+\varepsilon,\,-\varepsilon} .$$

For the special case $k_i = k$ (homogenous exponential chain), the following holds

$$F\xi_0 = (k - \varepsilon)\, \xi_1 , \qquad F\xi_1 = (k + \varepsilon)\, \xi_0 ,$$
$$\xi_0 = x_{0,\infty;\,-\varepsilon,\,+\varepsilon} \quad \text{and} \quad \xi_1 = x_{0,\infty;\,+\varepsilon,\,-\varepsilon} .$$

This is a hypercycle of order 2. This hypercycle admits a closed analytical solution, namely

$$\xi_0 = 2\varepsilon/\big(|k|\,(\exp(2\varepsilon t) - 1) + \varepsilon(\exp(2\varepsilon t) + 1)\big)\,,$$
$$\xi_1 = 2\varepsilon/\big(\varepsilon(\exp(-2\varepsilon t) + 1) - |k|\,(\exp(-2\varepsilon t - 1))\big)\,.$$

For small $\varepsilon$, both signals differ only a little from $1/(1 + |k|\,t)$, which represents a hyperbolically decreasing signal. Therefore, we could prove for case I (all parameters $K_i$ strictly positive) and case II (all parameters $K_i$ strictly negative) robust convergence against an exploding signal in case I and an extincting signal in case II.

III. Exponential chains with sign-changing parameters $K_i$.

The case $k_i = 0$ for all $i$ belongs to this type of behaviour. The following consideration gives only a rough impression of what occurs in this case. Because of

$$\exp(ax) > 1 + ax \quad \text{and} \quad F^{-1}(K.) > 1 + KI$$

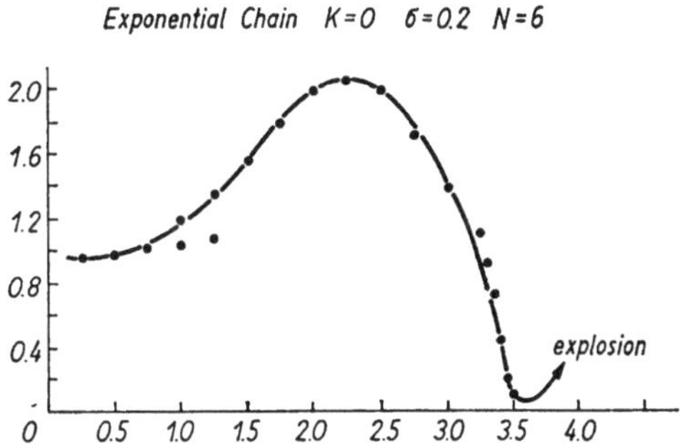


**Fig. 69** A random trajectory of a finite exponential chain with $K = 0$ under amplitude-bounded parameter fluctuations.

we have

$$x_{0,N} > \big(1 + IK_0(1 + IK_1(\ldots(1 + IK_N 1(t))\ldots))\big) =$$
$$= 1 + K_0 t + K_0 K_1 t^2/2! + \ldots + K_0 K_1 \ldots K_N t^{N+1}/(N+1)!\,.$$

The variables with the bar designate mean values of the corresponding original variables. For small values of the corresponding arguments, we have

$$\bar{x}_{0,N} \sim \sum_{i=0}^{N+1} \bar{a}_i t^i/i!\,, \qquad \bar{a}_i = \bar{K}_0 \bar{K}_1 \ldots \bar{K}_i\,,$$
$$\sigma^2_{x_{0,N}} = \sum_{i=0}^{N+1} \sigma^2_{a_i}(t^i/i!)^2\,.$$

For the case $\bar{a}_i = \bar{a}$ or $\bar{K}_i = \bar{K}$ the mean value $\bar{x}_{0,N}$ for small arguments differs from $\exp(\bar{K}t)$ only slightly. That means that for $\bar{K} > 0$, after some time the condition small arguments will be violated, and that for $\bar{K} < 0$, no definite conclusion is possible. For $\bar{K} = 0$, we see that the mean $x_{0,N}$ will be fixed at the value 1.

In all three cases, the variance is monotonously and quickly increasing. This means also that in the case $\bar{K} < 0$, the possibility to receive an exploding signal exists, and, in the case $\bar{K} = 0$, the instability of the state $x_{0,N} = 1$ exists. We see that the very beginning of this process will be similar to a stochastic process with independent increments.

# 3. Structure Design for Nonlinear and Instationary Processes

## 3.1. *System Concepts for Structure and Function Taken from Ecology*

From structures realised in ecology, we can derive the following information which exposes the framework of ecological structures:

- Subordination between species is widely distributed. One species depends upon another species, but the reverse is not necessarily the case. That means that we frequently meet polarized dependencies.
- Many of these subordination relationships are functionally realised through the principle of rate-coupling, meaning that the growth rate of the subordinated species $x$ depends on the state $x'$ of the dominating species in the sense

  $$dx/dt : x = f(x') .$$

- Frequently, the rate-coupling principle is realised in linear chains which can be sometimes closed to cycles

  $$dx_i/dt : x_i = K_i x_{i+1} .$$

- Sometimes the rule of immediate subordination is broken by feedback over several steps. The number of these feedback loops usually is relatively small.
- Chain structures and cycles exist in ecological systems as impressive substructures (clusters) which are relatively but not completely autonomous. There is usually a small number of connections (channels) between these substructures.

In this book, we deal mainly with model building for nonlinear dynamic systems from the point of view of ecology. In model building we are engaged not only with the simulation of a function, but we must always take into account the unity of the functional model with the structural model. We are deeply convinced that interaction mechanisms seen in ecological systems are of importance far beyond the boundaries of ecology. Therefore, we think that it is necessary to analyse the phenomena in the world by looking at them from the ecological point of view.

For structure models, we like to rely upon structural concepts met in ecology; for functional models we like to apply interaction mechanisms from ecology. To consider our world from the point of view of ecology does not mean that we believe blindly that this projection shows us the essential details we need, but at least we can hope to raise new and interesting questions and can sometimes make good use of the results which we derive under these conditions.

In modelling the functional aspect we will also borrow from ecology.

Very often, nature realises complex functions by using modular constructions from elementary modules. From sets of homogenous elements (cells, segments in plants and animals) and from using certain structural concepts (principles of architecture), complex systems for the realisation of highly specialised behavioural types are designed. In these cases it seems to make sense to use homogenous modules for filling a given system concept for structure with the functional content. This ecological approach leads us in the sense of a heuristic analogue conclusion to the *structure design principle* [18].

The structure of an interacting complex system should essentially be composed of chains and cycles as the most prevalent substructures. In the substructures, standard modules should be used as structural elements (nodes of a graph). The substructure should have only a relatively small number of feedback loops connected with it, and the number of connections between different substructures should also be relatively small. This picture of a structure which we will achieve with our structure design process is consistent with modern development in microelectronics. Also, in the structures of circuits in microelectronics, we frequently meet chains and cycles (for example, shift registers and counter chains). These basic structures (chains and cycles) contain many elementary modules such as diodes, flip-flops, etc. From these substructures, complex logical structures are designed. Because of the high degree of integration in microelectronics, solutions are preferred with a low number of interconnections and links to the environment.

Therefore, the picture of ecological structures seems to be also an interesting variant for the structure of microelectronical circuits, meaning for hardware realisations of nonlinear dynamic systems through electronic means. However, we must distinguish the aims of this book from those of microelectronics. We concentrate in this book mainly on analogous and not on discrete base modules with ordinary differential equations as the adequate form of description, instead of on the logical or automata models in microelectronics. However, we realize that some of the conceptional considerations in this book may lead to conclusions about nonlinear systems for impulse systems or about switching circuits.

## 3.2. Structure Design Rules and Consequences for the Structure Design with the Base Module Logarithmic Derivative

A unique base module $F$ is chosen (it should be a module realizing a differential operator, perhaps with some parameters for specification) for the dynamic elements of the structure under construction.

We shall describe the base module with an equation of the form

$$F(x) = x' .$$

For a given signal $x$ belonging to the definition area of the operator $F$, after application of $F$ a higher-level signal $x'$ will be generated. The operator $F$ shall have a pseudo-inverse operator $F^{-1}$ (similar to the Taylor operator construction from Section 2.5, but admitted also for the nonlinear case).

In structure design, we first think about the construction of chains with our base module $F$, described by equations of the form

$$F(x_i) = x_{i+1} , \qquad i = 0, 1, \dots ,$$

with state-variables $x_i$, in general, still unnormalised.

We speak of a chain design principle for a given signal $x(t)$ if we choose this signal as the bottom signal $x_0$ of a chain and construct that chain by successive applications of $F$ starting with $x_0$ with the description

$$F(x_i) = x_{i+1}, \qquad i = 0, 1, \ldots .$$

We speak of a network-design principle if we do not have the aim to generate $x_0$ using a pure chain. In this case we need a more flexible design approach. Let us assume we had already constructed in our design process an intermediate signal $x_i$ and now we are going to perform the next step.

For this purpose, we apply the base operator $F$ (after specification of the corresponding free parameters) on $x_i$ and analyse the corresponding generated signal $A_i$

$$F(x_i) = A_i .$$

We understand the produced signal to be a certain arithmetic expression $A_i$, consisting of signals which are contained in $A_i$ and linked by certain arithmetic operations. (In analogous systems, we meet as operations, preferably addition, subtraction and multiplication with constants, signal multiplication, signal division, and sometimes nonlinear signal transformations.)

In the arithmetic expression $A_i$, we first identify signals which are already known because they either were met already at earlier stages of the structure design process or they were generated by function generators. We mark all the signals thus identified and install connections between the signal $A_i$ and the places where these known signals were first met. Besides these already known signals, new signals in general may be contained in $A_i$. We have certain degrees of freedom within which we must agree which combinations of variables should be chosen for a new signal. That means we must fix which variable combinations we take foı the new signals and we must assign names to them because alle these signals will be future state-variables of the structure. These degrees of freedom in the fixation of the formerly unknown signals within an arithmetic expression $A_i$ are responsible for the fact that at the end of a structure design process we could receive not only one, but several different structure solutions, which, in a sense, are all equivalent to each other because they all should realise the original system's behaviour. Of course, our structure design is oriented not only toward one-dimensional problems but can be used also for multivariable problems.

If there is given a finite number of original signals with coupling relations between them, we must apply the structure design approach to all of these original signals, first separately and then taking into account the dependencies between these signals. In such a system known signals are all those which already were met at any place of the corresponding complex structure. By this process, links between the different chains or trees belonging to the different original signals can occur.

The structure design process ends successfully if, after a finite number of design steps, we arrive everywhere at arithmetic expressions $A_i$ consisting completely of already known signals. This gives us a closed structure. Fig. 70 schematically shows which kind of structures we get as the result of the described structure design process.

Obviously, every original signal which we generate by the structure under design is the root of a tree. Branches of the feedback loops grow bottom-up, from earlier design stages to later ones, and they stop there the corresponding design activity. Thus, the growth of the trees will be completely stopped at the end of the design process. Such branches can also grow between neighbouring trees. The taller the trees are and the less feedbacks or other connections we have, the more the structure solution approaches our ecological figures. The structure then consists mainly of chains and cycles.

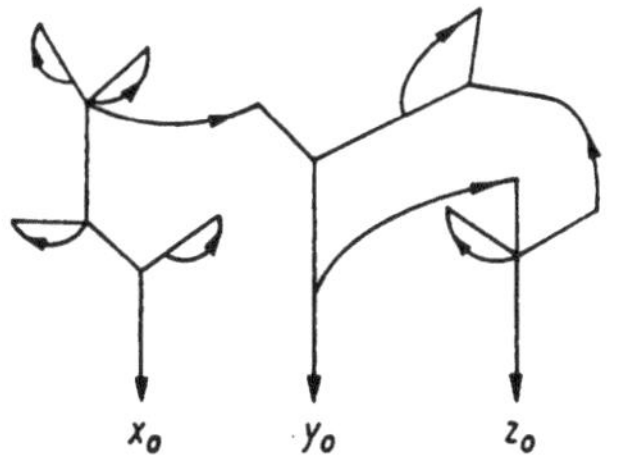


**Fig. 70** Schematic representation of the structure received as a result of the structure design process

It is suitable to introduce during the design process or at the end finite names $x_i$ for the output signals of the base modules $F$, taking them all as the dynamic state-variables in the structure which we have constructed. If we orient from the very beginning on normalised state-variables, we shall then use the standard nominations $\xi_i$. This is consistent with the tradition used in analogous computation techniques, where state-variables are always the output signals of the integrator modules. Of course, we may encounter the case where such a structure design process does not stop after a finite number of steps, in a situation where we could get a finite structure if we had skillfully performed the whole process. This takes place, for example, in the case of the design of a chain, where we take as a new signal $x_{i+1}$ every generated arithmetic expression $A_i$ without identifying known signals. In such a case, we had to stop this chain expression after a sufficiently large number of steps, and thus, we would get an approximated solution for which we must study the convergence behaviour.

Now we will study a special case which is of high interest for us, where we take as our base module the operator of logarithmic derivative

$$F = \mathrm{d}\ln/\mathrm{d}t$$

realising the rate coupling principle with the following equation

$$\mathrm{d}x/\mathrm{d}t = Kxx' \,.$$

We take into account the pseudo-inversion of the logarithmic derivative (as a nonlinear Taylor operator) using the initial value for normalisation purposes. Thus, we get the following solution of the module equation

$$x(t) = x(0)\,\mathrm{e}^{KIx'} = x(0)\,\xi(t) \quad \text{with} \quad \xi(0) = 1\,, \qquad I = \int_0^t .$$

Thus the normalised state-variable is then

$$\xi(t) = \mathrm{e}^{KIx'} .$$

If we, during the structure design process, meet parts of exponential chains

$$\mathrm{d}x_i/\mathrm{d}t = K_i x_i x_{i+1}$$

we always pass to a normalised description. Usually we use this normalisation everywhere in our structure as the result of the structure design process.

The consequence of doing this is that all initial values of state-variables will then be parameters of the structure, and the whole structure, parametrized in this way, can be considered as an initial automaton starting always with the normalised initial condition $\xi_i(0) = 1$. Obviously the normalised chain is given by the following equations:

$$d\xi_i/dt = K_i \xi_i \xi_{i+1}$$

after the substitution of the chain parameters by

$$K_i := K_i x_{i+1}(0) \ .$$

Now we describe the structure design process for this special case. All signals which we want to generate by the structure will first be transformed to normalised signals by division through $x(0)$. In the case that $x(0) = 0$ we must manipulate the given signal in such a way (for example by shifting) that we get a non-vanishing initial value.

On every normalised signal $\xi_0$ we now apply the operator of logarithmic derivation. Here, it must be assumed that $\xi_0(t)$ is differentiable, and it must be taken into account that the result of the operation is not defined in time points corresponding to the roots of $\xi_0(t)$. This property is responsible for parasitic poles and can be very dangerous for computer simulations, as will be demonstrated later on.

$F\xi_0 = A_0$ is a certain arithmetic expression. We identify in $A_0$ signals which we have already met (in the first step, this could be only $\xi_0$), make decisions about which combinations should be taken for new signals, assign names to the new signals and repeat for all newly introduced signals the step of applying the base operator $F = d \ln/dt$ after having normalised all new signals to the initial value $\xi(0) = 1$.

Of course it can always be a problem to secure during the design process to fulfill the initial value normalisation because we usually do not know what initial value about newly generated signals will have. Therefore, we shall frequently be forced to introduce for this purpose free parameters which we must specify during the simulation process.

Often we must be sure that all state-variables can assume only non-negative values to eliminate the danger of generating parasitic poles. We distinguish between the full parametrized structure design, where we take initial values completely into account, and all other conditions, with the aim of getting a structure, which we could immediately realize on a computer, and the design of only a structure graph (formal structure design). In the last case we are only interested in gaining a finite graph in which the desired trajectories can be imbedded, independent of possible bad properties arising from the additional state-variables.

We have still to stress an important specific property of the structure design based on the logarithmic derivative operator. This design offers the opportunity to remove all multiplication or division of signals because of the corresponding property of the logarithm function. This is achieved if we let ourselves be led by the following rules:

— Sum rule (Difference rule)

If the arithmetic expression $A$ can be decomposed into a sum

$$A = x_1 + x_2$$

then $x_1$ and $x_2$ will be introduced as unknown signals, and the structure design rules will be applied afterwards separately to $x_1$ and $x_2$.

— Product rule (Division rule)
If the arithmetic expression $A$ can be decomposed into a product

$$A = x_1 x_2$$

then we take $A$ itself as an unknown signal $x = x_1 x_2$ and apply the next design step immediately upon $x$ and get

$$Fx = Fx_1 + Fx_2 .$$

In this way, the second case (product, division) can be reduced to the first case (sum, difference).

This is obviously always possible if the arithmetic expression $A$ can be completely generated alone using the basic arithmetic operations. This gives us the important result that, by following the above formulated rules, the structure design process based on the logarithmic derivative operator always leads to structures which we realize with only the following operations:

— $F$ respectively $F^{-1} = e^I$,
— multiplication by constants,
— addition and subtraction.

If we agree to choose for the state-variables the output variables of the modules $F$ in the structure, we get at the end of our structure design process with the base operator $F = \mathrm{d}\ln/\mathrm{d}t$ a unified system description of the following form in the autonomous case

$$Fx_i = e_i + \sum_j g_{ij} x_j ,$$

in the nonautonomous case

$$Fx_i = e_i + \sum_j g_{ij} x_j + \sum_r h_{ir} y_r .$$

Here, $y_r$ are not necessarily the original input variables, but expressions depending only upon the input variables.

These are the basic equations of mathematical ecology. In special form, they were introduced early by Verhulst [10], later discussed by Lotka [39] and thoroughly studied by Volterra [1].

## 3.3. *Examples for the Illustration of the Structure Design Process Based upon the Operator Logarithmic Derivative*

### *1. Signal generator for an arbitrary polynomial time function*

#### *1.1 Polynomials in additive representation*

$$x(t) = x_0 = P_n(t) = \sum_{i=0}^{n} a_i t^i / i! ; \qquad a_i = P_n^{(i)}(0) .$$

We put

$$\xi_0 = x(t)/x(0) = P_n(t)/P_n(0) \quad \text{for} \quad P_n(0) \neq 0 .$$

Then we have

$$F\xi_0 = \mathrm{d}\ln \xi_0/\mathrm{d}t = P_n^{(1)}(t)/P(t) = \frac{P_n^{(1)}(0)}{P_n(0)}\xi_1 ,$$

$$F\xi_1 = P_n^{(2)}(t)/P_n^{(1)}(t) - \frac{a_1}{a_0}\xi_1 ,$$

$$F\xi_1 = \frac{a_2}{a_1}\xi_2 - \frac{a_1}{a_0}\xi_1 ,$$

. . . . . . . . . . . . . . . . . . . . . . . . .

$$F\xi_i = \frac{a_{i+1}}{a_i}\xi_{i+1} - \frac{a_i}{a_{i-1}}\xi_i .$$

Because

$$P_n^{(n)}(t) = a_n = \text{const} ,$$

the design stops on the $n^{\text{th}}$ level with the closing step

$$F\xi_n = -\frac{a_n}{a_{n-1}}\xi_n .$$

The closing step obviously corresponds to a hypercycle of order $r = 1$.

The design procedure is only then successful if, for all coefficients $a_i$ it holds, that they do not vanish, $a_i \neq 0$. We took into account in the design only the normalisation of the initial values. In the simulation process, we must expect further difficulties (poles) at all points in time where one of the introduced state-variables vanishes.

The equation we have derived can be written uniquely in the following form:

$$F\xi_i = \frac{a_{i+1}}{a_i}\xi_{i+1} - \frac{a_i}{a_{i-1}}\xi_i$$

Fig. 71 Structural scheme of a Riccati tower (beginning with the second plant)

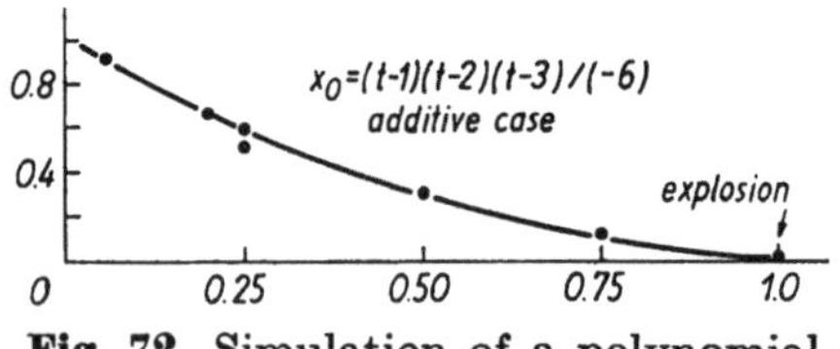


Fig. 72 Simulation of a polynomial of third order in additive representation after the structure design

if we agree about

$$a_{n+1} = 0 \quad \text{and} \quad a_{-1} = \infty .$$

Fig. 71 shows the corresponding structure. A tower of such a form, in which every elementary plant has an autocatalytic feedback (in our case, beginning with the variable $\xi_1$), we will call a Riccati tower. Every step is determined by an equation with the form of the Bernoulli-Riccati differential equation.

*1.2 Polynomials in multiplicative representation*

$$x(t) = P_n(t) = K \prod (1 - ta_i)^{k_i} \prod (1 + \lambda_j t + \vartheta_j t^2)^{l_j} .$$

This is the most general case of a polynomial in multiplicative representation with real roots and corresponding multiplicity for the linear factors, and conjugated complex roots with corresponding multiplicity for the quadratic factors.

We put

$$\xi_0 = P_n(t)/P_n(0) , \qquad P_n(0) = K .$$

We get

$$F\xi_0 = - \sum k_i a_i/(1 - ta_i) + \sum l_j(\lambda_j + 2\vartheta_j t)/(1 + \lambda_j t + \vartheta_j t^2) .$$

We introduce the following normalised additional state-variables

$$\xi_i = 1/(1 - ta_i) \quad \text{and} \quad \eta_j = (1 + 2\vartheta_j/\lambda_j t)\,(1 + \lambda_j t + \vartheta_j t^2) .$$

This is possible under the assumption $\lambda_j \neq 0$, meaning that imaginary roots must be excluded.

$$F\xi_i = a_i \xi_i ,$$

these are all hypercycles of order $r = 1$,

$$F\eta_j = 2 \frac{\vartheta_j/\lambda_j}{(1 + 2\vartheta_j/\lambda_j t)} - \lambda_j \eta_j .$$

We introduce further on the following normalised additional state-variables

$$\zeta_j = 1/(1 + 2\vartheta_j/\lambda_j t) .$$

Finally, we get

$$F\zeta_j = -2\vartheta_j/\lambda_j \zeta_j$$

which are again hypercycles of order $r = 1$. Fig. 73 shows schematically the structure of this design process.

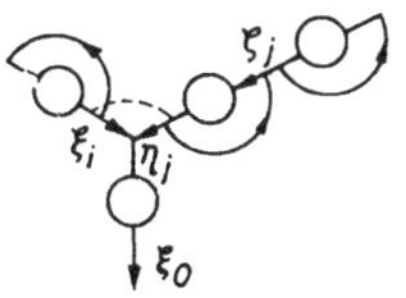


**Fig. 73** Structure of a polynomial in multiplicative representation

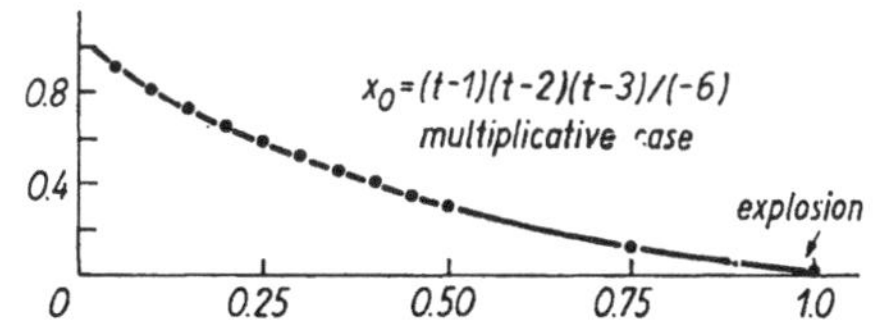


**Fig. 74** Simulation of a polynomial of third order in multiplicative representation after the structure design.

Obviously, in this case, all additional state-variables which we introduced are non-negative. Nevertheless, we must expect difficulties with the simulation as soon as we approach the first root of the polynomial.

*2. System with constant parameters (linear in the logarithmic derivatives, that means a special class of nonlinear systems)*

We consider the following differential equation

$$\sum_{i=0}^{n} a_i F^{n-i} x(t) = \sum_{j=0}^{m} b_j F^{m-j} y(t) \quad \text{with} \quad F = \mathrm{d}\ln/\mathrm{d}t .$$

This equation only implicitly defines the response $x(t)$ by the input signal $y(t)$. In principle, the initial values $x^{(i)}(0)$ can be chosen freely, but we must restrict ourselves in so far as we should demand $F^i x(0) \neq 0$ from reasons of normalisation.

First we construct an exponential chain for $x(t)$ up to level $n$. From the very beginning we use for this purpose normalised state-variables. Then we have

$$x(t) = x(0)\,\xi_0\,, \qquad F\xi_0 = Fx = F^1x(0)\,\xi_1\,,$$
$$F\xi_1 = F^2x = F^2x(0)\,\xi_2 \quad \text{and in general} \quad F\xi_i = F^{i+1}x(0)\,\xi_{i+1}\,.$$

After the initial value normalisation, the given equation changes its parameters but not the form. We obviously get

$$\sum_{i=0}^{n} a_i F^{n-i} x(0)\,\xi_{n-i} = \sum_{j=0}^{m} b_j F^{m-j} y\,.$$

This equation determines $\xi_n$ as a linear combination of state-variables of the chain which were constructed formerly. We can therefore express $\xi_n$ in the following way

$$\xi_n = \frac{1}{a_0 F^n x(0)} \Big(\sum_{j=0}^{m} b_j F^{m-j} y - \sum_{i=1}^{n} a_i F^{n-i} x(0)\,\xi_{n-i}\Big)\,.$$

Taking into account the corresponding feedback loop, obviously the structure design process stops in this case at the level $n - 1$.

Because of the introduced initial value normalisation, we can pass through the chain for the variables $\xi_i$ without any difficulty top-down, and thus generate at its bottom the desired variable $\xi_0 = x(t)/x(0)$. Fig. 75 shows the constructed structure

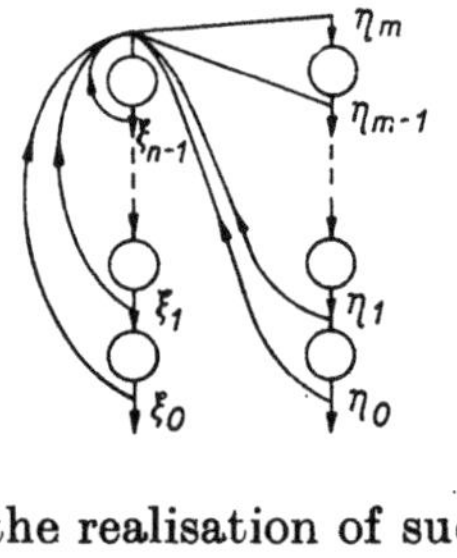

**Fig. 75** Structure of a (nonlinear) system with constant parameters

for the realisation of such a system. The whole state space description has now the following form

$$F\xi_i = F^{i+1}x(0)\,\xi_{i+1} \quad \text{for} \quad i = 0, 1, \dots, n-2\,,$$
$$F\xi_{n-1} = \frac{1}{a_0}\Big(\sum_{j=0}^{m} b_j F^{m-j} y - \sum_{i=0}^{n} a_i F^{n-i} x(0)\,\xi_{n-i}\Big)\,.$$

*3. Growth equation with power-product driving force/hyperlogistic evolution model*

This growth model is given by an ordinary differential equation of the following form

$$\mathrm{d}x/\mathrm{d}t = Kx^k(B - x^w)^l\,.$$

We normalise the growth indicator $x$ by

$$x(t) = x(0)\,\xi_0 \quad \text{and get for } \xi_0\,,$$
$$\mathrm{d}\xi_0/\mathrm{d}t = Kx(0)^{k-1+wl}\,\xi_0^k\big(B/x(0)^w - \xi_0^w\big)^l\,.$$

This means that the normalisation preserves the form of the equation and that the parameters change in the following way

$$K := Kx(0)^{k-1+wl}; \qquad B := B/x(0)^w .$$

With this variation, we again note

$$\mathrm{d}\xi_0/\mathrm{d}t = K\xi_0^k(B - \xi_0^w)^l$$

and, applying the structure design approach, we get the following sequence of equations

$$F\xi_0 = K\xi_0^{k-1}(B - \xi_0^w)^l = K(B-1)^l\,\xi_1 ,$$
$$F\xi_1 = (k-1)\,K(B-1)^l\,\xi_1 - wlK(B-1)^{l-1}\,\xi_2 ,$$
$$F\xi_2 = (k+w-1)\,K(B-1)^l\,\xi_1 - w(l-1)\,K(B-1)^{l-1}\,\xi_2 .$$

Obviously, we had to introduce the following additional state-variables

$$\xi_1 = \xi_0^{k-1}(B - \xi_0^w)^l/(B-1)^l , \qquad \xi_2 = \xi_0^{k+w-1}(B - \xi_0^w)^{l-1}/(B-1)^{l-1} .$$

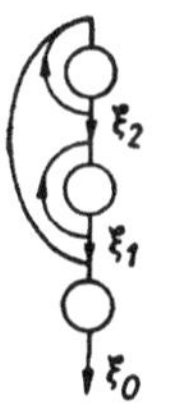


**Fig. 76** Lotka-Volterra structure of a hyperlogistic model

Fig. 76 shows the corresponding structure of a growth process of this kind. Without the specification of the parameters, we obtain structure equations of the following form

$$F\xi_0 = K_0\xi_1 ,$$
$$F\xi_1 = K_1\xi_1 + L_1\xi_2 ,$$
$$F\xi_2 = K_2\xi_2 + L_2\xi_1 .$$

The corresponding subsystem with the help state-variables $\xi_1$, $\xi_2$, is related to the well-known Lotka-Volterra system of the second order. The only difference is that the growth of the prey $\xi_1$ by a constant resource $e_1$ is now substituted by an autocatalytic growth with the rate $K_1$ and the constant death rate of the predators $\xi_2$ is now substituted by autocatalytic death rate $K_2$. If we meet these factors with the corresponding correct sign, then the prey population is growing much more intensively, but, meanwhile, death reduces the predator population more intensively (if we can assume a sufficient mass of these populations).

Let us now consider the special case of the basic harmonics $\sin t$ and $\cos t$, which also are solutions of a hyperlogistic differential equation, namely for the following set of parameters

$$K = 1 , \qquad k = 0 , \qquad B = 1 , \qquad w = 2 \quad \text{and} \quad l = 1/2 .$$

Of course we can meet some problems with normalisation here. Therefore, we introduce the following state-variables before the specification of the parameters

$$\eta_1 = (B-1)^l\,\xi_1 \quad \text{and} \quad \eta_2 = (B-1)^{l-1}\,\xi_2 .$$

Then we get

$$F\xi_0 = \eta_1 ,$$
$$F\eta_1 = -(\eta_1 + \eta_2) ,$$
$$F\eta_2 = (\eta_1 + \eta_2) .$$

It can be seen easily that in fact sin $t$ and cos $t$ obey this set of equations. Fig. 77 shows the corresponding structure of the generator for the basic harmonics.

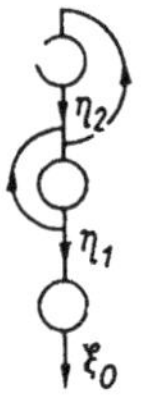


**Fig. 77** Structure representation for a basic harmonic sin $t$

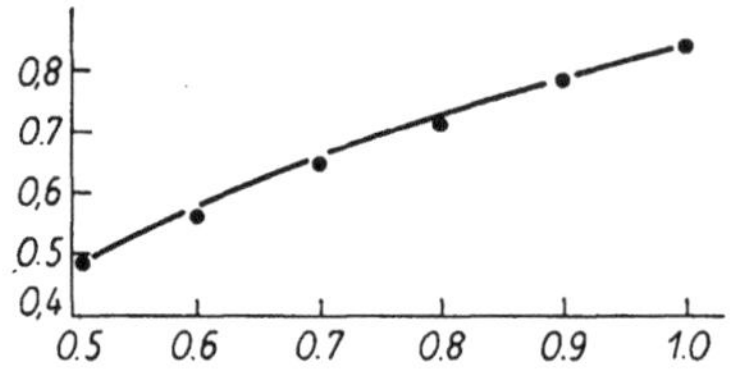


**Fig. 78** Simulation of the basic harmonics sin $t$ according to the structure in fig. 77

*4. Differential equation of the mathematical pendulum*

$$\mathrm{d}^2x/\mathrm{d}t^2 + D\,\mathrm{d}x/\mathrm{d}t + \gamma \sin x = 0 ; \qquad \gamma = g/l .$$

For the purpose of normalisation, we first put

$$x(t) = x(0)\,\xi_0 .$$

Then we obviously have

$$Fx = F\xi_0 = Fx(0)\,\xi_1 .$$

There is no problem with this if $x(0) \neq 0$ and $\dot{x}(0) \neq 0$.

The first condition can be always fulfilled if the pendulum does not rotate, but $\dot{x}(0) \neq 0$ excludes the reversal points of the motion. That means that, without additional measures, we can simulate at most half of a period of the pendulum oscillation

$$F\xi_1 = F^2x = x^{(2)}/x^{(1)} - Fx(0)\,\xi_1 .$$

Here we must substitute the equation of the pendulum motion and get

$$x^{(2)}/x^{(1)} = -D - \gamma \sin x/x^{(1)} .$$

Now we introduce the next additional state variable by

$$\sin x/x^{(1)} = \frac{\sin x(0)}{x^{(1)}(0)}\,\xi_2 .$$

We continue with

$$F\xi_2 = \frac{\cos x}{\sin x}\,x^{(1)} + D + \gamma\,\frac{\sin x(0)}{x^{(1)}(0)}\,\xi_2 .$$

We have

$$\frac{\cos x}{\sin x}\,x^{(1)} = L \cos x/\xi_2 \quad \text{with} \quad L = x^{(1)}(0)/\sin x(0) ,$$

$$\frac{\cos x}{\sin x}\,x^{(1)} = L \cos x(0)\,\xi_3 .$$

We have now to take into account the additional condition $\cos x(0) \neq 0$. Without any additional measures this condition restricts the simulation interval on one quarter of a full pendulum oscillation.

Further on we have

$$F\xi_3 = -\frac{\sin x}{\cos x} x^{(1)} - F\xi_2 .$$

We put

$$\frac{\sin x}{\cos x} x^{(1)} = \frac{\sin x(0)}{\cos x(0)} x^{(1)}(0)\, \xi_4 ,$$

$$F\xi_4 = \frac{\cos x}{\sin x} x^{(1)} + \frac{\sin x}{\cos x} x^{(1)} + x^{(2)}/x^{(1)} =$$

$$= L \cos x(0)\, \xi_3 + \frac{\sin x(0)}{\cos x(0)} x^{(1)}(0)\, \xi_4 - D - \gamma \frac{\sin x(0)}{x^{(1)}(0)} \xi_2 .$$

After introduction of corresponding structural parameters the equation for the realisation of the pendulum's motion assume the following form

$$F\xi_0 = K_0\xi_1 ,$$

$$F\xi_1 = -D + K_1\xi_1 + L_1\xi_2 ,$$

$$F\xi_2 = +D + K_2\xi_2 + L_2\xi_3 ,$$

$$F\xi_3 = -D - K_2\xi_2 - L_2\xi_3 + L_3\xi_4 .$$

$$F\xi_4 = -D + L_1\xi_2 + L_3\xi_3 - L_3\xi_4 .$$

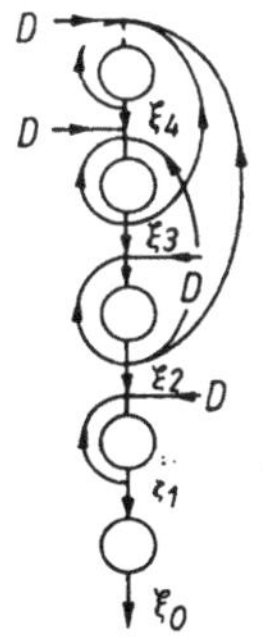


**Fig. 79** Structure representation for the mathematical pendulum

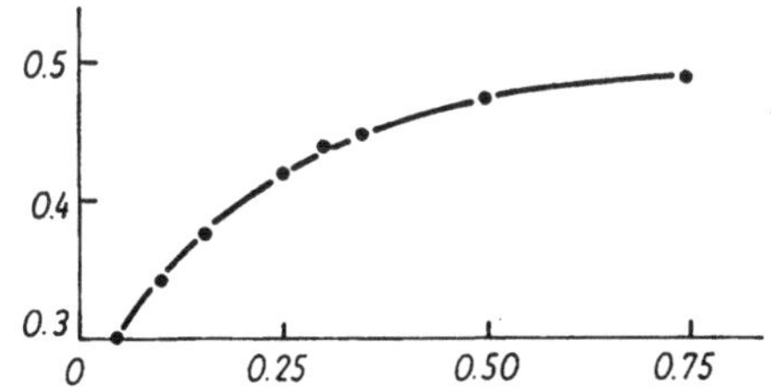


**Fig. 80** Simulation of the mathematical pendulum according to the structure in fig. 79

The corresponding Basic-like notation is

$$x(t) = x(0)\, \mathrm{X0} ,$$

$$\mathrm{FX0} = \frac{\dot{x}(0)}{x(0)} \mathrm{X1} = \mathrm{K0X1} , \qquad \mathrm{K0} = \dot{x}(0)/x(0) ,$$

$$\mathrm{FX1} = -D - \gamma \mathrm{K1X2} - \mathrm{K0X1} , \qquad \mathrm{K1} = \sin x(0)/\dot{x}(0) , \qquad \gamma = g/l ,$$

$$\mathrm{FX2} = \mathrm{D} + \gamma \mathrm{K1X2} + \mathrm{K2X3} , \qquad \mathrm{K2} = \dot{x}(0) \cot x(0) ,$$

$$\mathrm{FX3} = -\mathrm{D} - \gamma \mathrm{K1X2} - \mathrm{K2X3} - \mathrm{K3X4} \qquad \mathrm{K3} = \dot{x}(0) \tan x(0) ,$$

$$\mathrm{FX4} = \mathrm{K2X3} + \mathrm{K3X4} - \mathrm{D} - \gamma \mathrm{K1X2} .$$

Fig. 79 shows the corresponding structure as result of our design process for the case of the mathematical pendulum.

## 3.4. *The Universal Significance of Lotka-Volterra Equations* [1])

If we take into account the sum rule and product rule formulated above in the described structure design process, we can observe that the structure equations which we get at the end of our structure design process have the following form: *The logarithmic derivatives of all state variables are a linear combination of the state variables.* This could be seen in all examples given in Section 3.2. That means the following unified form of the structure equations can always be achieved

$$Fx_i = e_i + \sum_j g_{ij}x_j \quad \text{in the autonomous case,}$$

$$Fx_i = e_i + \sum_j g_{ij}x_j + \sum_r h_{ir}y_r \quad \text{in the nonautonomous case.}$$

Here the $y_r$ maybe higher logarithmic derivatives of the original input variables. With the introduction of the additional resource of a state variable

$$x_{n+1} = 1 \quad \text{with} \quad x_{n+1} = 0$$

we can always achieve the following homogenous representation of the structure equations

$$Fx_i = \sum_j G_{ij}x_j + \sum_r H_{ir}y_r \,.$$

It is astonishing that, on the basis of the theories available on ecological structures and on the postulatum of the rate-coupling principle, we, via our structure design principle for very general dynamic systems, derived the Lotka-Volterra equations, which are the basic equations of mathematical ecology. In restricted form, these equations were introduced as early as 1847 by Verhulst [10], were discussed first by Lotka in 1920 [39] and then analysed in greater detail by Volterra in 1931 [1].

Now it is time to say something about the range of the applicability of the structure design principle. The question is, under which conditions can a given set of ordinary differential equations be generated by a finite structure consisting of only the $F$-module and linear combinations. To span a framework for the applicability of our structure design principle, we introduce the notion of *arithmetic differential expression.* For this notion, we give a recursive definition by a notation similar to the Backus notation from the algorithmic languages in informatics [71]. With this notion, we introduce the *admissable differential equation* which determines the framework for the applicability of the structure design principle.

basic variable: = (variable name, type);
comment: variable name in our case always designates a time function.
constants are considered as special time functions.
type specifies the character of the corresponding time function.

[1]) We thank Dr. Biener, Sektion Mathematik der Humboldt Universität, Berlin, and Dr. Breitenecker, Institut für technische Mathematik der Technischen Universität, Wien, for simulation studies of some systems in Lotka-Volterra representation.

type: = input / output / parameter

comment: With the typenotation, we differentiate between input variables of the corresponding equation (meaning all exogenous influences), output variables (we agree that every equation should possess only one output variable), and parameters (which themselves can be represented by time functions or functions of other occurring variables consistent with the chosen function concept).

primary variable: = basic variable / d/d$t$ (primary variable) / function

comment: Within the framework of the primary variable notion, it is possible to build derivatives of arbitrary order for a given basic variable because of the recursive character of this definition.

term: = primary variable / term × primary variable / term: primary variable.

comment: Within the framework of the term notion, arbitrary power-products with integer exponents can be constructed. Of course, de facto arbitrary real exponents are admitted, but for this purpose we must apply a corresponding function concept.

elementary expression: term / elementary expression + term / elementary expression − term

comment: With the notion of elementary expression, we can get arbitrary linear combinations of terms.

arithmetic differential expression: elementary expression / arithmetic diff. expression +/−/×/: elementary expression

comment: With the notion of arithmetic differential expression, we can build all expressions which are possible by making use of the arithmetic operations +/−/×/: only.

Remark: Because of this kind of recursive generation, we can avoid putting brackets within the expression, which, without brackets, left-hand priority automatically holds.

Later on, we must still explain the notion of *function*. Especially important for the application of our structure design principle is the notion of *admissable differential equation*. For the one-dimensional case, we introduce this notion in the following way:

admissable differential equation :

$$= (x, \text{output})^{(n)} =$$
$$\text{arith.diff.expr.}\left((x, \text{outp.})^{(j)}, (y, \text{inp.})^{(k)}, (p, \text{parameter})\right)$$
$$\text{with } j = 0, 1, 2, \ldots, n-1, k \text{ unrestricted.}$$

For the multivariable case, the notion of *admissable differential equation* can be introduced similarly.

— We take a finite number of one-dimensional admissable differential equations for the corresponding output variables $x_j$, which attain the maximum order $n_j$.

— In the equation for $x_i$, the maximum order of $x_i$ on the right hand side should be $m_j < n_j$. That means that we assume that the whole system of differential equations is completely resolved after the derivatives of highest order for every output variable.

— Also, in the multivariable case, we use a function concept taken from the one-dimensional case, meaning that we admit only static nonlinearities which depend always on one single variable.

Now we must explain our understanding of the primary variable *function*.

function: $= f(x)$ with $f$ being a solution of an admissable differential equation of the form

$$f^{(n)}(x) = \text{arith.diff.expr.}\left(f^{(j)}(x); j = 0, 1, \dots, n-1, p(t), q(x)\right)$$

comment: Here, the independent variable $t$ (time) of the dynamic systems is substituted by the state-variable $x$, meaning that all derivatives must be taken after the variable $x$.

This function concept generalises the corresponding concept from the algorithmic languages. For example, we can now construct powers $x^k$ with arbitrary real exponent $k$, $e^{ax}$, $\sin x$, $\cos x$ etc. In the following, we present a sketch of the proof for the following main theorem.

Main Theorem of the Structure Design Principle:

All one-dimensional and multidimensional ordinary differential equations which are admissible according to the definitions introduced above can be (formally) represented by a finite structure using only the base module *Logarithmic Derivative* and linear combinations of the state-variables according to our structure design principle (with introduction of some additional help state-variables). To avoid a unnecessary and tremendous formalism, we give here the idea of the proof only for the case of a one-dimensional admissible differential equation taking into account that with these considerations, for every case a realising structure can be found immediately and may be later further simplifications.

For simplicity, we assume that we have only one input variable $y$ and also only one time-dependent parameter $p(t)$. The considerations will be a little more complicated for the case of some input variables $y_j$ and some parameters $p_k$ being either functions of time or functions of a corresponding state-variable $x$. Therefore, we are dealing with a differential equation of the following form

$$x^{(n)} = \text{ADE}\,(x^{(j)}, y^{(i)}, p)\,, \qquad j = 0, 1, \dots, n-1\,, \qquad i = 0, 1, \dots\,.$$

After application of rational operations on given variables, always we get expressions, which can be brought on the following form

$$\text{ADE} = \sum T_i / \sum \tilde{T}_j\,.$$

Here, $T_i$ and $\tilde{T}_j$ are terms according to the definition introduced above that meaning power-products in the basic variables $x^{(j)}$, $y^{(i)}$ and $p$, exposing the form

$$T_i = \prod_{j,r} x^{(j)k_{ij}} y^{(r)l_{ir}} p^{s_i}\,.$$

During the structure design process, ADE must be reduced via the reduction of all these terms to find a finite realising structure. First, we must provide chain structures for $x(t)$ as well as for $y(t)$ up to a certain level for the generation of all of the help state-variables we need to form the terms in the given ADE. Therefore, we start our design process with corresponding tower constructions for $x(t)$ and $y(t)$. Tower construction for $x(t)$: We begin with $x(t) = x_0$. Then we have

$$Fx = x^{(1)}/x = x_1\,, \qquad Fx_1 = x^{(2)}/x^{(1)} - x_1 = x_2 - x_1\,.$$

In general, we have

$$Fx_{i-1} = x_i - x_{i-1} \quad \text{with} \quad x_i = x^{(i)}/x^{(i-1)}\,.$$

We must continue this construction up to the level $i = n$ and get in finite terms

$$Fx_{n-1} = x_n - x_{n-1}\,.$$

Now we have the opportunity for the first time to make use of the given admissible differential equation. This part of a tower obviously has the form of a Riccati tower (see Section 3.3). This tower immediately presents us the following variables

$$x_i = Fx^{(i-1)} = x^{(i)}/x^{(i-1)}$$

for $i = 0, 1, \ldots, n$.

We can construct corresponding a Riccati tower up to an arbitrary level in the same way for the input signal $y(t)$, which gives us the following signals

$$y_i = Fy^{(i-1)} = y^{(i)}/y^{(i-1)}\,.$$

Now we must generate the signal $x_n$ by using the given admissible differential equation for $x$ by a suitable structure. We have

$$x_n = x^{(n)}/x^{(n-1)} = \frac{\text{ADE}}{x^{(n-1)}} = \frac{\sum T_i/x^{(n-1)}}{\sum \tilde{T}_j}\,.$$

A large reduction effect can be achieved if we adjoin for every summand

$$\frac{T_i/x^{(n-1)}}{\sum\limits_j \tilde{T}_j}$$

a corresponding additional state-variable.

We need not take into account separately the factor $x^{(n-1)}$, since this factor can be immediately included into the corresponding term $T_i$. Therefore we have to realise by a suitable structure the following variable:

$$\eta = \frac{\Pi\, x^{(j)k_{ij}}\, \Pi\, y^{(r)l_{ir}}\, p^{s_i}}{\sum \tilde{T}^t}\,.$$

Now we apply a design step to $\eta$ by determining the corresponding logarithmic derivative. In the result we get

$$F\eta = \sum x_{j+1}k_{ij} + \sum l_{ir}y_{r+1} + s_iFp - \frac{\sum \dot{\tilde{T}}_j}{\sum \tilde{T}_s}\,.$$

Because $j \leqq n - 1$ all variables $x_{j+1}$ and $y_{r+1}$ will be supplied by the corresponding towers already constructed. $F_p$ can be considered as an external variable and could be generated by a separate structure. Therefore, for the realisation of $\eta$, we still need a corresponding structure for the generation of the ratio

$$\frac{\sum \dot{\tilde{T}}_j}{\sum \tilde{T}_r}\,.$$

After the derivation of the terms in the counter, every term derivative produces a whole sum of terms. Therefore, the expression above assumes the following form

$$\tilde{\eta} = \frac{\sum T_s^*}{\sum \tilde{T}_r}.$$

The generation of $\tilde{\eta}$ is now quite similar to the realisation of $\eta$. We deal separately with every summand

$$\xi_s = T_s^* / \sum \tilde{T}_j .$$

We apply to $\xi_s$ a logarithmic design step and, with the same considerations as in the case of $\eta$, we get

$$F\xi_s = \sum x_{j+1} k_{ij}^s + \sum l_{ir}^s y_{r+1} + Y s_i^s F p - \tilde{\eta} .$$

This step stops the design process because $\tilde{\eta}$ now reproduces itself.

With this scheme of the proof we can be sure that all admissable differential equations can be represented by a finite structure at least formally. However this does not mean that it is always necessary to choose the way over a rational function at the initial stage of the design process. In principle, we can start with any representation (also implicit ones) of an ADE. In this principle are included the degrees of freedom in the structure design, which thus can lead to quite different representations in that the Lotka-Volterra equations will be in some sense, all equivalent to each other.

On the other hand the successful formal design of a corresponding structure does not guarantee that the result can be used for a good reproduction of the behaviour by simulation. On the contrary, in order to assure good simulation properties we must occupy ourselves deeply with the numerical properties of the Lotka-Volterra model we have derived. In special cases, this can lead to the necessity still to perform complicated transformations to get a regular numerical behaviour. At the end of this section, we will show how we should proceed if primary variables of the type *function* are contained in the ADE.

The function $f(x)$ must obey as a function depending on $x$, an admissable differential equation of that means

$$f^{(m)}(x) = \mathrm{ADE}\left(f^{(j)}(x), p(t), q(x)\right) , \qquad j = 0, 1, \dots , m - 1 .$$

Now we need a certain initial tower for the function $f(x)$. We get this tower for example in the following way

$$f_0 = f(x) ,$$

$$Ff_0 = \frac{f^{(1)}}{f} x^{(1)} = f_1 ,$$

$$Ff_1 = \frac{f^{(2)}}{f^{(1)}} x^{(1)} - f_1 + F(x^{(1)}) .$$

The general step is

$$Ff_i = f_{i+1} - f_i + F(x^{(1)}) \quad \text{with} \quad f_i = \frac{f^{(i)}}{f^{(i-1)}} x^{(1)} .$$

This initial tower will be constructed until we meet the signal

$$f_n = \frac{f^{(n)}}{f^{(n-1)}} x^{(1)} .$$

Now the reduction can start with the help of the given admissable differential equation for $f(x)$

$$f_n = \frac{\mathrm{ADE}\left(f^{(j)}(x),\, p(t),\, q(x)\right) x^{(1)}}{f^{(n-1)}}.$$

This reduction is performed similarly so that formerly described for differential equations over time $t$. We have only to take into account the additional influence of $x^{(1)}$ handled as an external variable. $q(x)$ will be realised in the same way as $f(x)$, meaning that means this structure design process leads to a hierarchical structure.

The considerations above can lead to the belief that the described structure design approach could be applicable without any restrictions. However, many difficulties may arise which may lead to great restrictions of practical use of this approach. On the one hand, the representation of a given differential equation in Lotka-Volterra form can be accompanied by a very large number of additional introduced state-variables which we encounter in a high-dimensional nonlinear problem, and also in cases where the original differential equation has a very low dimension. This is the case especially where the rate-coupling principle is not naturally given as a functional mechanism in the system's behaviour. In such a case an ecological viewpoint gives large distortions of the real phenomena.

On the other hand there are many of differential equations with great practical importance where we cannot assume that a sufficiently large number of derivatives of the trajectories exists. For example, in control theory we frequently meet nonlinear static characteristics (relays, hysteresis) with restricted differentiability in closed feedback loops. Here, our structure design approach can only give us approximate solutions if we approximate nondifferentiable characteristics by curves with a sufficiently large number of existing derivatives. A prediction of the consequences of such an approximation on the quality of the Lotka-Volterra representation is impossible. In spite of the demonstrated possible reduction of arbitrary ADE's by the structure design principle, the full range of applicability of this concept is still not identified.

The constructive considerations give us only sufficient conditions for imbedding of given ordinary differential equations into a finite structure. Therefore, we adjoin still more ideas for a future better understanding of this approach[1]).

We propose to consider this approach extended on complex variables in the Gaussian plane and concentrate the application of the structure design concept on so-called meromorphic functions. A meromorphic function $M(z)$ is any quotient of integer analytical functions $H(z)$ and $G(z)$

$$M(z) = H(z)/G(z)\,.$$

Integer analytical functions are characterized by the fact that they possess a Taylor series which converges everywhere

$$H(z) = \sum h_i z^i\,.$$

The necessary and sufficient condition for this convergence is the condition

$$(|h_i|)^{1/i} -> 0 \quad \text{for} \quad i -> \infty\,.$$

[1]) We thank Dr. Vogl, Institut für technische Mathematik, Technische Universität Wien, for some first ideas for studying the Lotka-Volterra structure design principle in the complex plane.

Obviously, the following operations map a given meromorphic function onto a meromorphic function

$$p = \mathrm{d}/\mathrm{d}t,\ +,\ /,\ -,\ \times,\ :$$

and most importantly

$$F = \mathrm{d}\ \ln/\mathrm{d}t\ .$$

This means that in the area of meromorphic functions, our structure design approach seems to be quite natural.

We should first restrict the structure design problem on function generators for time functions

$$M(t) = H(t)/G(t) \quad \text{with integer analytical functions } H(t) \text{ and } G(t)$$

with the hope that deriving a natural way to form the given function into a Lotka-Volterra representation may be into a relatively simple one (normal form idea). Because of

$$FM(t) = FH(t) - FG(t)$$

this problem reduces to the structure design for time functions their being specialisations of integer analytical functions. However we meet immediately the question of which conditions an integer analytical function $H(t)$ possesses a finite structure according our structure design approach; because it is clear that not every such function obeys an admissable autonomous ordinary differential equation. Perhaps in this way we can find counter-examples for our structure design approach showing that not every integer analytical function can be represented by a finite structure in our sense.

## 3.5. *Ordinary and Regular Volterra Representations*

On the different levels of the structure design process we meet various degrees of freedom in the introduction of additional state-variables. Therefore, we can get, as a result of the design process, different Volterra representations. These different representations may have a different dimension than the corresponding state-space.

We introduce the notion of *Volterra order* as the minimum number for a given state-space of the possible Lotka-Volterra representations corresponding to a given ordinary differential equation (general for all possible initial values configurations). It is very useful if we are able to restrict the behaviour of the trajectories on the positive cone

$$\xi_i \geqq 0$$

because then the occurrence of parasitic poles arising from roots of $f(t)$ or its derivatives can be avoided.

If we want always to take this condition into account during the design process we should introduce on all levels only those additional state-variables which are forbidden to assume negative values.

We call the minimum order of a Lotka-Volterra representation to a given ordinary differential equation under all such representations fulfilling the additional condition to be concentrated completely in the positive cone $\xi_i \geqq 0$ *regular Volterra order*. Obviously, the regular Volterra order cannot be less than the Volterra order. It is not clear whether every system with a finite Volterra order has at the same time always a finite regular Volterra order.

Let us give a simple example which shows the difference between an ordinary Lotka-Volterra representation and a regular Lotka-Volterra representation.

a) Formal structure design of a sine generator

$$x(t) = x_0 = \sin t ,$$

$$Fx_0 = \cos t/\sin t = x_1 ,$$

$$Fx_1 = -\sin t/\cos t - x_1 .$$

We put $x_2 = \sin t/\cos t$ and thus have

$$Fx_2 = x_1 + x_2 .$$

We get as a representation of $x(t) = \sin t$ the following Lotka-Volterra equations

$$Fx_0 = x_1 ,$$

$$Fx_1 = -(x_1 + x_2) ,$$

$$Fx_2 = (x_1 + x_2) .$$

Obviously $z = x_1x_2 = 1$ is a first integral of the corresponding motion.

The uniquely determined trajectory $x_0 = \sin t$ is the first coordinate of the corresponding state-vector after imbedding into the state-space $(x_0, x_1, x_2)$. Fig. 81 illustrates thé situation. It represents the projection of the motion in the plane $(x_1, x_2)$.

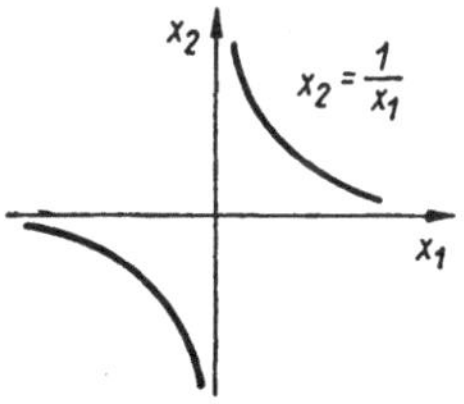


**Fig. 81** The projection of the motion of a nonregularized basic harmonics sin $t$ into a two-dimensional state-plane

Obviously, the trajectory is a curve on the cylindrical shells $x_1x_2 = 1$ in the third and first quadrants. The imbedded trajectory gives between two consecutive poles, $x_0$ respective $x_2$ with finite values of $x_0$, a part of the curve corresponding always to a quarter of a period of the sine signal. This is similar to what we have found in the case of the mathematical pendulum.

b) Design of a regular Lotka-Volterra representation of a sine signal

We must avoid during the design process any change of sign in the state-variables. Moreover we will now take the normalisation into account. A heuristic way to avoid the change of signs of corresponding signals is to shift them with a certain additive parameter. For this purpose we now put

$$x_0 = a + \sin t \quad \text{with} \quad a > 1 .$$

Then we have

$$\xi_0 = 1 + \alpha \sin t \quad \text{with} \quad \alpha = 1/a,$$

$$F\xi_0 = \alpha \frac{\cos t}{1 + \alpha \sin t} = \alpha \frac{(\cos t + b)}{1 + \alpha \sin t} - b \frac{\alpha}{1 + \alpha \sin t} \qquad \text{with } b > 1 \,,$$

$$F\xi_0 = \alpha(1 + b)\,\xi_1 - b\alpha\xi_2$$

with the help state-variables

$$\xi_1 = \frac{(\cos t + b)}{(1 + \alpha \sin t)\,(1 + b)}\,, \qquad \xi_2 = \frac{1}{1 + \alpha \sin t}\,.$$

Further on we have

$$F\xi_1 = -\frac{\sin t}{\cos t + b} - F\xi_0\,, \qquad F\xi_2 = -F\xi_0\,.$$

We have

$$\frac{\sin t}{\cos t + b} = \frac{1}{\alpha}\frac{(1 + \alpha \sin t)}{(\cos t + b)} - \frac{1}{\alpha\,(\cos t + b)} = \frac{1}{\alpha(1 + b)}\xi_3 - \frac{1}{\alpha(1 + b)}\xi_4\,.$$

We have introduced the help variables

$$\xi_3 = \frac{(\alpha \sin t + 1)}{(\cos t + b)}(1 + b)\,, \qquad \xi_4 = \frac{(1 + b)}{(\cos t + b)}\,.$$

Further on we have

$$F\xi_3 = F\xi_0 + \frac{1}{\alpha(1 + b)}\xi_3 - \frac{1}{\alpha(1 + b)}\xi_4\,,$$

$$F\xi_4 = \frac{1}{\alpha(1 + b)}\xi_3 - \frac{1}{\alpha(1 + b)}\xi_4\,.$$

We have now derived the following regular Lotka-Volterra representation for the basic harmonic oscillations

$$F\xi_0 = \left(\alpha(1 + b)\,\xi_1 - b\alpha\xi_2\right),$$

$$F\xi_1 = -\left(\alpha(1 + b)\,\xi_1 - b\alpha\xi_2\right) - \left(\frac{1}{\alpha(1 + b)}\xi_3 - \frac{1}{\alpha(1 + b)}\xi_4\right),$$

$$F\xi_2 = -\left(\alpha(1 + b)\,\xi_1 - b\alpha\xi_2\right),$$

$$F\xi_3 = \left(\alpha(1 + b)\,\xi_1 - b\alpha\right)\xi_2 + \left(\frac{1}{\alpha(1 + b)}\xi_3 - \frac{1}{\alpha(1 + b)}\xi_4\right),$$

$$F\xi_4 = \left(\frac{1}{\alpha(1 + b)}\xi_3 - \frac{1}{\alpha(1 + b)}\xi_4\right).$$

Obviously this extended system has the following three first integrals

$$\xi_0\xi_2 = 1\,, \qquad \xi_1\xi_0\xi_4 = 1\,, \qquad \xi_3 = \xi_0\xi_4\,.$$

Therefore, in this case we get a regular trajectory in the positive orthant $\xi_i > 0$. That means that we get a representation of the sine signal without an interruption by induced parasitic poles for the price of an increased dimension of the system, in comparison with the Lotka-Volterra representation without these conditions. At the same time we see that the resulting motion is bound to the intersection of the three hyperphases given by the three first integrals. Therefore, in fact the motion is again two-dimensional if we would use the freedom of initial values. A further reduction of the dimension is impossible, because the sine signal itself obeys an ordinary second-order differential equation.

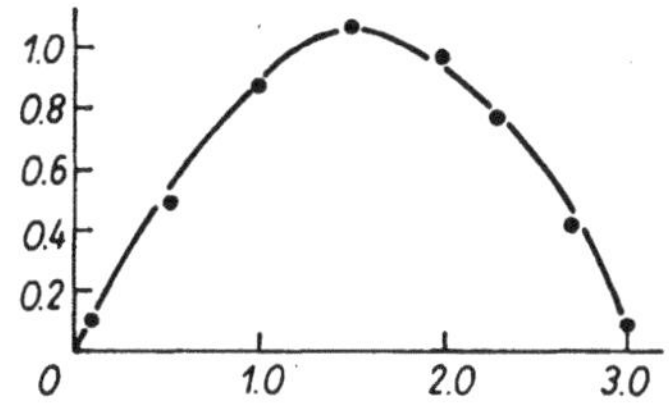


**Fig. 82**
Simulation of a regularized harmonic oscillation

## *3.6. Computer-Aided Design of Lotka-Volterra Representations*

We give here the general features of a possible computer aided design process of finite structures for the realisation of given sets of ordinary differential equations. The primary question is which steps of the structure design process should be delegated to the computer.

The a priori information of the design could be given in directly the form of analytical expressions for signals, or implicitly prescribing the behaviour by a set of ordinary differential equations using corresponding formulas to express the differential equations analytically. Here rational formulas play an important role, but, besides in these expressions functions given by their names can also occur. Therefore, we first need a suitable notation of these formulas for representation within the computer. For this purpose, algorithmic languages such as ALGOL, FORTRAN, PASCAL, etc. can be used. They offer two possibilities for formula notation.

Formulas can be expressed explicitly by an arithmetic expression or implicitly by a formula generator. However here we meet a difference from the common use of algorithmic languages for the expression of formulas. The representation of a formula by use of an algorithmic language has the aim of evaluating this formula.

On the contrary, we are not interested in computing values for these formulas; we are seeking a convenient notation of the corresponding formula. This difficulty can be easily overcome, however. We have only to interpret the value assignment statements as introductions of new names for right-hand expressions. The only danger here could be the usage of the same variable names on the left and the right sides of such a statement. This can be avoided without any problem. With this interpretation, we can use the formula expressions given by algorithmic languages as a basis for our aim, which is directed more toward something like formula manipulation than toward determination of values.

As we pointed out in Section 3.4, we need more degrees of freedom for the introduction of function names. This, however, is not a problem because for the most commonly functions used there must also be programs for the generation of these functions. Any of those programs can be interpreted as a generation of the function formula in the sense of formula manipulation. Because functions in our case must also obey admissable differential equations, the same regulations for their notations that we discussed previously can be applied. This is connected with a certain type of subprogram technique in formula manipulation software.

The basic operation in our structure design process is the application of the logarithmic derivative to a given expression, meaning

$$F = \mathrm{d}\ln/\mathrm{d}t \quad \text{respective} \quad \mathrm{d}\ln/\mathrm{d}x \qquad \text{(in case of a function).}$$

This is nothing but the computation of a new formula

$$\dot{A}/A$$

for a given formula $A$. This is a solved problem in formula manipulation. Within the system must be implemented the algebraic properties of *logarithmic derivation* such as

$$F(A \times B) = FA + FB\,, \qquad F(A : B) = FA - FB\,,$$

$$F(cA) = FA \qquad \text{if } c \text{ is a constant,}$$

$$F(A^k) = kFA\,,$$

$$F(A + B) = \frac{A}{A + B} FA + \frac{B}{A + B} FB\,.$$

We thus have the possibility of transforming given formulas in the computer in the same way that we used to do it by hand.

In every step of the design process after the application of $F$, we like to have the freedom to choose suitable subexpressions for which we like to introduce variable names, meaning that, we ourselves will determine which additional state-variables we should introduce. This demands that we use the formula manipulation system in an interactive mode. This should not exclude the possibility of applying standard procedures to formulas of a certain type, giving us more automation with which we derive a Lotka-Volterra structure.

The design process leads to a unified description by the Lotka-Volterra equations on one hand and to a finite structure on the other hand. It seems to be useful to represent, with the help of the computer, intermediate states, for example, to show a part of all equations which already have the Lotka-Volterra form, or to demonstrate that part of the structure which is already constructed. For the demonstration of that part of the structure already constructed it is useful to apply computer graphics. This is valuable information which gives some hints on the best way to continue the design process. The use of different colours for the representation of the structure can also be of interest. Because the whole design process has the character of an evolutionary tree we would propose that the same colour be used for all parallel applications of $F$, together with the resulting feedbacks in the graph. Parallel application of $F$ are recommended for all new variables occurring after a parallel design step. The *depth of the structure* would then be the number of different colours in the graph, if we use in the design process the kind of parallel processing considered above. The next cry for help from the side of a computer comes if we are not satisfied for other reasons, with the constructed structure.

Then the problem of transforming a given representation into another one more suitable for us arises.

We shall discuss this theme later on, after the introduction of some equivalence transformations for the Lotka-Volterra representations.

# 4. Systematisation and Unification of Growth Models with Lotka-Volterra Equations

## 4.1. *Hyperlogistic Growth as a Model for EVOLON*

Hyperlogistic growth is given by the $s$-shaped solutions of the hyperlogistic differential equation (power-product driving forces)

$$dx/dt = Kx^k(B - x^w)^l .$$

Because all considerations here are oriented toward practical applications, we are interested in a regular Volterra representation.

During the structure design process, we get the following sequence of expressions

$$x(t) = x(0) ,$$

$$F\xi_0 = K'(B' - 1)^l \xi_1$$

with

$$K' = Kx(0)^{k+w-1} , \qquad B' = B/x(0)^w , \qquad \xi_1 = \xi_0^{k-1}(B' - \xi_0^w)^l/(B' - 1)^l ,$$

$$F\xi_1 = (k - 1)\ K'(b' - 1)^l \xi_1 - wlK'(B' - 1)^{l-1} \xi_2$$

with

$$\xi_2 = \xi_0^{k-1+w}(B' - \xi_0^w)^{l-1}/(B' - 1)^{l-1} ,$$

$$F\xi_2 = (k + w - 1)\ K'(B' - 1)^l \xi_1 - w(l - 1)\ K'(B' - 1)^{l-1} \xi_2 .$$

With general parameter notations, we get the following structural equations for the structure, shown in Fig. 83

$$F\xi_0 = K_0\xi_1 ,$$

$$F\xi_1 = K_1\xi_1 + L_1\xi_2 ,$$

$$F\xi_2 = K_2\xi_2 + L_2\xi_1 .$$

The parameters of the representation have been chosen in such a way that we always fulfill the normalisation conditions

$$\xi_i(0) = 1 \quad \text{for} \quad i = 0, 1, 2 .$$

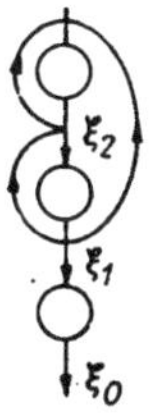


**Fig. 83** Structure representation of the power-product driving force model (hyperlogistic equation)

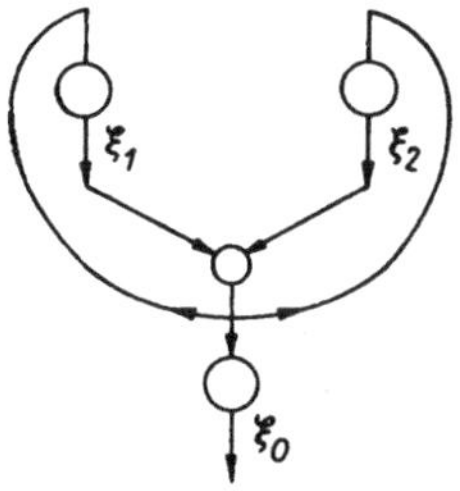


**Fig. 84** The structure representation for Bertalanffy's growth model

Of special interest is the case of the Bertalanffy growth model, given by the differential equation

$$dx/dt = ax^m - bx^n \quad \text{with} \quad m \geqq n\,.$$

Usually this is a special case of the hyperlogistic model, namely for $a$, $b > 0$ and $m > n$.

Here another structure realisation seems to be quite useful. We again put

$$x(t) = x(0)\,\xi_0$$

and get first

$$F\xi_0 = a'\xi_0^{m-1} - b'\,\xi_0^{n-1}$$

with $a' = ax(0)^{m-1}$ and $b' = bx(0)^{n-1}$.

We introduce two additional state-variables

$$\xi_1 = \xi_0^{m-1} \quad \text{and} \quad \xi_2 = \xi_0^{n-1}$$

and get then immediately the following Lotka-Volterra equations

$$F\xi_0 = a'\xi_1 - b'\xi_2\,,$$

$$F\xi_1 = (m-1)\,(a'\xi_1 - b'\xi_2)\,,$$

$$F\xi_2 = (n-1)\,(a'\xi_1 - b'\xi_2);$$

Fig. 84 shows the corresponding structure. It looks quite different from the structure in Fig. 83. Nevertheless, both structures can represent solutions of the Bertalanffy's growth model. This is therefore an example for the possibility to represent a given trajectory by different Lotka-Volterra structures.

## *4.2. Parameter Identification for the Hyperlogistic Growth Model*

In the following we describe an approach for the determination of the parameters $K_j$, $L_j$ of the hyperlogistic growth model in Lotka-Volterra representation, from which we then can determine the corresponding parameters $K$, $k$, $l$, $w$ and $B$ of the original hyperlogistic differential equation by using only a measured part of the trajectory of $\xi_0(t)$, meaning that we do not assume the measurability of the additional state-variables $\xi_1$ and $\xi_2$. At the same time this is the backward integration of the structural equations in this case, offering the possibility to determine all types of functions being represented by a Lotka-Volterra structure according to Fig. 85.

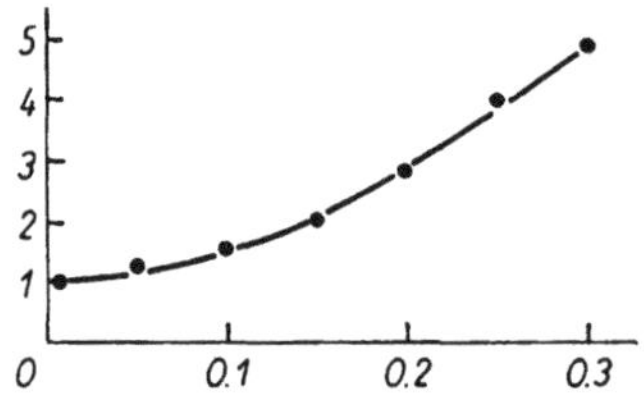


**Fig. 85** The simulation of Bertalanffy's growth model

At first we compute from the measured trajectory $\xi_0(t)$ the curve

$$w = F\xi_0 = \dot{\xi}_0/\xi_0\,.$$

From the viewpoint of numerical computation, some problems may arise, but parasitic poles cannot meet because, in this case $\xi_0$ cannot have roots.

After the substitution of $w = K_0\xi_1$ in the rest of equations, we get

$$Fw = \varkappa_1 w + q\,,$$

$$Fq = \varkappa_2 q + \lambda w\,.$$

Here we introduced the following parameters

$$\varkappa_1 = K_1/K_0\,, \qquad \varkappa_2 = K_2/L_1 \quad \text{and} \quad \lambda = L_2/K_0\,.$$

For the computation of $Fw$ we must divide by $\dot{\xi}_0$, but this cannot lead to great difficulties up to random disturbances.

With a fixed parameter $\varkappa_1$ which we afterwards modify adaptively, we introduce the signal

$$q(\varkappa_1) = Fw - \varkappa_1 w\,.$$

For the computation of $Fq$ we have to expect a root, because the EVOLON has a turning point. The neighbourhood of such a point must be removed from the given time series. However this can inflect different time intervals because it depends on the value of $\varkappa_1$. Finally, we can estimate from the last equation with the help of linear regression in dependence on the parameter $\varkappa_1$ the parameters $\varkappa_2(\varkappa_1)$ and $\lambda(\varkappa_1)$. Here, we can use the quadratic deviation $Q(\varkappa_1)$ as an objective function for the determination of the best value of $\varkappa_1$ by minimization. This must be done within a dominating seek cycle for the parameter $\varkappa_1$. Thus, we have obtained estimations for the parameters $\varkappa_1$, $\varkappa_2$ and $\lambda$. Further on, we have

$$K_0 = w(0)\,.$$

Therefore, we can get at once

$$K_1 = K_0\varkappa_1 \quad \text{and} \quad L_2 = K_0\lambda\,.$$

Then we have

$$Fw(0) = K_1 + L_1$$

which leads immediately to

$$L_1 = Fw(0) - K_1$$

and finally we get $K_2$ from

$$K_2 = L_1\varkappa_2\,.$$

Now we are able to determine principally the unknown parameters of the hyperlogistic differential equation from the following equations, which we get by immediate comparison

$$K_0 = K(B' - 1)^l\,, \qquad K_1 = (k-1)K_0\,, \qquad L_1 = wlK_0/(B'-1)\,,$$

$$L_2 = (k - 1 + w)\,K_0\,, \qquad K_2 = -w(l-1)\,K_0/(B'-1)\,.$$

From these equations, we immediately get the following expressions

$$k = 1 + K_1/K_2\,, \qquad w = (L_2 - K_1)/K_0\,, \qquad l = 1/(1 - K_2/L_1),$$

$$B' = 1 + |(L_2 - K_1)/(L_1 - K_2)|\,, \qquad K' = K_0\,|(L_1 - K_2)/(L_2 - K_1)|\,.$$

This method for determining the parameters has the disadvantage that it is oriented a priori toward only one of the possible function types defined by a structure according to Fig. 84, whereas the measured time series could lead to another function type also consistent with the given structure. Therefore, we present still another possibility for the determination of the parameters by backward integration of the given Lotka-Volterra model.

From the structural equations, we get immediately

$$\xi_2 = \frac{1}{L_1}\left(F\xi_1 - \frac{K_1}{K_0}F\xi_0\right) = \frac{1}{L_1}F(\xi_1/\xi_0^{\varkappa_1}) .$$

We substitute this expression into the last structural equation and get

$$F\xi_2 = \varkappa_2 F(\xi_0/\xi_0^{\varkappa_2}) + \lambda F\xi_0 .$$

If we take into account the initial value normalisation, we get at once from this equation the following first integral

$$\xi_2 = \xi_1^{\varkappa_2}\xi_0^{\lambda-\varkappa_1\varkappa_2} .$$

Therefore we get the following differential equation for the determination of $\xi_1$:

$$F(\xi_1/\xi_0^{\varkappa_1}) = L_1\xi_1^{\varkappa_2}\xi_0^{\lambda-\varkappa_1\varkappa_2} .$$

$\xi_0$ is a monotonous function oh time $t$. In order to simplify the integration, we now choose $\xi_0$ instead of time as the independent variable, meaning that we determine a state-space trajectory. For this transformation we must take into account

$$\mathrm{d}/\mathrm{d}t = \mathrm{d}/\mathrm{d}\xi_0 \cdot \mathrm{d}\xi_0/\mathrm{d}t = \mathrm{d}/\mathrm{d}\xi_0 \cdot K_0\xi_0\xi_1 .$$

With the notation

$$F_{\xi_0} = \mathrm{d}\ \ln/\mathrm{d}\xi_0$$

we get

$$F_{\xi_0}(\xi_1/\xi_0^{\varkappa_1}) = \frac{L_1}{K_0}\xi_1^{\varkappa_2-1}\xi_0^{m-1} = F_{\xi_0}\xi_1 - \varkappa_1/\xi_0$$

with

$$m = \lambda - \varkappa_1\varkappa_2 .$$

We get the following special cases:

1. $\varkappa_2 = 1$.

1.1. $m = 0$.

Then we have

$$\frac{\mathrm{d}}{\mathrm{d}\xi_0}\ln(\xi_1/\xi_0^{\varkappa_1}) = \frac{L_1}{K_0}\xi_0^{-1}$$

and, after integration

$$\xi_1 = \xi_0^{\varkappa_1 + L_1/K_0} .$$

If we put this result into the first Volterra equation, we get

$$\mathrm{d}\xi_0/\mathrm{d}t = K_0\xi_0^{1+\varkappa_1+L_1/K_0} .$$

Therefore, we get, in this special case, the normal autocatalytic differential equation

$$\mathrm{d}\xi_0/\mathrm{d}t = K_0\xi_0^k$$

and we get by comparison

$$k = 1 + (K_1 + L_1)/K_0 .$$

1.2. Now we assume $m \neq 0$.
In this case, we have

$$\mathrm{d}/\mathrm{d}\xi_0 \ln (\xi_1/\xi_0^{\varkappa_1}) = \frac{L_1}{K_0} \xi_0^{m-1} .$$

Now the integration leads to the result

$$\ln (\xi_1/\xi_0^{\varkappa_1}) = \frac{L_1}{K_0} \frac{(\xi_0^m - 1)}{m}$$

belonging to the following initial differential equation

$$\mathrm{d}\xi_0/\mathrm{d}t = K\xi_0^k \mathrm{e}^{w\xi_0^l} .$$

By comparison of the corresponding parameters, we get the following equation for the determination of the coefficients

$$K = K_0 \mathrm{e}^{-L_1/(K_0 m)} , \qquad k = \varkappa_1 + 1 , \qquad w = L_1/(K_0 m) , \qquad l = m = \lambda - \varkappa_1\varkappa_2 .$$

2. $\varkappa_2 \neq 1$.
Now we perform in the differential equation

$$F_{\xi_0}(\xi_1/\xi_0^{\varkappa_1}) = \frac{L_1}{K_0} \xi_1^{\varkappa_2-1}\xi_0^{m-1}$$

a transformation of the variables putting

$$\xi_1 = u^r$$

and get the result

$$r\, \mathrm{d}u/\mathrm{d}\xi_0 - \varkappa_1 u/\xi_0 = \frac{L_1}{K_0} u^{1+r(\varkappa_2-1)}\xi_0^{m-1} .$$

We choose

$$r = 1/(1 - \varkappa_2)$$

and get the following linear differential equation for $u$

$$\mathrm{d}u/\mathrm{d}\xi_0 - \frac{\varkappa_1}{r} u/\xi_0 = \frac{L_1}{K_0 r} \xi_0^{m-1} .$$

We solve this equation by the *variation of the constant* method. The general solution of the homogenous equation is

$$u(\xi_0) = c\xi_0^{1/r} .$$

From this the concept for the solution of the inhomogenous equation is

$$u(\xi_0) = c(\xi_0)\, \xi_0^{\varkappa_1/r} .$$

By substitution into the non-homogenous equation, we get the following differential equation for the determination of $c(\xi_0)$:

$$\mathrm{d}c(\xi_0)/\mathrm{d}\xi_0 = \frac{L_1}{K_0 r} \xi_0^{m-1-\varkappa_1/r} = \frac{L_1}{K_0 r} \xi_0^{\lambda-\varkappa_1-1} .$$

Now the solution process branches again.

2.1. $\lambda = \varkappa_1$, $L_2 = K_1$.
Here we have

$$c(\xi_0) = c(1) + \frac{L_1}{K_0 r} \ln \xi_0 .$$

With this result we get

$$\xi_1 = \xi_0 \left( c(1) + \frac{L_1}{K_0 r} \ln \xi_0^r \right)^r .$$

Because of the initial values normalisation $\xi_i(0) = 1$, we get $c(1) = 1$.

We must compare the result with the corresponding differential equation

$$d\xi_0/dt = K \xi_0^k (B - \ln \xi_0)^l .$$

For the case $L_1/(K_0 r) < 0$ we get the following equations for the determination of the coefficients (otherwise, the result would not belong to an EVOLON model)

$$K = K_0 \left| \frac{L_1}{K_0 r} \right|^r , \qquad k = \varkappa_1 + 1 \quad \text{and} \quad B = \left| \frac{K_0 r}{L_1} \right| , \qquad l = r .$$

2.2. Now $\lambda \neq \varkappa_1$ (the most general case).
In this case we get for $c(\xi_0)$ the result

$$c(\xi_0) = c(1) + \frac{L_1}{K_0 r(\lambda - \varkappa_1)} (\xi_0^{\lambda - \varkappa_1} - 1)^r .$$

Therefore, we get now (again with $c(1) = 1$)

$$\xi_1 = \xi_0^{\varkappa_1}(1) + \frac{L_1}{K_0 r(\lambda - \varkappa_1)} (\xi_0^{\lambda - \varkappa_1} - 1)^r .$$

By parameter comparison with the corresponding differential equation

$$d\xi_0/dt = K \xi_0^k (B - \xi_0^w)^l$$

we get the following equation for the determination of the coefficients

$$K = K_0 \left| \frac{L_1 - K_2}{L_2 - K_1} \right|^l , \qquad k = \frac{K_1}{K_0} + 1 , \qquad B = 1 + \left| \frac{L_2 - K_1}{L_1 - K_2} \right| ,$$

$$w = (L_2 - K_1)/K_0 , \qquad l = 1/(1 - K_2/L_1) .$$

## 4.3. *The Problem of Parameter Identification for the Pearson II Growth Model*

The system concept for the Pearson II growth model is

$$d\xi_0/dt = \varphi(t)$$

with $\varphi(t)$ obeying the following ordinary differential equation

$$d\varphi/dt = \varphi(t + a)/(b_0 + b_1 t + b_2 t^2) .$$

We control the structure design process in such a way that we need not know the positions of the roots, and that only non-negative state-variables $\xi_i \geqq 0$ arise with the initial values normalisation condition $\xi_i(0) = 1$. Then we get

$$F\xi_0 = \varphi(t)/\xi_0 = \varphi(0)\,\xi_1\,.$$

Without any restriction of generality, we can assume $\varphi(0) = 1$. Then we have

$$F\xi_1 = F\varphi - \xi_1\,.$$

According to Pearson's differential equation, we have

$$F\varphi = (t + a)/(b_0 + b_1 t + b_2 t^2) = \frac{a}{b_0}\xi_2\,.$$

We demand now $a \neq 0$ and $b_0 \neq 0$. Then we have

$$F\xi_2 = \frac{1}{(t+a)} - \frac{(b_1 + 2b_2)}{(b_0 + b_1 t + b_2 t^2)} = \frac{1}{a}\xi_3 - \frac{b_1}{b_0}\xi_4$$

with the additional state-variables

$$\xi_3 = \frac{a}{(t+a)}\,, \qquad \xi_4 = \frac{b_0}{b_1}\frac{(b_1 + 2b_2 t)}{(b_0 + b_1 t + b_2 t^2)}\,.$$

In the following we must assume $b_1 \neq 0$ and get

$$F\xi_3 = -\frac{1}{a}\xi_3\,, \qquad F\xi_4 = \frac{2b_2}{(b_1 + 2b_2 t)} - \frac{b_1}{b_0}\xi_4\,.$$

At last, we introduce

$$\xi_5 = b_1/(b_1 + 2b_2 t)$$

and get the equation

$$F\,\xi_5 = -2\,\frac{b_2}{b_1}\xi_5\,.$$

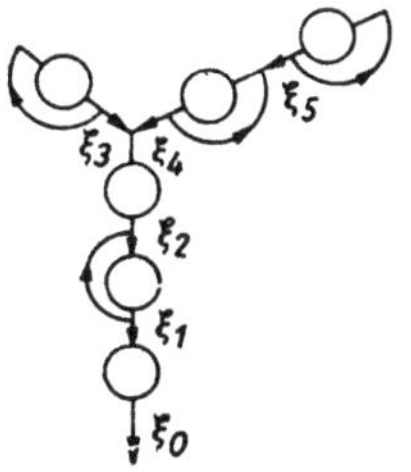


**Fig. 86** Lotka-Volterra structure of the growth models of Pearson II

The corresponding structure equation of the Lotka-Volterra representations are

$$F\xi_0 = K_0\xi_1\,,$$
$$F\xi_1 = K_1\xi_1 + L_1\xi_2\,,$$
$$F\xi_2 = M_1\xi_3 + M_2\xi_4\,,$$
$$F\xi_3 = K_3\xi_3\,, \qquad F\xi_4 = K_2\xi_4 + L_2\xi_5\,,$$
$$F\xi_5 = L_2\xi_5\,.$$

In order to identify parameters we now try to perform a backward integration of these structure equations. For this backward integration, and also for other equations, a certain base module plays an important role. We will call it *Riccati module.* This module is determined by the following equation

$$(F - \varkappa)\, x = Kx' \quad \text{with} \quad F = \mathrm{d}\ln/\mathrm{d}t\,.$$

The solution after $x$ for given input $x'$ can be computed by applying a suitable pseudo-inverse operator for $F - \varkappa$. If we again demand the normalisation condition $x(0) = 1$ here, then we get the following solution

$$x(t) = F\big(1 - \varkappa I F^{-1}(Kx')\big)^{-1/\varkappa}\,.$$

Of course, as a solution for the homogenous case with $x' = 0$, we get as eigen-solution of $F$ for the eigen-value $\varkappa$, again the hyperbola

$$x_h(t) = 1/(1 - \varkappa t)\,.$$

We shall now use these connections for the backward integration. First we obviously get

$$\xi_4(t) = F\big(1 - K_2 I F^{-1}(L_2\xi_5)\big)^{-1/K_2}\,.$$

From the equation for $\xi_2$ and this formula we get immediately

$$\xi_2(t) = \frac{1}{(1 - K_3 t)^{M_1/K_3}} \Bigg/ \left(1 - K_2 \int_0^t \frac{\mathrm{d}t}{(1 - L_3 t)^{L_2/L_3}}\right)^{M_2/K_2}.$$

From the second equation, we get

$$\xi_1(t) = F\big(1 - K_1 I F^{-1}(L_1\xi_2)\big)^{-1/K_1}$$

and, at last, for $\xi_0(t)$ follows from the first equation

$$\xi_0(t) = \frac{1}{(1 - K_1 I F^{-1}(L_1\xi_2))}\, K_0/K_1\,.$$

Up to this last quadrature we have got an explicit solution. We will now see how this result corresponds with the Pearson's differential equation. First we must for this purpose compute $\mathrm{d}\xi_0/\mathrm{d}t = \varphi(t)$ and get

$$\varphi(t) = \frac{K_0 F^{-1}(L_1\xi_2)}{(1 - K_1 I F^{-1}(L_1\xi_2))^{1 + K_0/K_1}}\,.$$

Then we get for the right-hand side of Pearson's differential equation

$$F\varphi = L_1\xi_2 + (1 + K_0 K_1) \Bigg/ \frac{K_1}{K_0} \frac{K_0 F^{-1}(L_1\xi_2)}{(1 - K_1 I F^{-1}(L_1\xi_2))}\,.$$

From this follows:
For $K_0 = -K_1$ (as it was the case in the derivation from Pearson II) we get

$$F = L_1\xi_2\,.$$

This is a generalisation of the originally given Pearson differential equation, which is reduced to the Pearson II case, if the following holds: $K_0 = -K_1 = 1$, $M_1 = -K_3 = 1/a$, $K_2 = M_2 = -b_1/b_0$, $L_2 = -L_3 = 2b_2/b_1$. If we knew the parameter of the corresponding Lotka-Volterra structure, we could determine the unknown coefficients of the Pearson II concept from the following conditions:

$$a = 1/M_1\,, \qquad b_0 = 1/(M_1L_1)\,, \qquad b_1 = -M_2/(M_1L_1)\,,$$
$$b_2 = -\tfrac{1}{2}\,(M_2L_2)/(M_1L_1)\,.$$

That means that we must solve the parameter estimation problem for the corresponding Lotka-Volterra structure. However this is already a relatively complicated problem of nonlinear regression. From a measured trajectory of $\xi_0(t)$ we get the condition

$$F\,\mathrm{d}\xi_0/\mathrm{d}t = (K_0/K_1 + 1)\,F\xi_0 + L_1\xi_2(t)\,.$$

Here $\xi_2(t)$ follows the system-concept

$$\xi_2(t) = \frac{1}{(1-K_3)^{M_1/K_3}\left(1 + \dfrac{K_2}{(L_2 - L_3)}\right)\left((1 - L_3t)^{-L_2/L_3+1} - 1\right)^{M_2/K_2}}\,.$$

With error adjustment in the mean square sense, we can expect to determine the following parameters

$$L_1,\ K_0/K_1,\ K_3,\ M_1,\ L_2,\ L_3,\ K_2 \text{ and } M_2\,.$$

For the separation of $K_0$ and $K_1$ we must take into account

$$\xi_0(t) = \frac{1}{\left(1 - K_1IF^{-1}(L_1\xi_2(t))\right)^{K_0/K_1}}$$

or the adjustment of $K_1$.

## 4.4. *The Problem of Parameter Identification for the Pearson I Growth Model*

After discussing the concept of Pearson I we now have

$$\mathrm{d}\xi_0/\mathrm{d}t = \varphi(\xi_0)$$

with $\varphi(\xi_0)$ now obeying Pearson's differential equation with $\xi_0$ as independent variable

$$F_{\xi_0}\varphi(\xi_0) = (\xi_0 + a)/(b_0 + b_1\xi_0 + b_2\xi_0^2)\,.$$

We have

$$F\varphi(\xi_0) = F_{\xi_0}\varphi(\xi_0)\,\varphi(\xi_0)\,.$$

Now we perform the structure design process on this growth concept under normalised conditions.

Without any restriction of generality, we can assume $\varphi(1) = 1$. Then we have

$$F\xi_0 = \varphi(\xi_0)/\xi_0 = \xi_1 = K_0\xi_1\,,$$
$$F\xi_1 = F\varphi - \xi_1 = \frac{(1+a)}{P_2(1)}\,\xi_2 - \xi_1 = K_1\xi_1 + L_1\xi_2$$

with

$$\xi_2 \sim \varphi(\xi_0 + a)/P_2(\xi_0)\,.$$

We must assume $1 + a \neq 0$, $P_2(1) = b_0 + b_1 + b_2 \neq 0$. Further on we get

$$F\xi_2 = \frac{1}{(\xi_0 + a)}\varphi(\xi_0) - \frac{(b_1 + 2b_2\xi_0)}{P_2(\xi_0)}\varphi(\xi_0) + F\varphi$$
$$= M_1\xi_3 + M_2\xi_4 + N_1\xi_2$$

with the additional state-variables

$$\xi_3 \sim \frac{1}{(\xi_0 + a)}\varphi, \qquad \xi_4 \sim \frac{(b_1 + 2b_2\xi_0)}{P_2(\xi_0)}\varphi,$$
$$F\xi_3 = -M_1\xi_3 + N_2\xi_2,$$
$$F\xi_4 = \frac{2b_2}{(b_1 + 2b_2\xi_0)}\varphi + K_2\xi_4 + N_3\xi_2, \qquad \xi_5 \sim \frac{\varphi(\xi_0)}{(b_1 + 2b_2\xi_0)},$$
$$F\xi_5 = L_3\xi_5 + N_4\xi_2.$$

Obviously we have

$$N_1 = N_2 = N_3 = N_4, \qquad K_2 = M_2.$$

Fig. 87 shows the corresponding Lotka-Volterra structure for the Pearson I growth model. We are surprised that this structure seems to be an extension of the structure

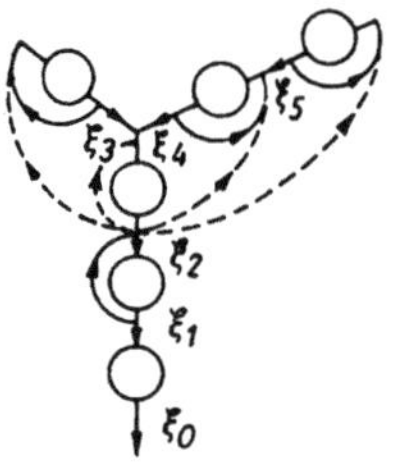


**Fig. 87** Lotka-Volterra structure of the growth model Pearson I

we have got for the Pearson II model with four additional feedback loops and the same amplification factor. This is something like a standard structure extension originating from the substitution of $\xi_0$ for $t$. However, because the structure gives more possibilities than the Pearson I model, from this structural result, it does not follow that the Pearson II model is a special case of the Pearson I model.

Now we give some first considerations to the problem of the parameter identification for the Pearson I model. We again choose the backward integration approach. From the structure equations, we get immediately with the help of the pseudo-inverse operator for $F - \varkappa$, the following equations[1])

$$\xi_5 = F\big(1 - L_3IF^{-1}(N_4\xi_2)\big)^{-1/L_3},$$
$$\xi_3 = F\big(1 - K_3IF^{-1}(N_2\xi_2)\big)^{-1/K_3},$$
$$\xi_1 = F\big(1 - K_1IF^{-1}(L_1\xi_2)\big)^{-1/K_1},$$
$$\xi_0 = \big(1 - K_1IF^{-1}(L_1\xi_2)\big)^{-K_0/K_1}.$$

---

[1]) We thank Dr. Jansen and Dr. Grote, ZKI, for contributions, to the problem of parameter identification of special classes of Lotka-Volterra systems.

Obviously, the variable $\xi_2$ has a central significance because of the large number of corresponding feedbacks. Further on we get

$$\xi_4 = F\left(1 - K_2 I\left(F^{-1}(N_3\xi_2)/(1 - L_3 I F^{-1}(N_4\xi_2))^{L_2/L_3}\right)\right).$$

At last we get for the important variable $\xi_2$ the following expression

$$\xi_2 = F(1 - N_1 I F^{-1}(M_1\xi_3)\, F^{-1}(M_2\xi_4))^{-1/N_1},$$

$$\xi_2 = F\Big(1 - N_1 I\Big((1 - K_3 I F^{-1}(N_2\xi_2))^{-M_1/K_3} \times$$

$$\times\ (1 - K_2 I\left(F^{-1}(N_3\xi_2)/(1 - L_3 I F^{-1}(N_4\xi_2))^{L_2/L_3}\right))\Big)\Big)^{-1/N_1}.$$

This is, for the given parameters, a complicated nonlinear iteration equation for the computation of $\xi_2(t)$. For the case of still unknown parameters, this is an implicit equation for the determination of the system concept for $\xi_2(t)$ according to the necessary adjustment by nonlinear regression in the following equation

$$F\, \mathrm{d}\xi_0/\mathrm{d}t = (1 + K_0/K_1)\, F\xi_0 + L_1\xi_2(t).$$

# 5. Equivalence Transformations for Lotka-Volterra Representations

## *5.1. Generalisation of the Lotka-Volterra Equations by Imbedding into the Class of Multinomial Differential Equations*

In the following, we shall assume that we are dealing with a regular Lotka-Volterra representation, meaning that the motion shall be concentrated into the positive cone $\xi_i \geqq 0$. In applying our structure design principle, we had, in every design step, to agree upon which combinations should be taken as new additional state-variables. Making use of these degrees of freedom, we get the result that we can, at the end of our design process, get some different solutions in the form of Lotka-Volterra equations all being able to imbed given trajectories. These different Lotka-Volterra representations may have the same number of state-variables, but they may also have different numbers of state-variables. Therefore we are led in a natural way to the problem of transforming a given Lotka-Volterra representation into another equivalent one, which also imbeds the given trajectory and, besides that, is more suitable according to our objectives. However it turns out that there is no direct way to get the desired wanted equivalence transformation of Lotka-Volterra representations. After imbedding the Lotka-Volterra equation into the broader class of multinomial differential equations, it can be shown, however, that we then have an interesting class of equivalence transformations for the multinomial differential equations which can be used for to achieve our aim, of getting equivalence transformations of Lotka-Volterra representations.

### *Multinomial Representation of Dynamic Systems*

We consider a set of $n$ state-variables $\xi_i$, $i = 0, 1, \dots, n-1$, with $\xi_i \geqq 0$. With term we designate every power product expression of the form

$$\Pi_r \xi_r^{a_r}$$

with arbitrary real exponents $a_r$. We call an arbitrary linear combination of terms multinomial

$$\sum_J A_J \, \Pi_r \xi_r^{a_{Jr}} \, .$$

For a better discrimination, we always choose big letters for term indices.

In the following discussion, we shall restrict ourselves to linear combinations consisting of a finite number of terms.

If we allow for driving forces to have the form of multinomials in a multivariable growth equation, we shall speak of a multinomial representation of the corresponding dynamic system. The corresponding ordinary differential equations we call multinomial differential equations. These are differential equations of the following form

$$F\xi_i = \sum A_{iJ} \prod_r \xi_r^{a_{Jr}} \quad \text{with} \quad F = \mathrm{d}\ln/\mathrm{d}t\,.$$

Obviously, the Lotka-Volterra equations are a special case of multinomial differential equations if and only if every exponent vector has exactly one component equal to 1 and all other components equal to zero. The following considerations shall lead us to a class of equivalence transformations for multinomial differential equations. We perform them in such a way that they hold also for the case of a generalised basic module following the generating equation

$$\mathrm{d}\Phi(x)/\mathrm{d}t = Kx'\,.$$

Taking into account the corresponding normalisation of the state-variables

$$\Phi(\xi_j(0)) = 0$$

the notion of a term can be generalised in the following way

$$\Phi^{-1}\big(\sum_r a_r \Phi(\xi_r)\big)\,.$$

A multinomial in this more general case is then an expression of the form

$$\sum A_J \Phi^{-1}\big(\sum a_{Jr} \Phi(\xi_r)\big)\,.$$

The corresponding analogue to multinomial differential equations is then

$$\mathrm{d}\Phi(\xi_i)/\mathrm{d}t = \sum A_{iJ} \Phi^{-1}\big(\sum a_{Jr} \Phi(\xi_r)\big)\,.$$

## 5.2. *Equivalence Transformations for Multinomial Representations*

For the state-variables of a dynamic system in multinomial representation we introduce the following (term-consistent) coordinate transformation

$$\eta_j = \Phi^{-1}\big(\sum t_{jr}^{-1} \Phi(\xi_r)\big)$$

with a regular $n,n$-matrix $T = (t_{jr})$.

The corresponding inverse transformation is then obviously

$$\xi_i = \Phi^{-1}\big(\sum t_{is} \Phi(\eta_s)\big)\,.$$

Because $T$ is a regular matrix and therefore transforms the 0-vector again into the 0-vector, the normalisation of the state-variables will be invariant under this transformation

$$\Phi(\xi_i(0)) = 0 \leftrightarrow \Phi(\eta_j(0)) = 0\,.$$

For the effect of this transformation on a multinomial differential equation, the following relations hold:

The initial multinomial differential equation in vectorial notation

$$\mathrm{d}\Phi(\xi)/\mathrm{d}t = A\Phi^{-1}\big(a\Phi(\xi)\big)$$

under the term-consistent coordinate transformation

$$\xi = \Phi^{-1}\big(T\Phi(\eta)\big)$$

is performed again into a multinomial differential equation, namely

$$d\Phi(\eta)/dt = T^{-1}A\Phi^{-1}(aT\Phi(\eta)) .$$

Every multinomial representation is uniquely determined by a matrix pair

$$(A, a) .$$

It is suitable to agree that $a$ contains rows which are all different from each other.

With $N$ being the number of all different terms on the right-hand side then $A$ is an $[n, N]$ and $a$ an $[N, n]$-matrix. Here $n$ is the number of state-variables and $N$ the number of different terms. If we apply all term-consistent equivalence transformations, then every matrix pair $(A, a)$, swells to an equivalence class of the form

$$(T^{-1}A, aT)$$

with $T$ being a regular $[n, n]$-matrix. However the fact that after the mapping by $T$ in the matrix $aT$ not necessarily all rows are different from each other. From this equivalence class of matrix pairs, we can choose a suitable representative depending on our concrete aims.

For multinomial differential equations, these equivalence transformations represent a valuable tool for simplifying the corresponding representations. It is necessary to study how these transformations inflect on the dynamic behaviour of the corresponding dynamic systems.

At this point it is useful to connect the transformations introduced above with the Kalman theory of multivariable linear systems [60]. In this case we meet the special case

$$\Phi(u) = u .$$

Obviously, in this case, the following initial values normalisation condition holds

$$\xi_i(0) = 0 .$$

The term-consistent coordinate transformations here have the form

$$\eta = T^{-1}\xi \quad \text{resp.} \quad \xi = T\eta .$$

Terms here are the linear forms

$$\sum a_r\xi_r ,$$

and for the multinomials, we get

$$\sum A_J \sum a_{Jr}\xi_r .$$

That means, that in this case, the multinomials reduce themselves again on linear forms. In this case, the multinomial differential equations are not a generalisation of the linear differential equations. Therefore, the multinomial representation of an autonomous multivariable linear system has the form

$$d\xi/dt = Aa\xi .$$

We always meet here the product $Aa$ of the system matrix. Therefore the term-consistent equivalence transformations are given in this cas eby the following equivalence class of system matrices

$$T^{-1}ST \quad \text{with} \quad S = Aa$$

which coincides with the very well known equivalence transformations of multivariable linear systems.

Let us now return to the general multinomial differential equations.[1]) From the equivalence transformation law derived above, the conclusion immediately follows that all elements of the matrix $aA$ are invariants under the equivalence transformations of multinomial differential equations because it holds

$$aTT^{-1}A = aA .$$

However we can, of course, not state that all of these invariants are independent of each other.

## 5.3. *Discussion of Some Properties of Term-Consistent Equivalence Transformations Using Linear Algebra*

To make the changes under the equivalence transformations as transparent as possible, we introduce some suitable notations. Let $B = (b_{ij})$ be an arbitrary $[m, n]$-matrix. The row vectors we note as $b_{i.}$ and the column vectors, $b_{.j}$. The product matrix $\times$ vector is then

$$Bb = \big((b_{i.}, b)\big) .$$

Here, $(a, b)$ is the scalar product of two vectors $a$ and $b$. For the product matrix $\times$ matrix we get

$$BC = \big((b_{i.}, c_{.j})\big) .$$

With these notations, we can characterize the equivalence class of matrix pairs introduced above in the following way

$$\big(((t_{i.}^{-1}, A_{.J})), ((a_{J.}, t_{.j}))\big) .$$

Now we use the notion of biorthogonal bases. Let $e_j = t_{.j}$ and be an arbitrary (linearly independent) vector base in the $n$-dimensional space. The corresponding and uniquely determined (dual) base $f^i$ is then given by the following condition of biorthogonality

$$(f^i, e_j) = \delta_j^i = \begin{cases} 1 & \text{for} \quad i = j , \\ 0 & \text{for} \quad i \neq j . \end{cases}$$

Obviously, the elements of the dual base $f^i$ are just the rows of the inverse matrix $T^{-1}$

$$f^i = t_i^{-1} .$$

Therefore, an equivalence class of matrix pairs can be characterized as follows, with the help of biorthogonal bases

$$\big((f^i, A_{.J}), (a_{J.}, e_j)\big) .$$

The degrees of freedom for the choice of representative out of this equivalence class is therefore given by the possibility to choose an arbitrary vector base in the $n$-dimensional space. If we make use of these possibilities, we can follow different means of simplifying the representation of a multinomial system.

---

[1]) We thank Prof. Dr. Hinrichsen, Forschungsschwerpunkt Dynamische Systeme, Universität Bremen, for helpful discussions on the problem of equivalence transformations of multinomial systems.

The following considerations can be understood as an empirical introduction to a possible normal form theory for multinomial systems.[1])

*A-normal forms*

*I. Full rank of A : Rank (A) = n*

$A_{.J}$ are $N$ (number of terms) $n$-dimensional vectors. The number of linearly independent columns $A_{.J}$ cannot be larger than $n$. According to our assumption Rank $(A) = n$, we have $n$ linearly independent columns, and a set of $n$ independent columns might be chosen in quite different ways.

Let us choose a set of $n$ linearly independent columns $A_{.J}$ and take these columns as base in the $n$-dimensional space

$$A_{.J} = e_j\,, \qquad J = j = 0, 1, \dots, n-1\,.$$

Then every column of $A$ must be a certain linear combination of the vectors of the chosen basis

$$A_{.J} = \sum l_{Jj} e_j\,, \qquad J = 1, 2, \dots, N$$

($l_{Jj} = E$ for $J, j = 0, 1, \dots, n-1$ can be easily achieved if we bring, by permutation of the columns of $A$, the vectors of the chosen basis on the left-hand first $n$ places). With the help of the dual basis $f^i$ corresponding to the chosen $e_j$ we get

$$\tilde{A}_{iJ} = (f^i, a_{.J}) = l_{Ji}\,.$$

After performing such a transformation, the multinomial differential equation obviously assumes the following form:

- There are main terms $J = 0, 1, \dots, n-1$, which correspond to the chosen columns of $A$ and may be assigned to exactly one equation of the derived multinomial representation because these terms are met in only one of the equations.
- For the case $N > n$ there are still other terms (secondary terms) which can occur in more than one of the state equations. They correspond to term indices $J \geqq n$.

Of great interest and practical importance is the special case

$$\text{Rank}\,(A) = n \quad \text{and} \quad N = n\,.$$

In this case, no secondary terms will be left and the multinomial differential equation assumes the following form after performing the equivalence transformation

$$\mathrm{d}\Phi(\xi_i)/\mathrm{d}t = K_i \Phi^{-1}\left(\sum \tilde{a}_{ij} \Phi(\xi_j)\right)\,.$$

This means, that in the case of $\Phi(u) = \ln u$, we get a representation of the following form

$$F\xi_i = K_i \prod \xi_j^{a_{ij}}\,.$$

If in such a way all driving forces have such a power-product form, we call that representation a Riccati-representation.[1])

---

[1]) We thank the aspirant Dipl. Math. Nguyen Hai Bao for basic ideas in connection with the robust rank, and for the elaboration of efficient algorithms for the determination of the robust rank.

*II. Incomplete rank of A, Rank* $(A) = s < n$

In this case the columns $A_{.J}$ offer only $s$ linearly independent vectors which can be used for the construction of a part of a vector base in the $n$-dimensional space. In one of the possible ways, we choose $s$ linearly independent columns of $A$ and bring them into the first left-hand places in $A$ by permutation of columns. Every column $A_{.J}$ is then a linear combination of the chosen columns

$$A_{.J} = \sum_{j=0}^{s-1} l_{Jj} e_j \quad \text{with} \quad A_{.J} = e_j\,, \qquad J = j = 0, 1, 2, \ldots, s-1\,.$$

Now we adjoin to the chosen columns $e_j$, $j = 0, 1, \ldots, s-1$, a number of $n - s$ additional linearly independent vectors to construct a full base of the $n$-dimensional vector base and determine the correspondent dual vector base $f^i$. Now we get the following result for the transformed matrix $\tilde{A}$

$$\tilde{A}_{iJ} = (f^i, A_{.J}) = \begin{cases} l_{Ji} & \text{for} \quad i = 0, 1, 2, \ldots, s-1\,, \\ 0 & \text{for} \quad i = s, s+1, \ldots, n-1\,. \end{cases}$$

We now encounter a similar situation to that in case I. Again, we have got main terms for $i = 0, 1, \ldots, s-1$ which enter in only one of the equations, and we may have secondary terms which may occur in more than one of the equations. But the most important result here is that after the transformation, the driving forces of the equations $s, s+1, \ldots, n-1$ vanish identically. That means that the new state-variables

$$\eta_i = \Phi^{-1}\big((f^i, \Phi(\xi)\big) = \text{const} \quad \text{for} \quad j = s, \ldots, n-1$$

will be first integrals. The trajectory of the motion is then necessarily contained in the intersection of the corresponding $n - s$ hyperfaces.

In the case of ordinary multinomial differential equations, the corresponding first integrals won by our transformation (besides other first integrals which might exist) have the power-product form

$$\eta_i = \Pi\, \xi_r^{f_{ir}}$$

called *a-normal form.*

*III. Full rank of the exponent matrix, Rank* $(a) = n$

Under this assumption, there exist within the set of all rows $a_{J.}$ $n$ linearly independent vectors which can be chosen in different ways in the case for $N > n$. We choose a set of $n$ linearly independent rows $a_{J.}$ and take them as dual base $f^i$ in the $n$-dimensional vector-space

$$f^i = a_{J.} \quad \text{for} \quad i = J = 0, 1, \ldots, n-1\,.$$

Again, this kind of indexing is realised by permutation of the rows. By this means, we have chosen from the set of all terms $n$ terms with linearly independent exponent vectors.

In the sense of a biorthogonality relation to the dual base $(f^i)$ corresponds a uniquely determined primary base $(e_j)$ of the $n$-dimensional vector space. After performing the corresponding term-consistent coordinate transformation in the state-space of the multinomial differential equation, we get the following new exponent matrix

$$\tilde{a} = \left((a_{J.}, e_{i})\right) .$$

However all rows of $a$ are linear combinations in our dual base, meaning that

$$a_{J.} = \sum_{s=0}^{n-1} l_{Js} f^{s} .$$

Therefore, we have for the transformed matrix

$$\tilde{a}_{Ji} = (a_{J.}, e_i) = l_{Ji}$$

and for $J = 0, 1, \dots, n-1$ we get $l_{Ji} = \delta_{Ji}$. Therefore, all chosen terms will be linear after the transformation, meaning that they are identical with the corresponding state variables $\eta_i$, $i = J$. The other terms preserve their form of power-products according to the corresponding linear combination coefficients $l_{Ji}$.

For the special case $N = n$, if the number of terms coincides with the dimension of the state-space and rank $(a) = n$, then the original multinomial differential equation is transferred by our transformation into a Lotka-Volterra representation.

### *IV. Incomplete rank of the exponent matrix a, rank* $(a) = s < n$

From all $N$ terms we can now choose in different ways a set of $s$ linearly independent terms and bring, by corresponding permutation, the exponent vectors of these terms to the first upper places of the matrix $a : J = 0, 1, \dots, s-1$. In a suitable way, we adjoin to these vectors $n - s$ additional ones and take the set of $n$ linearly independent vectors thus constructed as a dual base $f^i$ in the $n$-dimensional vector space. On the base of the biorthogonality condition, we then construct for the dual base $f^i$ the corresponding primary base $e_j$. Again we have the linear dependencies

$$a_{J.} = \sum_{j=0}^{s-1} l_{Jj} f^{j} .$$

According to the chosen basis, we now perform the equivalence transformation and get for the transformed exponent matrix $\tilde{a}$ the following result

$$\tilde{a}_{Ji} = (a_{J.}, e_i) = \begin{cases} l_{Ji} , & i = 0, 1, \dots, s-1 , \\ 0 , & i = s, \dots, n-1 . \end{cases}$$

Therefore, the important result follows that, in the transformed multinomial system, exactly $s$ terms will be linear and are identical with the transformed variables $\eta_i$ for $i = 0, 1, \dots, s-1$, whereas the variables $\eta_i$ with $i = s, \dots, n-1$ do not enter into the transformed terms. That means the variables $\eta_i$ do not occur in the driving forces. Besides this, there might exist terms which preserve their power-product form in the variables $\eta_i$ for $i = 0, 1, \dots, s-1$.

The four cases of transformations described reflect some possible aims in simplifying a given multinomial differential equation, namely to try maximally transform into a diagonal form either the matrix $A$ or the matrix $a$. The equivalence transformations can be used to get suitable *normal forms* for multinomial differential equations. Indirectly, this holds also for Lotka-Volterra representations because every multinomial differential equation can immediately be transformed into a Lotka-Volterra representation if we introduce names for all terms and deal with the terms as new additional state-variables. The problem of which kind of normal forms can be achieved by this class of transformations according to our aims to simplify a given multinomial representation arises. Below we list some of possible aims for simplification. One big problem is open, namely what the consequence is of such a transformation in regard to the state-space behaviour of the considered dynamic system.

1. The matrix $A$ should have, as much as possible, zero elements in order to minimize the number of terms occurring in the driving forces.
2. The matrix $a$ should have, as much as possible, zero elements in order to minimize the number of variables occurring in the terms.
3. The matrix $A$ should have, as much as possible, zero rows, because every zero row means that the corresponding transformed state-variable is a first integral of the motion with a power-product form. This is obviously one possibility to reduce the system's dimension.
4. The matrix $a$ should have, as much as possible, zero columns, because every zero column corresponds to a state-variable not occurring in all of the driving forces. That means that there is a state-variable which could be eliminated. This is another possibility to reduce the system's dimension.
5. The matrix $a$ should have, as much as possible, rows equal to each other, because these rows belong to the same terms, and they can be combined with each other. This is obviously another possibility to reduce the number of different terms in the driving forces.

If we are interested in following some of these aims, at the same time, we must solve a multiobjective decision-making task. At our control disposal is only one regular matrix $T$ in the $n$-dimensional vector space for solving such a vector-optimization task.

One way to get efficient solutions in the sense of vector optimization can be characterized as follows. We first use our resource $T$ in such a way that we reach a certain normal form, for example in the sense of objective 3. Then we consider the subgroup within the group of all regular affine transformations $T$ for which the achieved normal form of type 3 is invariant, and use the rest of the degrees of freedom in our resource (using this subgroup) to optimally follow another aim, for example, objective 1. It turns out that this way favors the first chosen objective very much.

## 5.4. *Examples for the Equivalence Transformation of Multinomial Systems*

According to the possibilities for term-consistent equivalence transformations for multinomial systems, we can propose the following *multinomial structure design rule*: We apply the structure design approach described in Chapter 3 only until the corresponding right-hand sides of the differential equations assume a multinomial form. After that, we can use the possibilities of equivalence transformations of multinomial systems instead of introducing a large number of additional state-variables. After having optimally used the equivalence transformations, we then introduce names only for the rest of nonlinear terms.

1. *Regular representation of* $x_0 = 1 + \alpha \sin t$ *with* $0 < \alpha < 1$, $b > 1$ *as regularisation parameters.*

We have shown that a regular sine generator is given by the following Lotka-Volterra equations on the positive cone $x_i \geqq 0$.

$$Fx_0 = \left(\alpha(1+b)\, x_1 - b\alpha x_2\right),$$

$$Fx_1 = -\left(\alpha(1+b)\, x_1 - b\alpha x_2\right) - \left(\frac{1}{\alpha(1+b)}\, x_3 - \frac{1}{\alpha(1+b)}\, x_4\right),$$

$$Fx_2 = -\left(\alpha(1+b)\, x_1 - b\alpha x_2\right),$$

$$Fx_3 = \left(\alpha(1+b)\, x_1 - b\alpha x_2\right) + \left(\frac{1}{\alpha(1+b)}\, x_3 - \frac{1}{\alpha(1+b)}\, x_4\right),$$

$$Fx_4 = \left(\frac{1}{\alpha(1+b)}\, x_3 - \frac{1}{\alpha(1+b)}\, x_4\right).$$

Obviously to use the resource $T^{-1}$ for transforming $A$ into $T^{-1}A$ means that it is possible for us to perform the addition of any linear combination of rows of the matrix $A$ to each other. Therefore, by the following form of $T^{-1}$

$$T^{-1} = \begin{pmatrix} 1 & 0 & 0 & 0 & 0 \\ 1 & 1 & 0 & 0 & 1 \\ 1 & 0 & 1 & 0 & 0 \\ -1 & 0 & 0 & 1 & -1 \\ 0 & 0 & 0 & 0 & 1 \end{pmatrix}$$

we can achieve for the transformed matrix $T^{-1}A$ the following form

$$T^{-1}A = \begin{pmatrix} 0 & \alpha(1+b) & -b\alpha & 0 & 0 \\ 0 & 0 & 0 & 0 & 0 \\ 0 & 0 & 0 & 0 & 0 \\ 0 & 0 & 0 & 0 & 0 \\ 0 & 0 & 0 & \dfrac{1}{\alpha(1+b)} & \dfrac{1}{\alpha(1+b)} \end{pmatrix}.$$

The new terms are given by the transformed matrix $a$

$$aT = T = \begin{pmatrix} 1 & 0 & 0 & 0 & 0 \\ -1 & 1 & 0 & 0 & -1 \\ -1 & 0 & 1 & 0 & 0 \\ 1 & 0 & 0 & 1 & 1 \\ 0 & 0 & 0 & 0 & 1 \end{pmatrix} \qquad \begin{matrix} z_0 \\ z_0^{-1} z_1 z_4^{-1} \\ z_0^{-1} z_2 \\ z_0 z_3 z_4 \\ z_4 \end{matrix} \quad .$$

The new state-variables $z_i$, expressed by the old ones $x_i$, are given by the following coordinate transformations

$$z_i = \Pi x_r^{t_{ir}^{-1}} .$$

Therefore, we get the following expressions for the new state-variables

$$z_0 = x_0 , \qquad z_1 = x_0 x_1 x_4 , \qquad z_2 = x_0 x_2 , \qquad z_3 = x_0^{-1} x_3 x_4^{-1} , \qquad z_4 = x_4 .$$

Obviously, the new coordinates $z_1$, $z_2$, $z_3$ are first integrals, meaning that they are constant along their trajectories. Therefore, we have reduced the dimension of the differential equation to 2, but we had to pay for it because we now must deal with a multinomial differential equation. It has the following form

$$Fz_0 = (1 + b)\, z_0^{-1} z_1 z_4^{-1} - b z_0^{-1} z_2 ,$$

$$Fz_4 = \frac{1}{\alpha(1 + b)} z_0 z_3 z_4 - \frac{1}{\alpha(1 + b)} z_4 .$$

By assigning names to all terms different from each other we can easily transform the multinomial differential equation into a Lotka-Volterra representation. Therefore, we get the following equivalent Lotka-Volterra representation in comparison with the system of equations originally given

$$Fy_0 = -\alpha(1 + b)\, y_0 + b\alpha y_1 - \frac{1}{\alpha(1 + b)} y_2 + \frac{1}{\alpha(1 + b)} y_3 ,$$

$$Fy_1 = -\alpha(1 + b)\, y_0 + b\alpha y_1 ,$$

$$Fy_2 = (1 + b)\, y_0 - b\alpha y_1 + \frac{1}{\alpha(1 + b)} y_2 - \frac{1}{\alpha(1 + b)} y_3 ,$$

$$Fy_3 = \frac{1}{\alpha(1 + b)} y_2 - \frac{1}{\alpha(1 + b)} y_3 .$$

We recognize at once another first integral $y_0 y_2 = \text{const}$.

After applying the equivalence transformation, we arrived at a new Lotka-Volterra representation with the dimension reduced by about 1 in comparison with the original system.

*2. Hyperlogistic differential equation*

The hyperlogistic EVOLON model is given by the following differential equation

$$\mathrm{d}x/\mathrm{d}t = K x^k (B - x^w)^l .$$

With the substitution

$$u = B - x^w$$

we get immediately

$$Fx = K x^{k-1} u^l ,$$

$$Fu = -wK x^{k+w-1} u^{l-1} .$$

This is already a multinomial differential equation with the matrices

$$A = \begin{pmatrix} K & 0 \\ 0 & -wK \end{pmatrix}, \qquad a = \begin{pmatrix} k - 1 & l \\ k + w - 1 & l - 1 \end{pmatrix}.$$

The corresponding matrix of invariants is

$$aA = \begin{pmatrix} K(k-1) & -wlK \\ K(k+w-1) & -(l-1)\,wK \end{pmatrix}.$$

That means that we have four invariants for five parameters, with the consequence that, in the sense of our equivalence representations, one parameter is redundant. The derived multinomial differential equation is already a diagonal form for $A$, meaning that it is a Riccati normal form.

Now we intend to pass to a diagonalisation of $a$ to determine a corresponding Lotka-Volterra representation. We have

$$D = \det a = (k-1)(l-1) - l(k+w-1).$$

For $D \neq 0$ the exponent matrix $a$ has the full rank, and it is therefore possible to diagonalize $a$. For this purpose we choose

$$f^1 = \big((k-1), l\big) \quad \text{and} \quad f^2 = \big((k+w-1), (l-1)\big).$$

Then we have

$$e_1 = \frac{1}{D}\begin{pmatrix} l-1 \\ -(k+w-1) \end{pmatrix} \quad \text{and} \quad e_2 = \frac{1}{D}\begin{pmatrix} -l \\ (k-1) \end{pmatrix}.$$

The corresponding coördinate transformation is then

$$z_j = \Pi\, x_r^{f_r^i} \quad \text{with} \quad x_1 = x \quad \text{and} \quad x_2 = u$$

or

$$z_1 = x^{k-1}u^l \quad \text{and} \quad z_2 = x^{k+w-1}u^{l-1}.$$

For the transformed matrices $aT$ and $T^{-1}A$, we get

$$aT = \begin{pmatrix} 1 & 0 \\ 0 & 1 \end{pmatrix}, \qquad T^{-1}A = \begin{pmatrix} K(k-1) & -wlK \\ K(k+w-1) & -w(l-1)\,K \end{pmatrix},$$

and for the corresponding Lotka-Volterra equations, we get

$$Fz_1 = K(k-1)\,z_1 - wlz_2\,,$$
$$Fz_2 = K(k+w-1)\,z_1 - w(l-1)\,Kz_2\,.$$

For $D = 0$, we get a similar result. In this case we have

$$k + w - 1 = \varkappa(k-1) \quad \text{and} \quad l - 1 = \varkappa l\,.$$

We now construct the dual base in the following way

$$f^1 = \big((k-1), l\big) \quad \text{and arbitrary} \quad f^2 = (1, 0)\,.$$

For this choice we get the following primary base

$$e_1 = \begin{pmatrix} 0 \\ 1/l \end{pmatrix} \quad \text{and} \quad e_2 = \begin{pmatrix} 1 \\ -(k-1)/l \end{pmatrix}.$$

The corresponding coordinate transformation is then

$$z_1 = x^{k-1}u^l \quad \text{and} \quad z_2 = x\,.$$

For the transformed matrices $aT$ and $T^{-1}A$, we get

$$aT = \begin{pmatrix} 1 & 0 \\ \varkappa & 0 \end{pmatrix} \quad \text{and} \quad T^{-1}A = \begin{pmatrix} 0 & K \\ -wK/l & wK(k-1)/l \end{pmatrix}.$$

Therefore, we get as the result of the equivalence transformation in this case

$$Fz_1 = Kz_1^{\varkappa},$$

$$Fz_2 = -\frac{wK}{l} z_1 + \frac{wK}{l}(k-1)\, z_1^{\varkappa}.$$

Obviously, $z_2$ is no longer a control variable in the driving forces.

To pass to a Lotka-Volterra equation we have to introduce an additional state-variable

$$z_3 = z_1^{\varkappa}$$

and get the end-result

$$Fz_1 = Kz_3,$$

$$Fz_2 = -\frac{wK}{l} z_1 + \frac{wK}{l}(k-1)\, z_3,$$

$$Fz_3 = \varkappa K z_3.$$

*3. Two-body problem (Kepler dynamics)*

Let $M$ be the mass of the sun, $m$ the mass of the earth and $\gamma$ Newton's gravity constant. According to Newton's attraction principle, the following equations of motion hold

Motion of earth: $\mathrm{d}^2\vec{r}/\mathrm{d}t^2 = M\gamma(\vec{R} - \vec{r})/||\vec{R} - \vec{r}||^3$,

Motion of sun: $\mathrm{d}^2\vec{R}/\mathrm{d}t^2 = m\gamma(\vec{r} - \vec{R})/||\vec{R} - \vec{r}||^3$.

We take the position of the sun as reference point and introduce the radius vector $\vec{\varrho}$ with

$$\vec{\varrho} = \vec{r} - \vec{R}.$$

For the relative vector $\vec{\varrho}$ we get the following equation of motion

$$\mathrm{d}^2\vec{\varrho}/\mathrm{d}t^2 = -(M + m)\,\gamma\vec{\varrho}/||\vec{\varrho}||^3.$$

If we know the trajectory $\vec{\varrho}(t)$, then the motion $\vec{R}(t)$ of the sun can be easily determined

$$\vec{R}(t) = \vec{R}(0) + (\mathrm{d}\vec{R}(0)/\mathrm{d}t)\, t - \frac{m}{(M+m)}\left(\vec{\varrho}(t) - \vec{\varrho}(0) - \mathrm{d}\vec{\varrho}(0)/\mathrm{d}t \cdot t\right).$$

First, we will derive a multinomial differential equation corresponding to the Kepler dynamics (but without introducing special measures to guarantee a regular system which can be immediately simulated on a computer). In space-coordinates, we have

$$\vec{\varrho} = (\xi, \eta, \zeta).$$

Then we have

$$F\xi = \xi_1, \qquad F\eta = \eta_1 \quad \text{and} \quad F\zeta = \zeta_1.$$

Further on we have

$$F\xi_1 = -(\xi_1 + K\varkappa/\xi_1),$$
$$F\eta_1 = -(\eta_1 + K\varkappa/\eta_1),$$
$$F\zeta_1 = -(\zeta_1 + K\varkappa/\zeta_1)$$

with the following notations

$$K = (M + m)/\gamma \quad \text{and} \quad \varkappa = 1/||\varrho||^3 .$$

Obviously $\varkappa$ is the only coupling between the state-variables introduced up to now. For the coupling variable $\varkappa$, we get the following equation

$$F\varkappa = -3\varkappa^{2/3}(\xi^2\xi_1 + \eta^2\eta_1 + \zeta^2\zeta_1) .$$

Thus, we have already arrived at a multinomial differential equation for Kepler's two-body system. That means that, to a given system with six mechanical degrees of freedom, there corresponds a multinomial representation with only seven degrees of freedom.

We now pass to the notation for the corresponding matrices $A$ and $a$ of the system. For this purpose we introduce the following sequence of terms

$$\xi_1, \eta_1, \zeta_1, \varkappa/\xi_1, \varkappa/\eta_1, \varkappa/\zeta_1, \varkappa^{2/3}\xi^2\xi_1, \varkappa^{2/3}\eta^2\eta_1, \varkappa^{2/3}\zeta^2\zeta_1 ,$$

and for the variables we note the following order

$$\xi, \eta, \zeta, \xi_1, \eta_1, \zeta_1 .$$

Then we get obviously the following system matrices

$$A = \begin{pmatrix} 1 & 0 & 0 & 0 & 0 & 0 & 0 & 0 & 0 \\ 0 & 1 & 0 & 0 & 0 & 0 & 0 & 0 & 0 \\ 0 & 0 & 1 & 0 & 0 & 0 & 0 & 0 & 0 \\ -1 & 0 & 0 & -K & 0 & 0 & 0 & 0 & 0 \\ 0 & -1 & 0 & 0 & -K & 0 & 0 & 0 & 0 \\ 0 & 0 & -1 & 0 & 0 & -K & 0 & 0 & 0 \\ 0 & 0 & 0 & 0 & 0 & 0 & -3 & -3 & -3 \end{pmatrix},$$

$$a = \left(\begin{array}{ccccccc} 0 & 0 & 0 & 1 & 0 & 0 & 0 \\ 0 & 0 & 0 & 0 & 1 & 0 & 0 \\ 0 & 0 & 0 & 0 & 0 & 1 & 0 \\ 0 & 0 & 0 & -1 & 0 & 0 & 1 \\ 0 & 0 & 0 & 0 & -1 & 0 & 1 \\ 0 & 0 & 0 & 0 & 0 & -1 & 1 \\ 2 & 0 & 0 & 1 & 0 & 0 & 2/3 \\ 0 & 2 & 0 & 0 & 1 & 0 & 2/3 \\ 0 & 0 & 2 & 0 & 0 & 1 & 2/3 \end{array}\right) \begin{array}{l} \\ \\ f^1 \\ f^2 \\ f^3 . \\ f^4 \\ f^5 \\ f^6 \\ f^7 \end{array}$$

In the process of trying to get a Lotka-Volterra representation, we will try to get something like a diagonal form for $a$. Obviously, we must deal with full rank of $a$, because we find seven linearly independent rows of $a$. Therefore, after our theory, seven of the nine terms can be transformed into linear terms.

If we choose a dual base according to our proposal, we get the corresponding primary base $e_i$ at the right-hand side of the matrix $a_i$

| $e_1$ | $e_2$ | $e_3$ | $e_4$ | $e_5$ | $e_6$ | $e_7$ |
|---|---|---|---|---|---|---|
| −5/6 | −1/2 | 0 | −5/6 | 1/2 | 0 | 0 |
| −5/6 | 0 | 1/2 | −5/6 | 0 | 1/2 | 0 |
| −5/6 | 0 | 0 | −1/3 | 0 | 0 | 1/2 |
| 1 | −1 | 0 | 1 | 0 | 0 | 0 |
| 1 | 0 | −1 | 1 | 0 | 0 | 0 |
| 1 | 0 | 0 | 0 | 0 | 0 | 0 |
| 1 | 0 | 0 | 1 | 0 | 0 | 0 |

After applying the corresponding equivalence transformation with $T = (e)$, we get for the new exponent matrix the following result

$$aT = \begin{pmatrix} 1 & -1 & 0 & 1 & 0 & 0 & 0 \\ 1 & 0 & -1 & 1 & 0 & 0 & 0 \\ 1 & 0 & 0 & 0 & 0 & 0 & 0 \\ 0 & 1 & 0 & 0 & 0 & 0 & 0 \\ 0 & 0 & 1 & 0 & 0 & 0 & 0 \\ 0 & 0 & 0 & 1 & 0 & 0 & 0 \\ 0 & 0 & 0 & 0 & 1 & 0 & 0 \\ 0 & 0 & 0 & 0 & 0 & 1 & 0 \\ 0 & 0 & 0 & 0 & 0 & 0 & 1 \end{pmatrix}.$$

The transformed multinomial differential equation system then has the following form

$$Fz_1 = -z_1 - Kz_4\,,$$
$$Fz_2 = z_1z_4/z_2 + Kz_2 - 3z_5 - 3z_6 - 3z_7\,,$$
$$Fz_3 = z_1z_4/z_3 + Kz_3 - 3z_5 - 3z_6 - 3z_7\,,$$
$$Fz_4 = z_1 + Kz_4 - 3z_5 - 3z_6 - 3z_7\,,$$
$$Fz_5 = z_1z_4/z_2 - Kz_2 - 2z_5 - 2z_6 - 2z_7\,,$$
$$Fz_6 = z_1z_4/z_3 - Kz_3 - 2z_5 - 2z_6 - 2z_7\,,$$
$$Fz_7 = z_1 - Kz_4 - 2z_5 - 2z_6 - 2z_7\,.$$

We can pass to a Lotka-Volterra system by introduction of two additional state-variables

$$u = z_1z_4/z_2 \quad \text{and} \quad v = z_1z_4/z_3$$

with the additional equations

$$Fu = -u - Kz_2\,, \qquad Fv = -v - Kz_3\,.$$

## 5.5. *Some Numerical Aspects of the Normal Form Problem*[1])

We concentrate these considerations up on the case where we are interested in determining $A$ normal forms, meaning that we try to find such a regular matrix $T$ that the transformed matrix $T^{-1}A$ has a maximum number of zeros.

In Section 5.3, we proposed the following approach to solve this problem:

1. First, we determine the rank $(A) = a \leqq n$
2. Out of the whole set of $N$ columns $A_{.J}$ of $A$, we choose a set of $s$ linearly independent columns and take them as part of a base $e_j$, $j = 1, \dots, s$, of the $n$-dimensional vector space. This can be done in several different ways.
3. We adjoin to the partial base $e_j$, $j = 1, 2, \dots, s$, additional $n - s$ independent vectors $e_j$, $j = s + 1, \dots, n$. This can be done in different ways. Now every column $A_{.J}$ can be uniquely expressed by
$$A_{.J} = \sum_{j=1}^{s} l_{Jj} e_j\,.$$
4. For the primary base $e_j$ we determine uniquely the corresponding dual base $f^i$ and agree to construct $T$ in the following way
$$T = (t_{ij}) \quad \text{with} \quad t_{.j} = e_j$$
resp.
$$T^{-1} = (t_{ij}^{-1}) \quad \text{with} \quad t_{i}^{-1} = f^i\,.$$
5. After applying $T^{-1}$ to $A$, we get the result expressed as a structure
$$T^{-1}A = \begin{array}{r} \\ \\ \\ \text{s} \end{array}\!\!\left\{\begin{matrix} 1 & 0 \dots 0 \\ 0 & 1 \dots 0 \\ \cdot\; \cdot & \cdot\; \cdot\; \cdot \\ 0 & 0 \dots 1 \end{matrix}\right. (e_{jJ}^T)\,,$$
$$n - s \left\{\begin{matrix} 0 & 0 \dots 0 & 0 \dots 0 \\ 0 & 0 \dots 0 & 0 \dots 0 \\ \cdot\;\cdot & \cdot\;\cdot\;\cdot & \cdot\;\cdot\;\cdot \\ 0 & 0 \dots 0 & 0 \dots 0 \end{matrix}\right. .$$

The problem arises how we can use our degrees of freedom in such way that the rest matrix $l_{jJ}^T$, $j = 1, \dots, s$; $J = s + 1, \dots, N$ (if $N > s$), contains as many zeros as possible. In trying to find an answer to this problem, we must ask the following questions:

We define the group of all regular transformations $K$, preserving in the transformation $K\tilde{A}$ the structure achieved in Step 5 above. We use the possibilities of choosing such a $K$ to produce in $l_{jJ}^T$ as many zeros as possible. After our studies, the effect which

---

[1]) Most of the ideas of Section 5.5 were found in cooperation with aspirant Dipl. Math. Nguyen Hai Bao, Centre of Sciences, Hanoi, SRV

can be additionally achieved using the operations is that $K$ is small, meaning that the main effect lies in choosing $s$ linearly independent vectors $e_i$ out of the columns of $A$. From the numerical point of view, a component $l_{jJ}$ with a value which is small enough can be handled like a zero. How can we validate the measure of putting all these matrix components with small values to "zero" and most importantly, what will the consequences be of such small values on the dynamic behaviour of the corresponding multinomial differential equation ? This question is connected with the problem of parameter robustness of multinomial systems.

We can consider any multinomial differential equation as a reliable model of a real system only if it possesses the parameter robustness in the sense that amplitude bounded fluctuations of all its parameters (which are physical variables) have no essential influence on the type of the motion of the system. That means that if any matrix element is not identical to zero because of the given physical structure of the system, but has a small value, then two cases are possible. Either, despite the small value, the corresponding influence in the driving force is essential and the corresponding element should not be dropped and should be robust, or this variable is itself the effect of the system's perturbation. If it is the effect of a perturbation, then either it can be dropped if the rest of the system is robust against this perturbation, or the assumed model is not relevant.

This robustness problem already begins earlier in our approach, with the determination of the rank ($A$). The rank $(A) = s$ is robust if small fluctuations of the elements of $A$ cannot change the value $s$. It makes no sense to base our approach on a non-robust rank.

We can try to determine a robust rank ($A$) in the following way:

— We first normalise all non-zero columns $A_{.J}/||A_{.J}|| = n_J$.
— A system of $s$ vectors $n_J$, $J = 1, 2, \ldots, s$, has the robust $\varepsilon$-rank if the orthogonal distance of any $n_J$ from the subspace spanned by all $n_I$, $I \neq J$, has a length larger than or equal to $\varepsilon$.

There are different approaches to determine a set of $s$ linearly independent columns $A_{.J}$ which realize the robust rank. Obviously, the robust rank as a function of $\varepsilon$ is a monotonously decreasing step function which for $\varepsilon = 0$ gives $s = \text{rank}\ (A)$ and for $\varepsilon \to \infty$ runs to zero. The longer the system "lives", the better is the whole robustness picture (meaning the broader this step function is) for a matrix $A$.

Let us assume that we have demanded a certain $\varepsilon$ and have determined the corresponding robust rank $(A) = s'$. We can then achieve a structure similar to that we have discussed above, but if we construct $s_j$ for $j = s' + 1, \ldots, s = \text{rank}\ (A)$ in such a way that we reduce the value of $\varepsilon$, this structure shows us $s - s'$ rows with matrix values being very small, meaning that they may be handled as zeros. Now, though we have returned to the question we raised above as to whether the system is robust in respect to these quasi-zeros. If we put all these rows to zero, then we have hyperfaces of the form

$$\eta_j = \Pi\, z_r^{f_r^j} = \text{const.}$$

With the quasi-zeros in these rows, we meet certain deviations of the system's state from these hyperfaces. If, in a certain neighbourhood of these hyperfaces, they are stable attractors, then the system itself will try to cause the right-hand sides of the corresponding equations to vanish, meaning that, in this case, the system will be robust against these quasi-zeros.

Thus, we can reduce one of these robustness problems of linear algebra in connection with our systems dynamics to a problem which can be solved by analysis of the qualitative behaviour of such systems. We will describe some results of studies of the qualitative behaviour in chapter 8.

## *5.6. Software Concept for Continuing the Structure Design Approach Using Equivalence Transformations*

We have described in Section 3.6 the principal way in which we can use a certain special language in dialogue mode to transform a concrete differential equation into the Lotka-Volterra form. We should now integrate into this approach the possibility of applying our structure design rule with the aim that, after a small number of steps, we will achieve the form of a multinomial differential equation. Here we must pass to numerical calculations using the concrete form of the matrix pair $(A, a)$ which contains all the parameters of the multinomial system, and follow different aims as they were described above in connection with the corresponding normal form structures. For this purpose, we should use only robust algorithms and start with a robust multinomial differential equation (how we can make sure of this ?). This will insure that the result, as a normal form with the reduction of the number of state variables and (or) the number of right-hand side control variables, is again a robust multinomial differential equation. If this step did not yet lead us to a Lotka-Volterra representation, we can now agree upon additional state-variables for the rest of nonlinear terms, and thus achieve the Lotka-Volterra description. The robustness of the Lotka-Volterra description itself is a very big problem. We shall discuss it further in Chapter 8.

# 6. The Riccati Representation and Its Applications

## 6.1. *Elimination of Primary Autocatalytic Effects from the Lotka-Volterra Equations*

We want to remove from the Lotka-Volterra representation

$$Fx_i = \sum G_{ij}x_j$$

the autocatalytic term $G_{ii}x_i$, meaning that we remove feedback influences of every state-variable $x_i$ onto itself. This can be done in the following way. We introduce a new state-variable $\xi_i$ by the demand

$$F\xi_i = Fx_i - G_{ii}x_i$$

which, in the normalized case $x_i(0) = 1$, coincides with

$$\xi_i = x_i/F^{-1}G_{ii}x \quad \text{with} \quad F^{-1} = \exp(I)\,, \qquad I = \int_0^t .$$

If we wish to determine from this substitution the original state-variable $x_i$, we must solve a special case of the Riccati differential equation. In this case, a closed analytical solution exists, namely

$$x_i = F(1 - G_{ii}I\xi_i)^{-1/G_{ii}}\,.$$

We recognize for the non-autocatalytic case with $G_{ii} = 0$ that this transformation reduces to the identical transformation

$$x_i = \xi_i\,.$$

If we put this transformation into the original Lotka-Volterra equations we get the following integral equation, which we will call integral *Riccati representation* because we made use of the Euler-Riccati differential equation

$$\xi_i(t) = \prod_{j \neq i} (1 - G_{jj}I\xi_j)^{-G_{ij}/G_{jj}}\,.$$

This integral representation appears to be a key to the study of the qualitative properties of such systems.

For the case $G_{ii} = 0$, the corresponding factor in the integral Riccati representation obviously reduces to exp $(G_{ij}I\xi_j)$. By a second equivalence transformation, we can easily return to a representation in the form of a multinomial differential equation. For this purpose, we put

$$u_j = 1 - G_{jj}I\xi_j \quad \text{for} \quad G_{jj} \neq 0\,, \qquad j \in A\,,$$

$$u_j = \exp(I\xi_j) \quad \text{for} \quad G_{jj} = 0\,, \qquad j \notin A\,.$$

Thus we get for the newstate-variables $u_j$ the following multinomial differential equation, which we call *Riccati representation*

$$j \in A\,, \qquad \dot{u}_j = -G_{jj} \prod_{r \in A-j} u_r^{-G_{jr}/G_{rr}} \prod_{s \in \bar{A}} u_s^{G_{js}}\,, \qquad x_j = \frac{1}{u_j} \prod_{r \in A-j} u_r^{-G_{jr}/G_{rr}} \prod_{s \in \bar{A}} u_s^{G_{js}}\,,$$

$$j \notin A\,, \qquad \dot{u}_j = u_j \prod_{r \in A} u_r^{-G_{jr}/G_{rr}} \prod_{s \in \bar{A}-j} u_s^{G_{js}}\,, \qquad x_j = \prod_{r \in A} u_r^{-G_{jr}/G_{rr}} \prod_{s \in \bar{A}-j} u_s^{G_{js}}\,.$$

Obviously, the normalisation condition is preserved, meaning that from $x_i(0) = 1$ follows $u_i(0) = 1$ and vice versa. However it is worth mentioning that by being a little more general, we can pass from the Lotka-Volterra representation to an equivalent Riccati representation by applying any substitution of the following form

$$x_i = K_i F u_i \quad \text{with} \quad K_i \neq 0\,.$$

Most important for the use of the Riccati representation is the property that if we restrict the motion of the original system on the positive cone $x_i \geqq 0$, then all state-variables $u_i$ are monotonous functions and, therefore, every original state-variable $x_i$ can be represented as a power-product of monotonous functions.

If the normalisation condition is fulfilled, the more general substitution leads to

$$F u_i = \frac{1}{K_i} \prod u_j^{G_{ij} K_j}$$

where the original state-variables $x_i$ are represented by

$$x_i = \prod u_j^{G_{ij} K_j}\,.$$

From this representation, we see a close connection to the term-consistent equivalence transformation, namely, if

$$\tilde{T}_{ij} = G_{ij} K_j$$

is a regular matrix, then the Riccati transformation can be considered as a term-consistent equivalence transformation.

## 6.2. *Generalisation of the Riccati Transformation for the Case of a General Base Module*

We consider, in the case of a general base module, the following analogy to the Lotka-Volterra equations

$$\mathrm{d}\Phi(x)/\mathrm{d}t = Gx\,.$$

Here we substitute

$$x = K\,\mathrm{d}\Phi(u)/\mathrm{d}t$$

with an arbitrary regular matrix $K$.

Conversely, we get (for the normalised case)

$$u = \Phi^{-1}(IK^{-1}x) \quad \text{with} \quad I = \int_0^t\,.$$

That means that we are indeed dealing with an equivalence transformation.

After substitution of our transformation on the right-hand side of the original differential equation we get

$$\mathrm{d}\Phi(x)/\mathrm{d}t = GK\,\mathrm{d}\Phi(u)/\mathrm{d}t$$

and integrating this relation taking into account the normalisation condition we get

$$x = \Phi^{-1}(GK\Phi(u)) \,.$$

For the case with a regular matrix $GK$, this transformation is obviously a term-consistent equivalence transformation. In general, however, we get for $u$ the following multinomial differential equation

$$\mathrm{d}\Phi(u)/\mathrm{d}t = K^{-1}\Phi^{-1}(GK\Phi(u)) \,.$$

Only for the case where $K$ is a regular diagonal matrix do we get from this result the corresponding Riccati representation.

## 6.3. *Basic Modes of Behaviour of Nonlinear Systems in Riccati Representation*

In general, it follows from the Riccati representation that we are dealing with a multiplicative interaction within the driving forces between the different state-variables. The elementary factors in this interaction model have all the power form

$u^k$ with

$u = 1 - GI\xi$ in the autocatalytic case $G \neq 0$,

$u = e^{I\xi}$ in the nonautocatalytic case $G = 0$.

We are interested only in the behaviour in the positive cone, which is preserved under the transformation leading from $x$ to $\xi$, and under the transformation leading from $\xi$ to $u$. We discuss the typical modes of the motion under the normalisation condition, meaning

$$x_i(0) = \xi_j(0) = u_i(0) = 1 \,.$$

We differentiate between the following cases:

$G = 0$: $u^k = e^{kI\xi}$ is quasi-exponentially increasing for $k > 0$ and quasi-exponentially decreasing for $k < 0$.

$G > 0$: In this case, $u$ is necessarily decreasing.
If $\xi$ very quickly approaches zero, then $u$ can stabilize on a value smaller than 1 and therefore $u^k$ stabilizes on a value smaller than 1 for $k > 0$ and greater than 1 for $k < 0$. Nevertheless, it can occur that $\xi$ needs an infinite time to approach 0, but $I\,\xi$ is converging so that $GI\xi$ stays smaller than 1. In this case, we meet quasi-logistic decrease for $k > 0$, and quasi-logistic increase for $k < 0$. However, it is also possible that $GI\xi$ approaches a constant smaller than 1 in a finite time. In this case, $u^k$ shows a quasi-parabolic saturation to a value smaller than 1 for $k > 0$, and to a value greater than 1 for $k < 0$.
Maybe $GI\xi$ approaches 1 in finite time. In this case $u^k$ approaches 0 in a finite time for $k > 0$ (we call this quasi-parabolic decrease), and $u^k$ approaches $\infty$ in a finite time for $k < 0$ (we call it quasi-hyperbolic increase). However, the

case may also exist in which an infinite time is necessary for $GI\xi$ to reach the value 1. In this case, we meet quasi-exponential decrease for $k > 0$, and quasi-exponential increase for $k < 0$.

$G < 0$: In this case, $u$ increases.
If here again $\xi$ reaches 0 in a finite time, we again meet quasi-parabolic increase of $u^k$ for $k > 0$, and quasi-parabolic decrease for $k < 0$. If $\xi$ needs an infinite time to reach 0, however, we meet quasi-logistic increase of $u^k$ for $k > 0$, and quasi-logistic decrease of $u^k$ for $k < 0$. If the $I\xi$ is not converging, then $u^k$ approaches infinity for $k > 0$. If this process lasts only a finite time, we meet quasi-hyperbolic increase, but if it needs an infinite time for it, we meet quasi-parabolic increase. If $k < 0$, then $u^k$ can approach 0 in a finite time, meaning quasi-parabolic decrease. If it needs an infinite time to approach 0, we meet quasi-hyperbolic decrease.

It is necessary to mention that we cannot differentiate in this rough qualitative comparison, between quasi-parabolic increase and quasi-exponential increase on one hand, and, on the other hand, between quasi-hyperbolic decrease and quasi-exponential decrease.

It is necessary to use always the attribute *quasi* because the concrete functions will differ from the well-known parabola, hyperbola or exponentials. A part of such influences are very well known from linear system theory. We get there the basic modes, for example, the decomposition into so-called partial fractions of the corresponding transfer functions [61]. There we generally only meet power functions in time $t$ and exponentials. Therefore, in linear system theory, we meet only the following cases:
quasi-parabolic (or quasi-exponential) increase,
quasi-hyperbolic (or quasi-exponential) decrease,
quasi-logistic increase and quasi-logistic decrease.
Therefore, it follows, that extinction (quasi-parabolic decrease) or explosion (quasi-hyperbolic increase) are specific features of non-equilibrium processes in nonlinear systems.

We will now construct simple examples for the demonstration of all these basic modes of nonlinear systems.

## 6.4. *Examples for the Representation of Nonlinear Systems in Riccati Form*

*1. Hyperlogistic evolution model*

$$\mathrm{d}x/\mathrm{d}t = Kx^k(B - x^w)^l \,.$$

The corresponding Lotka-Volterra representation was

$$Fx_0 = Kx_1 \,,$$
$$Fx_1 = (k-1)\,Kx_1 - wlKx_2 \,,$$
$$Fx_2 = (k-1+w)\,Kx_1 - w(l-1)\,Kx_2 \,,$$

with

$$x_1 = x_0^{k-1}(B - x^w)^l \,, \qquad x_2 = x_0^{k-1+w}(B - x^w)^{l-1} \,.$$

For the Riccati representation (not necessarily normalized), we get

$$\mathrm{d}u_0/\mathrm{d}t = u_0 u_1^{-1/(k-1)}\,, \qquad x_0 = u_1^{-1/(k-1)}\,,$$
$$\mathrm{d}u_1/\mathrm{d}t = -K(k-1)\, u_2^{-l/(l-1)}\,, \qquad x_1 = u_1^{-1} u_2^{-l/(l-1)}\,,$$
$$\mathrm{d}u_2/\mathrm{d}t = Kw(l-1)\, u_1^{-(k-1+w)/(k-1)}\,, \qquad x_2 = u_2^{-1} u_1^{-(k-1+w)/(k-1)}\,.$$

*2. A function generator for $x(t) = At^k\,\mathrm{e}^{at}$*

The Lotka-Volterra equations are

$$x(t) = x_0\,, \qquad Fx_0 = kx_1 + ax_2\,,$$
$$Fx_1 = -x_1\,,$$
$$Fx_2 = 0\,.$$

The result of the transformation into Riccati form is

$$\mathrm{d}u_0/\mathrm{d}t = u_0 u_1^k u_2^a\,, \qquad x_0 = u_1^k u_2^a\,,$$
$$\mathrm{d}u_1/\mathrm{d}t = 1\,, \qquad x_1 = 1/u_1\,,$$
$$\mathrm{d}u_2/\mathrm{d}t = u_2\,.$$

*3. A function generator for the Tschebyscheff polynomials*

$$x_0 = T_n(t) = \cos\left(n \arccos\,(2t-1)\right)\,, \qquad 0 \leqq t \leqq 1\,, \qquad \text{with } n \text{ integer}$$

and

$$y_0 = \frac{1}{\sqrt{t(1-t)}}\,.$$

$y_0$ can easily be realised by a Lotka-Volterra structure. For the Lotka-Volterra equations we get

$$Fx_0 = nx_1\,,$$
$$Fx_1 = -nx_1 - nx_2 + Fy_0\,,$$
$$Fx_2 = nx_1 + nx_2 + Fy_0\,.$$

After transformation into the Riccati representation we get

$$\mathrm{d}u_0/\mathrm{d}t = u_0 u_1\,, \qquad x_0 = u_1\,,$$
$$\mathrm{d}u_1/\mathrm{d}t = nu_2 y_0\,, \qquad x_1 = u_2 y_0/u_1\,,$$
$$\mathrm{d}u_2/\mathrm{d}t = -nu_1 y_0\,, \qquad x_2 = u_1 y_0/u_2\,.$$

*4. Regularized sine generator*

For the signal $x_0 = 1 + \dfrac{\sin t}{a}$ with the following additional state-variables

$$x_1 = \frac{a}{(b+1)}\,\frac{\cos t + b}{\sin t + a}\,, \qquad x_2 = \frac{a}{b}\,\frac{1}{\sin t + a}\,,$$
$$x_3 = \frac{(b+1)}{a}\,\frac{\sin t + a}{\cos t + b}\,, \qquad x_4 = (b+1)\,\frac{1}{\cos t + b}$$

we get the following regularized Volterra equations

$$Fx_0 = \left(\frac{(b+1)}{a}x_1 - \frac{b}{a}x_2\right),$$

$$Fx_1 = \left(-\frac{a}{(b+1)}x_3 + \frac{a}{(b+1)}x_4\right) - \left(\frac{(b+1)}{a}x_1 - \frac{b}{a}x_2\right),$$

$$Fx_2 = -\left(\frac{(b+1)}{a}x_1 - \frac{b}{a}x_2\right),$$

$$Fx_3 = \left(\frac{(b+1)}{a}x_1 - \frac{bx_2}{a}\right) - \left(-\frac{a}{(b+1)}x_3 + \frac{a}{(b+1)}x_4\right),$$

$$Fx_4 = -\left(-\frac{a}{(b+1)}x_3 + \frac{a}{(b+1)}x_4\right).$$

From this, we get the following Riccati representation

$$du_1/dt = \frac{(b+1)}{a}u_2^{-1}u_3u_4\,, \qquad x_1 = u_1^{-1}u_2^{-1}u_3u_4\,,$$

$$du_2/dt = -\frac{b}{a}u_1^{-1}\,, \qquad x_2 = u_2^{-1}u_1^{-1}\,,$$

$$du_3/dt = -\frac{a}{(b+1)}u_1u_2u_4^{-1}\,, \qquad x_3 = u_1u_2u_3^{-1}u_4^{-1}\,,$$

$$du_4/dt = +\frac{a}{(b+1)}u_3^{-1}, \qquad x_4 = u_4^{-1}u_3^{-1}\,,$$

$$du_0/dt = u_0u_1u_2\,, \qquad x_0 = u_1u_2\,.$$

*5. The Brusselator (Generator of a Hopf bifurcation)*

This example demonstrates something like a Robinson-Friday interaction.

Robinson: $\quad dx_1/dt = -ax_1 + K + L_1x_1^\alpha x_2^\beta$

($ax_1$ is Robinson's investment in the resource Friday, $K$ is a central resource (perhaps nature). Robinson gains, by exploiting Friday $L_1x_1^\alpha x_2^\beta$).

Friday: $\quad dx_2/dt = bx_1 - L_2x_1^\alpha x_2^\beta$

(Friday uses from the investment in him $ax_1$, the part $bx_1$, therefore, normally $b < a$ should hold because of dissipation and returns the amount of $L_2x_1^\alpha x_2^\beta$ to Robinson).

At first we transform the given equations into Lotka-Volterra form

$$Fx_1 = -a + K/x_1 + L_1x_1^{\alpha-1}x_2^\beta\,,$$

$$Fx_2 = bx_1/x_2 - L_2x_1^\alpha x_2^{\beta-1}\,.$$

This is obviously a multinomial differential equation in which normally two of the nonlinear terms can be transformed into linear terms. We decide for the following possibility

$$w_1 = 1/x_1\,, \qquad w_2 = x_1/x_2\,, \qquad w_3 = x_1^{\alpha-1}x_2^\beta\,, \qquad w_4 = x_1^\alpha x_2^{\beta-1}\,.$$

With this substitution, we get the following Lotka-Volterra equations

$$\begin{aligned}
Fw_1 &= +a - Kw_1 - L_1 w_3 ,\\
Fw_2 &= -a + Kw_1 + L_1 w_3 - bw_2 + L_2 w_4 ,\\
Fw_3 &= -a(\alpha - 1) + K(\alpha - 1)\, w_1 + L_1(\alpha - 1)\, w_3 + b\beta w_2 - L_2\beta w_4 ,\\
Fw_4 &= -a\alpha + K\alpha w_1 + b(\beta - 1)\, w_2 - L_2(\beta - 1)\, w_4 + L_1\alpha w_3 ,\\
Fw_5 &= 0 \qquad (w_5 = 1 \text{ for the resource constant}) .
\end{aligned}$$

By passing to the Riccati representation we get

$$\begin{aligned}
du_1/dt &= K u_3^{-L_1/K} u_5^{a} , & x_1 &= u_1^{-1} u_3^{-L_1/K} u_5^{a} ,\\
du_2/dt &= b u_1^{K/b} u_3^{-1/(\alpha-1)} u_4^{1/(\beta-1)} u_5^{-a} , & x_2 &= \frac{1}{b} F u_2 ,\\
du_3/dt &= -L_1(\alpha - 1)\, u_1^{\alpha-1} u_2^{\beta} u_4^{-\beta/(\beta-1)} u_5^{-a(\alpha-1)} , & x_3 &= \frac{-1}{L_1(\alpha - 1)} F u_3 ,\\
du_4/dt &= L_2(\beta - 1)\, u_1^{\alpha} u_2^{\beta-1} u_3^{-\alpha(\alpha-1)} u_5^{-a\alpha} , & x_4 &= \frac{1}{L_2(\beta - 1)} F u_4 ,\\
du_5/dt &= u_5 \quad \text{that means} & x_5 &= 1 .
\end{aligned}$$

We assume $\alpha > 1$ and $\beta > 1$. Under these conditions, only $u_3$ decreases, all other variables can only increase. Obviously we meet in the driving force of $u_4$ a competition between $u_3$ and $u_5$ which could lead to a bifurcation if $u_3$ does not approach a constant too quickly.

*6. Simple sine generator*

Here we will demonstrate that parasitic poles occurring in the Lotka-Volterra representation can be removed by passing to the Riccati representation

$$x_0(t) = A \sin \omega_0 t .$$

We get immediately the following nonregular Lotka-Volterra representation

$$\begin{aligned}
Fx_0 &= \omega_0 x_1 ,\\
Fx_1 &= -\omega_0(x_1 + x_2) ,\\
Fx_2 &= \omega_0(x_1 + x_2) .
\end{aligned}$$

The corresponding Riccati representation, in this case, is

$$\begin{aligned}
du_0/dt &= u_0 u_1 , & x_0 &= u_1 ,\\
du_1/dt &= \omega_0 u_2 , & x_1 &= u_2/u_1 ,\\
du_2/dt &= -\omega_0 u_1 , & x_2 &= u_1/u_2 .
\end{aligned}$$

The Lotka-Volterra representation leads to a parasite pole every quarter of a period (at the roots of sine function and cosine function), whereas, obviously, in the corresponding Riccati representation, no parasitic poles are present. The Riccati representation resembles the result in Section 3 for the Tschebyscheff polynomials. We have to interpret $2ny_0$ as circle frequency $\omega_0$, meaning that the Tschebyscheff polynomials are, in this interpretation, harmonics with time-variant frequency.

*7. Langevin differential equation*

For the laser (see Haken, H. [121]), the Langevin differential equation

$$\mathrm{d}q/\mathrm{d}t = \alpha q - \beta q^3$$

plays an important role. Without restricting the generality we can assume $q(0) = 1$. This is an example where, with help of the Riccati representation, a closed analytical solution can be easily determined.
With the following additional state-variables

$$q_1 = q_0^2, \qquad q_2 = 1$$

we get the Lotka-Volterra representation

$$Fq_0 = \alpha q_2 - \beta q_1 ,$$
$$Fq_1 = 2\alpha q_2 - 2\beta q_1 ,$$
$$Fq_2 = 0 .$$

The transformation to the Riccati representation yields

$$\mathrm{d}u_0/\mathrm{d}t = u_0 u_1^{-1/2} u_2^{\alpha} , \qquad x_0 = u_1^{-1/2} u_2^{\alpha} ,$$
$$\mathrm{d}u_1/\mathrm{d}t = 2\beta u_2^{2\alpha} , \qquad x_1 = u_1^{-1} u_2^{2\alpha} ,$$
$$\mathrm{d}u_2/\mathrm{d}t = u_2 , \qquad x_2 = 1 .$$

We can get in this case the closed analytical solution without needing to pick our way through the decomposition into partial fractions. With the initial condition $u_i(0) = 1$, we get in the following sequence

$$u_2 = \mathrm{e}^t , \qquad u_1 = 1 + \frac{\beta}{\alpha}(\mathrm{e}^{2\alpha t} - 1)$$

and therefore

$$x_0 = q_0 = \mathrm{e}^{\alpha t} \Big/ \sqrt{1 + \frac{\beta}{\alpha}(\mathrm{e}^{2\alpha t} - 1)} .$$

*8. Symmetric Bautin model for the Shabotinski reaction* [48]

$$\mathrm{d}x/\mathrm{d}t = ay + E_1 x - x^3 - xy^2 ,$$

$$\mathrm{d}y/\mathrm{d}t = -ax + E_2 y - y^3 - yx^2 .$$

We assume $x(0) \neq 0$ and $y(0) \neq 0$. Under this condition, we transform the equation with certain changes of the coefficients in such a way that the normalisation condition $x(0) = y(0) = 1$ is fulfilled.

We get

$$\mathrm{d}x/\mathrm{d}t = a_1 y + E_1 x - Mx^3 - Nxy^2 ,$$
$$\mathrm{d}y/\mathrm{d}t = a_2 x + E_2 y - Ny^3 - Myx^2 .$$

First, we transform these equations into the Lotka-Volterra representation

$$Fx_0 = a_1 y_0/x_0 + E_1 - Mx_0^2 - Ny_0^2 ,$$
$$Fy_0 = a_2 x_0/y_0 + E_2 - Mx_0^2 - Ny_0^2 .$$

We introduce the following already normalized additional state-variables

$$x_1 = y_0/x_0 , \qquad y_1 = x_0/y_0 , \qquad x_2 = x_0^2 , \qquad y_2 = y_0^2 , \qquad z = 1 .$$

With these variables we get the following Lotka-Volterra equations

$$Fx_0 = a_1 x_1 + E_1 z - Mx_2 - Ny_2 ,$$
$$Fx_0 = a_2 y_1 + E_2 z - Mx_2 - Ny_2 ,$$
$$Fx_1 = a_2 y_1 - a_1 x_1 + (E_2 - E_1) z , \qquad Fx_2 = 2a_1 x_1 + 2E_1 z - 2Mx_2 - 2Ny_2 ,$$
$$Fy_1 = a_1 x_1 - a_2 y_1 + (E_1 - E_2) z , \qquad Fy_2 = 2a_2 y_1 + 2E_2 z - 2Mx_2 - 2Ny_2 ,$$
$$Fz = 0 .$$

The transformation into the Riccati representation leads to

$$\mathrm{d}u_0/\mathrm{d}t = u_0 u_1 u_2^{-1/2} v_2^{-1/2} w^{E_1} , \qquad x_0 = u_1 u_2^{-1/2} v_2^{-1/2} w^{E_1} ,$$
$$\mathrm{d}v_0/\mathrm{d}t = v_0 v_1 u_2^{-1/2} v_2^{-1/2} w^{E_2} , \qquad y_0 = v_1 u_2^{-1/2} v_2^{-1/2} w^{E_2} ,$$
$$\mathrm{d}u_1/\mathrm{d}t = a_1 v_1 w^{(E_2 - E_1)} ,$$
$$\mathrm{d}v_1/\mathrm{d}t = a_2 u_1 w^{(E_1 - E_2)} ,$$
$$\mathrm{d}u_2/\mathrm{d}t = 2Mu_1^2 v_2^{-1} w^{2E_1} ,$$
$$\mathrm{d}v_2/\mathrm{d}t = 2Nv_1^2 u_2^{-1} w^2 E^{2E_2} ,$$
$$\mathrm{d}w/\mathrm{d}t = w .$$

Because of the high redundancy of state-variables, this system transparently exposes ways for finding a closed analytical solution.

We get the following sequence of consequences

$$w = \mathrm{e}^t$$

and from this

$$\mathrm{d}u_1/\mathrm{d}t = a_1 v_1 \, \mathrm{e}^{(E_2 - E_1)t} , \qquad \mathrm{d}v_1/\mathrm{d}t = a_2 u_1 \, \mathrm{e}^{-(E_2 - E_1)t}$$

and finally

$$\mathrm{d}^2 u_1/\mathrm{d}t^2 = a_1 a_2 u_1 + (E_2 - E_1)\, \mathrm{d}u_1/\mathrm{d}t ,$$
$$\mathrm{d}^2 v_1/\mathrm{d}t^2 = a_1 a_2 v_1 + (E_1 - E_2)\, \mathrm{d}v_1/\mathrm{d}t .$$

That means that both state variables $u_1$ and $v_1$ obey a second-order linear differential equation with constant parameters. The structure of the solution is then given by

$$u_1 = A \, \mathrm{e}^{at} + \bar{A} \, \mathrm{e}^{\bar{a}t}$$

with $a$ and $\bar{a}$ being the roots of the so-called characteristic equation

$$a^2 = a_1 a_2 + (E_1 - E_2)\, a .$$

Because of the normalisation condition $u_1(0) = 1$, we have

$$A + \bar{A} = 1 .$$

Further on we find

$$v_1 = \frac{1}{a_1} e^{-(E_2-E_1)t} (Aa\, e^{at} + \bar{A}\bar{a}\, e^{\bar{a}t}) .$$

Because of $v_1(0) = 1$ we have

$$Aa + \bar{A}\bar{a} = a_1 .$$

Therefore $A$ and $\bar{A}$ are uniquely determined.

Now we continue the considerations without explicit notation of the formulas. From the last two equations follows

$$d(u_2v_2)/dt = \text{Lin Exp} \qquad \text{(a linear combination of exponentials)}.$$

By integration follows

$$u_2v_2 = 1 + I \text{ Lin Exp} .$$

Therefore, we get for $x_0$ and $y_0$ the following form of the solutions

$$x_0 = \text{Lin Exp}/\sqrt{1 + I \text{ Lin Exp}} , \qquad y_0 = \text{Lin Exp}/\sqrt{1 + I \text{ Lin Exp}} .$$

### 9. *Model of a bimolecular chemical reaction (reaction in a fluid solution)*

A solid component may dissolve in water after the law [77]

$$x(t) = x_0(1 - e^{-at}) .$$

Here, $x_0$ is the initial amount of the solid component, and $x$ that part which is already in liquid solution. Two components of the part which is already in liquid solution may interact according to a bimolecular reaction. The amount of the secondary product already built from the two components may be $y$. Still in solution are the parts $x_1(t)$ and $x_2(t)$. We demand that the following conditions should hold: $x_2 = lx_1$ with $l$ being the same relation between the two components as it is already realised in the solid state; further on let $x_1 = \varkappa x$.

Therefore, $x_1(t)$ as well as $x_2(t)$ are functions with a similar trajectory as $x(t)$. Still free for reaction with each other in the bimolecular reaction are only the quantities $x_1 - y$ and $x_2 - y$. Therefore, we get the following kinetic reaction equation for the synthesis of the reaction product $y$.

$$dy/dt = K(x_1(t) - y)\,(x_2(t) - y) = Kx_1x_2 - Kx_2y + Ky^2 .$$

We start the process with $y(0) = 1$, and derive the corresponding Lotka-Volterra equations

$$Fy_0 = Kx_1x_2/y_0 - K(x_1 + x_2) + Ky_0 .$$

After introduction of the additional state-variable $y_1 = x_1x_2/y_0$ we get

$$Fy_0 = Ky_1 - K(x_1 + x_2) + Ky_0 ,$$
$$Fy_1 = Fx_1 + Fx_2 - Ky_1 + K(x_1 + x_2) - Ky_0 .$$

Then we get the following Riccati representation

$$du_0/dt = -Ku_1\, e^{-KI(x_1+x_2)} , \qquad du_1/dt = Ku_0x_1x_2\, e^{KI(x_1+x_2)} .$$

Here $u_0$ and $u_1$ obey second order linear differential equations with time-varying coefficients.

## 6.5. *Numerical Aspects and Advantages of Riccati Representation*

The most important feature of the Riccati representation consists of the fact that all state-variables $u_i$ are monotonous functions in time, at least if we restrict the consideration on the positive cone $u_j \geqq 0$. The consequence of this property is that all state-variables of the original Lotka-Volterra system are power-products of monotonous functions. This can be an advantage because of the relative simplicity of the representation, but it can, at the same time, be a source of difficulty. This is because all state-variables $u_i$ now expose a tendency to generate either a computer zero or to run into overflow. The consequence of these properties is a restriction of the time in which it is in fact possible to simulate the systems behaviour with a computer.

If we encounter these phenomena, we can try to avoid them by rescaling operations in the $u$-space, preserving the behaviour of those $x$-components in which we are interested. We will demonstrate this possibility of rescaling for the simple example of the two-dimensional Lotka-Volterra system

$$Fx_1 = A - kx_2 , \quad Fx_2 = lx_1 - B , \qquad x_i(0) = 1 , \qquad i = 1, 2 ,$$

with the Riccati representation

$$\mathrm{d}u_1/\mathrm{d}t = u_2^{-k} \mathrm{e}^{At} , \qquad x_1 = u_2^{-k} \mathrm{e}^{At} ,$$
$$\mathrm{d}u_2/\mathrm{d}t = u_1^{l} \mathrm{e}^{-Bt} , \qquad x_2 = u_1^{l} \mathrm{e}^{-Bt} .$$

In this case, both state variables $u_1$ and $u_2$ are monotonously increasing, but, because of the competition of both driving forces with the restricted resources $A$ resp. $B$, we know that the consequence is a quasi-cyclic motion in the $x$-plane. In the simulation experiment however, $u_1$ or $u_2$ necessarily run into overflow. Here, a very simple rescaling is possible which repairs this disaster successfully. In other cases, similar, but more specific, repair measures are possible.

We rescale, in this case, in the following way

$$t = t_0 + \Delta$$

and get

$$x_1 = (u_2^{-1} \mathrm{e}^{At_0/k}) \mathrm{e}^{A\Delta} , \qquad x_2 = (u_1 \mathrm{e}^{-Bt_0/l})^l \mathrm{e}^{-B\Delta} .$$

We know that, in the course of our structure design principle, roots of expanded signals or roots of their derivatives necessarily lead to parasitic poles in the Lotka-Volterra representation, with the consequence of very restricted possibilities for the computer simulation of the systems behaviour in the Lotka-Volterra representation. It turns out that we get rid of these parasitic poles if we pass from the Lotka-Volterra representation to the Riccati representation. This could be seen already in Section 6.4, in the example of a sine generator. The following considerations shall demonstrate the generality of this favourable property.

Let us assume that we are going to expand a signal of the following form

$$x_0(t) = y_0(t)\,\xi(t)$$

with $y_0(t_0) = 0$ and $\xi(t)$ without any roots.

Then we get

$$Fx_0(t) = Fy_0(t) + F\xi(t)\ .$$

We assume that $\xi(t)$ now offers difficulties in the structure design process. We see that, arising from the root $t_0$ of $y_0(t)$, we get a parasite pole in the Lotka-Volterra description.

If $x_0(t)$ has no autocatalytic influence then $x_0 = Fu_0$ or

$$u_0 = \mathrm{e}^{I(y_0\xi)}$$

with no parasitic pole at $t = t_0$.

If, on the other hand, $x_0(t)$ has an autocatalytic influence, then we can decompose in the following way

$$Fx_0 = Fy_0 + \eta(t) + G_{00}x_0$$

and

$$Fu_0 = -G_{00}x_0 = \eta + F(y_0/x_0) \quad \text{or} \quad u_0(t) = F^{-1}\eta + 1/\xi(t)$$

with also no parasitic pole at $t = t_0$.

The redundancy of state-variables $u_i$ in the Riccati representation can often be used successfully to determine closed analytical solutions. This could be demonstrated in Section 6.4, in examples 7 and 8.

The multiplicative character of the interaction in the driving forces of the Riccati representation seems to be a key to influence the qualitative behaviour of a nonlinear system in this description because we could distinguish in Section 6.3 basic modes of behaviour and can hope that, by controlling some parameters in the Riccati representation (in the sense of master variables), we have the possibility to produce qualitative effects like bifurcations. If we have to interpret our findings in the $u$-space in the $x$-space of the original Lotka-Volterra system, we must take into account the scaling invariances. We shall always get the results in the $x$-space by applying nonlinear mappings given by the term expressions of the $x_i$ by the $u_j$.

Because it can be shown that it is possible to generate reliable function generators with the Riccati representation, and because the Riccati representation offers a rather regular systematic handling of ordinary differential equations, it seems to be useful to imagine a hardware concept for the realisation of a system's behaviour in Riccati representation supported by a suitable hardware architecture. In the case where all exponents in the Riccati representation are integers, we need not forbid the system to pass the boundaries of the positive cone $u_i \geqq 0$. That means that, in this case, sign-changing can give us some benefits for the simulation of the system. It should be mentioned, however, that the usual realisation of the power functions on microcomputers $u$ excludes the possibility of negative sign of $u$ even in the case of an integer exponent $k$, meaning that we must use a programmed power-function in this case.

One benefit of admitting changing of signs of the $u$-variables on the system trajectories is connected with a graph method for analysing roughly the qualitative behaviour which stems from the process logic. We shall describe this interesting possibility in the next Section 6.6.

## *6.6. The Use of Oriented Graphs for the Description of Some Qualitative Properties of Riccati Representations with Integer Exponents* [1])

The graph method which we will present here can be applied for any Riccati representation under the assumption that all exponents in the driving forces are integers, and that we, in general, do not restrict the motion to only the positive cone $u_i \geqq 0$ in the $u$-space.

We first describe the scheme of this method. Let

$$\mathrm{d}u_i/\mathrm{d}t = K_i \prod u_r^{k_{ir}}, \qquad i = 1, 2, \dots, n \qquad \text{and all } k_{ir} \text{ being integer.}$$

Using the ideas of kinetic logic [87] we introduce *sign states* and *growth states*. A sign state is an admissible vector $(v_1, v_2, \dots, v_n)$ with $v_i = \operatorname{sign} u_i$. Here, we understand the sign function in the following sense

$$\operatorname{sign} u = \begin{cases} +1 \text{ (shortly } +) & \text{for} \quad u > 0\,, \\ 0 & \text{for} \quad u = 0\,, \\ -1 \text{ (shortly } -) & \text{for} \quad u < 0\,. \end{cases}$$

*Admissable vector* means that possibly not all sign states are allowed, if we take into account the possible restrictions on $x_i$ or on the original Lotka-Volterra system.

A growth state is an admissable vector $(w_1, w_2, \dots, w_n)$ with $w_i = \operatorname{growth} u_i$. We understand the growth function in the following sense

$$\operatorname{growth} u = \begin{cases} +1 \text{ (shortly } \uparrow) & \text{if} \quad \mathrm{d}u/\mathrm{d}t > 0\,, \\ 0 \text{ (shortly } \emptyset) & \text{if} \quad \mathrm{d}u/\mathrm{d}t = 0\,, \\ -1 \text{ (shortly } \downarrow) & \text{if} \quad \mathrm{d}u/\mathrm{d}t < 0\,. \end{cases}$$

*Admissable growth state* means that, because of the concrete constellation in the driving forces in the Riccati representation, possibly not all combinations of $\uparrow$, $\downarrow$ and $\emptyset$ can occur.

For a more clearer construction, we interpret the set of admissable sign states as the set of states of a graph game, and the set of admissable growth states as the set of states for another graph same. Both graph games are obviously coupled with each other by the concrete equations in the Riccati representation.

In any state of a graph game, we must define the set of possible steps. In our case, the possible steps (the elementary grammar of the graph game) are completely specified by the given Riccati representation. We shall now describe the procedure in general terms.

Let us start in an arbitrary admissable sign state. Because of the integer exponents $k_{ir}$ in the Riccati representation, the sign state and the power product driving forces on the right-hand side define a uniquely determined growth state. In this growth state, we find, based on the past sign state, the corresponding possible new sign states, taking into account the possible running of races between the different variables. Therefore, in every state of both graph games, the set of possible steps is completely specified. Thus, we get two coupled graphs containing as nodes sign states respective to growth states. Sometimes it may be useful to consider only the subgraphs consisting either only of sign states or of growth states. These graphs can offer useful information about the

---

[1]) The ideas for this approach originated from intensive discussions with Prof. Dr. P. Schuster, Prof. Dr. K. Sigmund and Dr. D. Hofbauer, Universität Wien, in Spring, 1983

qualitative behaviour of a Riccati representation. The problem then arises, however, as to which kind of information we can obtain from these graphs about the qualitative behaviour of the original Lotka-Volterra equation or the original given ordinary differential equation.

We will demonstrate the application of this graph approach for the qualitative behaviour of the hypercycle of order $n = 2$.

$$Fx_i = K_i x_{i+1} \quad \text{with} \quad i \bmod 2 \, .$$

If we admit arbitrary initial values, we can remove the parameters $K_i$ by the substitution

$$y_i = K_{i-1} x_i \, , \qquad i \bmod 2 \, .$$

Therefore, we must discuss the following Riccati representation,

$$\mathrm{d}y_0/\mathrm{d}t = y_0 y_1 \, , \qquad \mathrm{d}y_1/\mathrm{d}t = y_1 y_0 \, ,$$

We quickly find the following inferences in the sign graph or the growth graph, respectively,

$$(+, +) \to \quad (+, +) \qquad \text{(stable explosive mode)}$$

$$(+, -) \to \quad (0, -) \to (0, -)$$

$$(-, +) \to \quad (-, 0) \to (-, 0)$$

$$(-, -) \to \begin{cases} (0, -) \to (0, -) \\ (-, 0) \to (-, 0) \\ (0, 0) \to (0, 0) \end{cases}$$

Fig. 88 shows the sign graph together with possible transitions under disturbances, which we have assumed only influence the sign state 0. We observe that $(+, +)$ is the only stable focus, the other foci (containing always at least one 0) are unstable, since under disturbances, they expose the behaviour of a saddle.

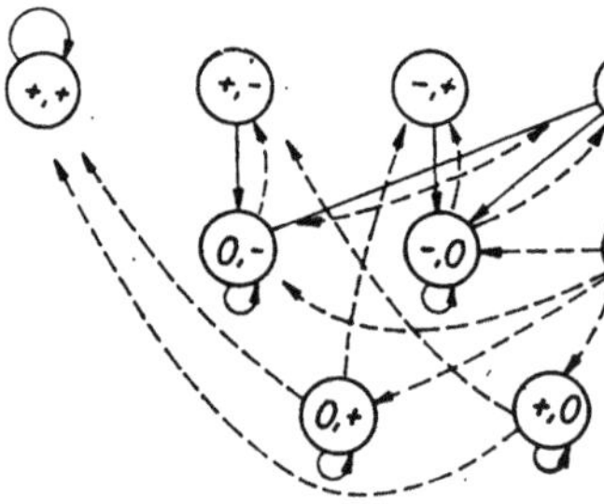


**Fig. 88** The sign-graph of an exponential hypercycle of order $r = 2$ together with transitions arising from perturbations

## 6.7. *Multiobjective Problem of the Introduction of Additional State Variables and the Use of Local Riccati Models*

In Section 2.13, we discussed the possibility of replacing the right-hand sides in a system of ordinary differential equations

$$\mathrm{d}z_i/\mathrm{d}t = f_i(z_1, z_2, \dots, z_n) \, , \qquad i = 1, 2, \dots, n,$$

by local power-product models in the neighbourhood of a reference point of time $t_0$. In the corresponding point $z^0$ in the state-space, we then get the following approximation of the dynamic behaviour

$$dz_{iM}/dt = f_i(z^0)\, \Pi\, (z_i/z_j^0)^{a_{ij}} .$$

This approximation obviously has the form of our Riccati representation. Taking this into account, we are now able to propose the following procedure for a combination of our structure design method and the method of determination of local Riccati models in order to get suitable approximations for the dynamic behaviour of a nonlinear system using a differential equation description. We propose to solve the following polyoptimization task:

- Apply several of the first steps of the structure design process. This leads us to a certain simplification on one hand (for example exposing some term summands), and, at the same time, to the introduction of some additional state-variables.
- Interrupt the structure design process and use the differential equation thus obtained as the object for determination of local Riccati models as described in Section 2.13.

As is usually the case, if we use local models, we must determine a number of $m$ local models with certain interface condition for a reliable description of the trajectories of the original differential equation. On the other hand, we must adjoin a number of $r$ state-variables for the first steps of our structure design process. Thus both numbers $Q_1 = m$ and $Q_2 = r$ are competing with each other. When $r$ is small, $m$ will be high and vice versa. This multiobjective situation should be solved by a rational compromise between the objectives connected with these two criteria. Depending upon the concrete task we will solve this problem with the considered ordinary differential equation.

## 6.8. *The Concept for the System Architecture of a Special Processor for the Integration of Ordinary Differential Equations*

In Section 6.1. we have shown that a huge class of ordinary differential equations can be transformed into a so-called Riccati representation, meaning into a state-variable description of the following form

$$du_i/dt = K_i \prod_r u_r^{k_{ir}} , \qquad i = 1, 2, \dots , n .$$

The original given state-variables $x_i$ and the interesting output variables among them are power-products of the Riccati state-variables $u_i$

$$x_i = \prod_r u_r^{l_{ir}} .$$

We agree that for $i = 1, 2, \dots , s$ we get those state-variables $x_i$ in which we are interested (for example, the state-variables of the original given differential equations).

This unified concept for the system description of ordinary differential equations allows for a standard realisation by an integrated module in VLSI-technique (Very Large System Integration).

In the following we shall describe a possible system architecture for such a module based on the Riccati representation with an orientation on parallel processing. First we report, however, on some ideas how to handle the dynamic properties of such a system description. Afterwards we give the scheme of a possible system architecture taking into account the dynamic properties and using the possibilities of parallel processing with a concept similar to the data-flow technique.

### *Assumptions and Handling of the Dynamic Properties*

We demand that all state-variables $u_i$ and $x_i$ are restricted on the positive cones $u_i \geqq 0$, respectively $x_i \geqq 0$ to avoid unwanted phenomena in connection with a change of the sign of the state-variables. Because of the concrete form of the Riccati representation it follows that all $u_i$ are monotonous functions on time $t$, namely, increasing for $K_i > 0$ and decreasing for $K_i < 0$. This holds not necessarily for the $x_i$, because they are given by multiplicative interaction of powers of the $u_i$'s. Therefore, we meet the danger, that $u_i$ either runs into overflow or into computer zero. In both cases the computation process has to be stopped. This might occur already after a short time, meaning, after having computed only a relatively small part of the interesting trajectories for the $x_i$, $i = 1, 2, \dots, s$.

We should try to avoid this bad phenomenon by the operation *rescaling*. In general we understand under the notion rescaling the following operation:

We transform the state-variables of the Riccati representation in the following way hoping to get rid of the overflow danger or computer zero

$$u_r = v_r \varkappa_r .$$

For the new state-variables $v_i$ we then get the following equations

$$\mathrm{d}v_i/\mathrm{d}t = K_r' \prod v_r^{k_{ir}}$$

with the new velocity amplifications

$$K_i' = K_i \prod \varkappa_r^{k_{ir}} / \varkappa_i .$$

But we are not allowed to use for this purpose arbitrary sets of factors $\varkappa_r$; we must assure, that the values of all interesting original variables $x_i$, $i = 1, 2, \dots, s$, are not influenced by this measure. That means we have to fulfill the following invariance conditions

$$\prod \varkappa_r^{l_{ir}} = 1 , \quad \text{for} \quad i = 1, 2, \dots, s .$$

We cannot be sure, that this rescaling operation leads always to a success. Let us study now this problem in more detail. For the security of computations we have to demand generally

$$\mathrm{EPS} \leqq u_i \leqq \mathrm{OFLOW}$$

and

$$\mathrm{EPS} \leqq v_i \leqq \mathrm{OFLOW}$$

for the old variables $u_i$ and also for the new variables $v_i$. Let us assume that $u_{i_0}$ has reached its admitted upper limit OFLOW. Then it seems to be feasible to transform this variable in the following way

$$v_{i_0} = u_{i_0} / \mathrm{OFLOW}$$

meaning that $\varkappa_{i_0} = \text{OFLOW}$ and $v_{i_0}$ starts with the value 1. We have then to distribute this measure for $u_{i_0}$ on the other state-variables by choosing suitable factors $\varkappa_r$, $r \neq i_0$. For this purpose we must take into account the invariance conditions

$$\sum l_{ir} \ln \varkappa_r = -l_{i i_0} \ln \text{OFLOW} , \qquad i = 1, 2, \dots , s$$

and also the security conditions

$$\text{EPS} \leqq u_i/\varkappa_i \leqq \text{OFLOW}$$

taking into account the actual configuration of the state-variables $u_i$. That means we have to determine a feasible solution,

$$\xi_r = \ln \varkappa_r , \qquad r \neq i_0 ,$$

of the following linear program

$$\sum_{r \neq i_0} l_{ir} \xi_r = -l_{i i_0} \ln \text{OFLOW} ,$$

$$\ln u_r - \ln \text{OFLOW} \leqq \xi_r \leqq \ln u_r - \ln \text{EPS} .$$

If a feasible solution of this linear program does not exist, then the computation process must be stopped. If there are some feasible solutions we can try to choose the "best" of them under some useful objectives. We can proceed similarly if $u_{i_0}$ has reached the admitted lowest level EPS. In this case we put

$$v_{i_0} = u_{i_0}/\text{EPS}$$

meaning that we choose $\varkappa_{i_0} = \text{EPS}$ and start $v_{i_0}$ with the value 1. Now we get the following linear program

$$\sum_{r \neq i_0} l_{ir} \xi_r = -l_{i i_0} \ln \text{EPS} ,$$

$$\ln u_r - \ln \text{OFLOW} \leqq \xi_r \leqq \ln u_r - \ln \text{EPS} .$$

If a feasible solution of this linear program does not exist, then the computation process must be stopped. Now we inform about the ideas for the architecture of the planned processor to get an applicable concept for the division of labour within the integrated system.

The integrated system should consist of some parallel working ROM's (**R**ead **O**nly **M**emory), each one designated to process one equation of the form

$$\mathrm{d}u_i/\mathrm{d}t = K_i \prod u_r^{k_{ir}} .$$

Of course, it is possible to combine some of these equations in one ROM or, opposite to this solution, to split every such right-hand sum into a row of more elementary modules orientating to use a matrix array for the realisation of the whole structure.

For simplicity we shall for the following considerations assume that for every equation of this type corresponding to the state-variable $u_i$ we provide just one $\text{ROM}_i$. By this $\text{ROM}_i$ in every access tact an operation of the form

$$\text{LET } ui = ui + D * Ki * u_r^{k_{ir}}$$

is performed. Herein $D$ is the step width. We restrict here on Euler forward integration. Of course, also other more qualified integration procedures like trapez-rule or Runge-Kutta integration should be offered by the concrete realisation of such an integrated system. Besides these decentralized ROM's there should exist one or more centralized modules for the realisation of the OS (operation system), the communication of the initial values, the generation of the wanted output signals, the storage of all state-variables $u_i$ and last not least for the performance of the rescaling operation included the feasibility test of the linear programs.

Let us first continue the consideration of the function of the $ROM_i$. Within such a ROM the power function $u^k$ must be realised and additionally the multiplication of these power products and the performance of an integration step according to the Riccati form of the driving force. The module $ROM_i$ gets from outside the informations $D$, $ui(0)$, $Ki$ and $k_{ir}$.

All these values should be stored within the module $ROM_i$. In the cell for $u_i$ first $u_i(0)$ will be stored and afterwards the actual value $u_i$. In every working tact the actual values $u_r$ for all $r \neq i$ will be communicated from outside to the $ROM_i$. This communication could be done by a logical network immediately from all other $ROM_r$ or from a centralised puffer monitoring the actual state-variables $u_r$. Necessarily every $ROM_i$ executes event-oriented a first step of a rescaling operation, namely,

$$\text{IF } ui \leqq \text{EPS THEN LET } ui = ui/\text{EPS},$$

$$\text{IF } ui \geqq \text{OFLOW THEN LET } ui = ui/\text{OFLOW}.$$

After that $ROM_i$ gives a rescaling report signal to the centralized part of the system responsible for the execution of the global rescaling operation. $u^k$ is naturally realised in form of a software module within every $ROM_i$. The different powers are computed sequentially one after the other and multiplied in a fast multiplier immediately after the execution of all such power operations.

If we would apply instead of such a $ROM_i$ a row of smaller ROM's we had the possibility to perform the computation of all necessary powers in parallel processing.

As a closing step now the integration operation follows. All these operations need a certain time resource within a $ROM_i$ and we have to expect that they cannot be executed synchronously in the different ROM's. Therefore, we orientate on an asynchronous data flow concept for the interaction of all these ROM's with the central part of the system. We will give an idea how we think this communication could be realised. During the active work of $ROM_i$ a "red" light is shining at its output. Having completed the work of one step as described above, the "red" light is switched off informing thus the central part of the system that $ROM_i$ is ready for a next execution.

At the next central tact when the central part has observed that all ROM's have switched off their "red" lights, the central part can do its work with the computed information and initiate the next global computation step. (Such an asynchronous control of a network we published in [119]). In this phase the central part registrates rescaling reports from the ROM's. Before activation the next global computation step the rescaling operation must be executed by the central part.

In a global computation step without the necessity to perform a rescaling operation the central part takes over all the computed state-variables $u_i$ into a central buffer memory and communicates these values back into the ROM's for the next tact.

If rescaling operation is necessary, in a central part the feasibility test must be executed, the linear program must be solved, the rescaling operation must be performed with the actual content of the central buffer memory, and now the rescaled state-variables must be communicated to the different ROM's and, what is most important, the different ROM's must get now new values of their velocity amplifications $K_i$. The computation of new values $K_i'$ of the $K_i$ can be done within the ROM's itself within a separate rescaling tact.

Thus, before communicating the new rescaled state-variables to the ROM's we tell them the values of all $\varkappa_i$ and compute the values of $K_i'$ using only the computation of powers and power products within the ROM's. Which additional tasks must be solved by the central part of the system? We assume that the whole system is provided by the user with a system of differential equations already transformed by the user himself into a Riccati representation. The user will then introduce the parameters $K_i'$ and $k_{ir}$ into the system together with the initial values $xi(0)$, $i = 1, 2, \ldots, s$, of those original state-variables $x_i$ in the computation of which the user is interested. The user also communicates the parameters $l_{ir}$ for the connection between the Riccati state-variables $u_r$ with the original (Volterra) state-variables $x_i$.

We assume that all initial values for the $x_i$, $i = s + 1, \ldots, n$, are normalised, meaning

$$xi(0) = 1 \quad \text{for} \quad i = s + 1, \ldots, n .$$

Then firstly the central part of the system has to compute the initial values of the Riccati state-variables $u_r$ solving the following $n \cdot n$ system of linear equations

$$\ln x_i(0) = \sum l_{ir} \ln u_r(0) , \qquad i = 1, 2, \ldots, n .$$

Therefore, the central part of the system needs the following software tools for:

— computation of ln ( )
— computation of exp ( )
— solution of a $n \cdot n$ linear system of equations

To avoid difficulties in our discussion now we shall assume that the matrix

$$L = (l_{ir})$$

is regular, what, in general, need not be the case. The central part must have an OS (operation system) handling the inputs and outputs and the distribution of informations to the ROM's and gathering the informations from the ROM's.

After having determined the initial values $u_r(0)$ the central part of the system can communicate these values to the ROM's and initiate then the parallel work of all these ROM's. If all the ROM's have completed one computation step from the lights of the ROM's is formed a logical signal (conjunction) informing the central part that all ROM's are ready to perform the next computation step.

As already above described the central part gathers the computed values $ur(k)$ into a central buffer and distributes these values back to all ROM's and initiates afterwards the ROM's for the performance of the next computation step. If the central part has at the end of a computation step received from one of the ROM's a boundary report (OFLOW or EPS) it has to perform the corresponding rescaling operation. This could be done by another software-tool within the central part of the system — solving of a linear program.

It follows the transmission of the computed scaling factors $\varkappa_r$ to the ROM's and a help computation step must be inserted to compute corrected values of the velocity amplification parameters $K_i'$ within the ROM's. Afterwards the rescaled state-variables $u_r$ are transferred to the ROM's and the next global computation step can be initiated.

In such a way the work will be continued for a given number of global computation steps or until a logical test will be fulfilled. The step number or the organisation of the corresponding logical test should be also included into the task profile of the central part. May be it is not senseful to include into the central part a whole software tool for the solution of a linear program.

Then instead of solving linear programming tasks some simple heuristice could be applied. But then it might occur that the computation stops because none of the included heuristics will be applicable further on despite of the possibility that a linear program would possess a feasible solution and could thus continue the computation process.

In the following considerations we will give some examples for simplifying heuristics instead of solving a linear program. Let us assume that the central part has got from the $\mathrm{ROM}_{i_0}$ an OFLOW signal. Then we have to find a solution of the following equations

$$\sum l_{ir}\xi_r = 0\,, \qquad i = 1, 2, \dots, s\,,$$

$$\ln u_r - \ln \mathrm{OFLOW} \leqq \xi_r \leqq \ln u_r - \ln \mathrm{EPS}\,.$$

Now we prescribe threshold factors $k_u$ and $k_l$ and check which of the $u_r$ fulfills the following condition

$$u_r \geqq k_u\,\mathrm{OFLOW} \quad \text{with} \quad k_u = 1/m \text{ and } m \text{ an integer.}$$

The number of these $u_r$ may be $N_1$. At the same time we check which $u_r$ obey the condition

$$u_r \leqq k_l\,\mathrm{EPS} \quad \text{with} \quad k_l = m' \text{ and } m' \text{ an integer.}$$

The number of the corresponding $u_r$ may be $N_2$. We agree now that we will modify by rescaling only those $u_r$ fulfilling one of both conditions and preserve the values of all other $u_r$ which obviously still are not in danger. Then the equations which we have now to fulfill are simplified in the following way

$$\sum_{r\in I_1} l_{ir}\xi_r + \sum_{s\in I_2} l_{is}\xi_s = 0\,,$$

$$\ln u_r - \ln \mathrm{OFLOW} \leqq \xi_r \leqq \ln u_r - \ln \mathrm{EPS}\,.$$

We havé the idea that all $\varkappa_r$ are something like $\mathrm{OFLOW}/q$ with a small integer number $q$, and that all $\varkappa_s$ are something like EPS $p$ with a small integer number $p$.

Then $\xi_r \sim \ln \mathrm{OFLOW} - \ln q$, that means $\xi_r$ is a relatively large positive number and $\xi_s \sim \ln \mathrm{EPS} + \ln p$, that means $\xi_s$ is a relatively small negative number. We have then to solve the following equations with small integer numbers $q_r$ and $p_s$

$$-\sum l_{ir} \ln q_r + \sum l_{is} \ln p_s = -\sum l_{is} \ln \mathrm{EPS} - \sum l_{ir} \ln \mathrm{OFLOW}\,.$$

The inequality conditions are in this case automatically fulfilled. By varying the integers $m$ and $m'$ we must secure that $N_1 + N_2 = s$, so that we have now to solve only an $s \cdot s$ regular linear system of equations. In this case we could use a software-tool which we necessarily have to provide within the central part of the system for the computation of the initial values of the Riccati state-variables $u_r(0)$. If it occurs, that $N_1 + N_2 > s$, we could orientate on some equal factors $q_r$ respectively $p_s$.

Frequently, we are interested in two state-variables $x_1$ and $x_2$ (for example, position and velocity for a mechanical system). Then we could try to find a homogenous solution, meaning that all $q_r = q$ and all $p_s = p$. Thus, we have to solve only a system of two equations

$$- \sum l_{1r} \ln q + \sum l_{1s} \ln p = - \sum l_{1s} \ln \mathrm{EPS} - \sum l_{1r} \ln \mathrm{OFLOW}\,,$$

$$- \sum l_{2r} \ln q + \sum l_{2s} \ln p = - \sum l_{2s} \ln \mathrm{EPS} - \sum l_{2r} \ln \mathrm{OFLOW}\,.$$

# 7. Structure-Building and Lotka-Volterra Equations

We understand evolution as a unity of growth and structure-building. The autonomous growth of natural systems is driven by the tendency to continuously increase the growth rate. This is possible if, in the background of the systems phenomena, an aim-orientated structure-building takes place, forming a cooperative structure in a short time to make use of all visible resources. In the saturation phase, the system meets a scarcity of resources or strong competition with other systems, and it has great difficulties in supplying the complex structure. Then a structural decomposition begins, the system decomposes into smaller structural parts, with reduced supply problems and a chance to synthetisize, by recombination, a new structure better fitted for the next evolutionary step. Thus, dynamic changes of structure and quantitative growth phenomena are strongly connected with each other in the unity of an EVOLON model. All these phenomena which we qualitatively characterized are consequences of the dialectic polarity of cooperation and competition.

If we consider from this point of view the Lotka-Volterra equations

$$Fx_i = \sum G_{ij}x_j$$

we can state that they contain a set of two-particle interactions (Robinson $x_i$, Friday $x_j$), with cooperation of $x_j$ into the direction of $x_i$ if $G_{ij} > 0$, competition of $x_j$ against $x_i$ if $G_{ij} < 0$, and we can say that not necessarily is $G_{ij}$ symmetric, meaning that cooperation in one direction can coexist with competition in the opposite direction.

The structure design principle can also be interpreted as a real possibility in a huge class of systems to reduce the complex interaction structure to a set of two-particle interactions. This interpretation seems to be interesting also for macroeconomy and sociology, as far as the dynamic phenomena there can be reliably described by sets of ordinary differential equations.

We deal in this chapter primarily with the formalisation of structures using the cluster technique. This is a restriction from the point of view of the role of structures in systems dynamics, but it is suitable for the Lotka-Volterra representation of systems. We will first discuss an effective fuzzy approach for cluster building, which is applicable to the competition matrix $G_{ij}$ of a Lotka-Volterra description, and which is at the same time, a good background for cluster building as a consequence of dynamic interaction. Afterwards, we will derive different notions of structure immediately from the equilibrium behaviour of dynamic systems.

## *7.1. Solution of Nonlinear Assignment Problems and Structure-Building Procedures on the Basis of the Fuzzy Approach*[1])

Assignment problems are of increasing importance in many different fields, as, for example, in industrial applications (microelectronic circuit design, allocation problems for large-manufacturing lines), in social sciences (formation of social groups, growth of settlements and their decomposition in substructures), and in ecology (decomposition of a global ecosystem into biotopes). All these problems can be formally described as cluster building processes.

Every structure is built up of elements (individuals, atoms, species, singletons, etc.) on the basis of a relationship between pairs of elements which must be examined objectively. Depending upon the character of the concrete assignment problem, we meet quite different forms of binary relations (number of connections, competition coefficients, similarity measure, metric distance). Using this binary relation, the set of all elements can be rationally decomposed into clusters which are not necessarily disjointed. Thus, we meet, in a natural way, fuzzy membership correspondences of the elements to the different clusters. On one level of consideration we find a fuzzy partition of the set of all elements into clusters. However, we can also consider the structure-building process as a hierarchical process. On the next higher level of consideration, we accept the clusters already constructed as aggregated elements and try to build up superclusters from them.

It is important to mention that cluster-building is a dynamic process at the end of which we find a stabilized cluster decomposition into the given set of elements. Therefore, in some sense, a fixed cluster decomposition is something like an equilibrium. For this process, a suitable interaction of global and local objectives is essential; it is similar to problem-solving in general or to chess playing especially. Therefore, it seems useful to consider the final result of a dynamic cluster-building process as the equilibrium point of a dynamic process, or as a solution of a (mostly nonlinear) fixed-point equation. Here, we study two quite different fuzzy approaches to cluster-building. The first is a generalisation of the cluster-building procedures from Späth and Bezdek, respectively. The second is an application of the replicator equation (Schuster and others [11], [14]) for cluster building purposes. Both approaches were checked on large examples and showed quite satisfactory properties.

General features of a fuzzy cluster-building process are as follows. The elements of the considered universe are designated by the indices $1, 2, \ldots, n$. Variable elements are denoted by $i$ and $j$, variable clusters by $k$ and $l$ and variable superclusters by $\varkappa$ and $\lambda$. The basic relation between any pair of elements $i$ and $j$ is given by a (deterministic) matrix of coefficients $S(i, j)$ with real elements not necessarily being a symmetric matrix. Sometimes negative elements are also admitted (for example, in the case of competition coefficients of species in an ecosystem).

The correspondence of an element $i$ to a cluster $k$ is modelled by a membership coefficient $U(i, k)$. These coefficients take over the role of state-variables in the dynamic cluster-building process.

In order to construct superclusters, we need an interaction matrix $\tilde{S}(k, l)$ between each pair of clusters $k$ and $l$, and a fuzzy correspondence $\tilde{U}(k, \varkappa)$ of the clusters $k$ to

---

[1]) We thank Dr. Richardt, Institute für Nachrichtentechnik, Berlin, for providing us with his ideas for nonlinear assignment problem-solving.

the superclusters $\varkappa$. The following notions are specific for our approach to fuzzy cluster-building. We introduce first an affinity measure $A(i, k)$ for the tendency of the element $i$ to belong to the cluster $k$, and define it in the following way

$$A(i, k) = \sum_j S(i, j)\, U(j, k)\,.$$

The idea for this measure stems from deterministic clustering, in which $U(j, k)$ can assume only the values 0 and 1 : 1 for all elements $j$ belonging to the cluster $k$ and else 0. Then $A(j, k)$ is the number of connections of the element $i$ into the cluster $k$. Our measure $A(i, k)$ obviously is a fuzzification of the deterministic case. It seems to be quite natural to introduce a similarity measure for any pair of clusters $k$ and $l$ by

$$\tilde{S}(k, l) = \sum_{i,j} U(i, k)\, S(i, j)\, U(j, l)$$

which is obviously a fuzzy generalisation of the number of connections of elements $i$ belonging to the cluster $k$ and to elements $j$ belonging to the cluster $l$.

In the course of the cluster-building process, we must change in an aim-orientated manner the values of the memberships $U(i, k)$ $\left(\text{or } \tilde{U}(k, \varkappa)\right)$ meaning that the cluster partition is always contained in the matrix $U(i, k)$. For the dynamic cluster-building process bilateral relations must always be taken into account, namely

$$\sum_k U(i, k) = U(i) \quad \text{the mass of the particle } i,$$

$$\sum_i U(i, k) = U(k) \quad \text{intermediate mass of the cluster } k$$

respectively in the two-level hierarchical process

$$\sum_\varkappa \tilde{U}(k, \varkappa) = U(k) \quad \text{intermediate mass of the cluster } k,$$

$$\sum_k \tilde{U}(k, \varkappa) = U(\varkappa) \quad \text{intermediate mass of the supercluster } \varkappa.$$

In general, we shall assume that the masses $U(i)$ of the elements $i$ are given. The first idea for cluster-building would be to apply a linear iteration procedure. We change the fuzzy memberships $U(i, k)$ according to their affinities $A(i, k)$ to the classes with the help of the following iterative procedure

$$F(i, k) := A(i, k)\,.$$

We interpret $F(i, k)$ as an attraction force and, taking into account the normalisation condition; we get

$$U(i, k) = U(i)\, F(i, k) / \sum_k F(i, k)\,.$$

This obviously corresponds to a Markov chain procedure. It turns out that it was not very successful to expose a structure in the course of iteration. Therefore, we passed to an essential nonlinear procedure by introducing master variables and taking into account the weight of the different intermediate masses $U(k)$ of the clusters.

We introduce the following concept

$$F(i, k) := A(i, k)^{\mathrm{EV}} / U(k)^{\mathrm{EG}}$$

with the master variables EV and EG, and we then apply the normalisation condition as usual:

$$U(i, k) = U(i) F(i, k) / \sum_k F(i, k) .$$

We set EG = EV $\cdot \varrho$ with a slowly adapting variable $\varrho$. In many cases we used a fixed value for $\varrho$. The master variable EV $> 0$ is a very important parameter. Obviously, a small value for EV transforms a given distribution $A(i, k)$ into a very flat one, and a very high value for EV generates a peak-like distribution. This means that, by increasing the value of EV, we approach more and more a deterministic correspondence of the elements $i$ to the clusters $k$.

We started every seeking for useful cluster configurations process with a uniform distribution of the elements to all clusters (the number of clusters must be prescribed), which we had to disturb slightly because the uniform distribution is always a solution of the corresponding nonlinear fixed-point equation. There are two ways to reduce the number of clusters in the course of the seeking process. If we meet two clusters with about the same correspondences for all elements, we can smelt them together and continue with only one of them. It may turn out that, for each element, the correspondence to a given cluster is very small, and this cluster can thus be removed.

EV is something like a measure of the fuzziness of our correspondence decision. In all analyse cases, we could find useful cluster decompositions by a minimum increase of the value of EV compared with the initial value. (We call this the principle of minimum decrease of fuzziness.) In some cases, we found oscillation phenomena, for example, oscillations of the correspondence amplitudes or switching of one structure decomposition into another and returning after one iteration immediately to the first one (period 2!). In some cases, we could eliminate instabilities from the structure-building process only by a trick, using the diagonal elements $S(i, i)$ as additional control parameters of the process.

For elements in many cases, $S(i, i)$ is zero because an element $i$ normally has no connections with itself. This condition is not valid in the hierarchical case. Here $\tilde{S}(k, k)$ is obviously the number of inner connections between elements within a given cluster $k$. Nevertheless, in all these cases we artificially manipulate the values of $S(i, i)$, meaning that we introduce something like pseudo-connections. The physical meaning of such a manipulation is obvious. The higher the value of $S(i, i)$ is, the higher will be the autocatalytic weight of the element $i$. This increases the tendency to separate the part from the whole. If we use the same concept for $\tilde{U}(k, \varkappa)$, together with the connection relation $\tilde{S}(k, l)$, we get a two-level hierarchical approach, and meet the poly-optimization problem that in finding the clusters $k$, we use the tendency to separate two clusters $k$ and $l$ from each other as well as possible. This implicates the aim to reach small interaction coefficients $\tilde{S}(k, l)$ as the basis for cluster-building on the next higher level. That means that if we want to structure simultaneously on both levels, we need a certain compromise between the local and the more global aim. This is a clustering procedure mainly for the case of non-vectorial objects. It can, however, be applied successfully in the case of vectorial objects also, if the similarity matrix $S(i, j)$ is derived from a relevant matrix in the corresponding feature space.

## 7.2. *The Replicator Form of Lotka-Volterra Equations and Its Role for Structure Building Phenomena*[1])

The Lotka-Volterra equations describe the dynamics of ecosystems. In concrete ecological systems, we can observe structure-building phenomena leading to structures consisting of subsystems (biotopes) in different levels of a hierarchical structure. For simplicity, we shall not take into account the distribution of species over the space of a certain area here, or in connection with that space dynamics like phenomenon of immigration and emigration. We will get information about possible biotopes without the space dimension, in the sense of possible clusters which might coexist if an area offers sufficient specific niches.

The question arises whether it is necessary to identify, for the dynamics of structuring in ecosystems, new first laws, or whether the Lotka-Volterra equations themselves already determine the structure-building in an ecosystem. We are convinced that, for this purpose, no equations other than the Lotka-Volterra equations are necessary because cooperation and competition as quantitative effects are completely contained in the Lotka-Volterra equations, and, therefore, discontinuous effects of decomposition in cluster configurations should be a consequence of the dynamic interaction structure.

The following considerations are meant as a contribution to the structuring of ecosystems as a first approximation. It should be mentioned that, usually, a given Lotka-Volterra equation shows, on one hand, a drift tendency towards infinity because of the explosive force in the interaction structure. Opposing this tendency, nevertheless, a cluster building process takes place. If we are going to make a special study of the cluster-building process, we must get rid of the drift tendency of the system as a whole. For this purpose, we apply, following Schuster and others [11], [14], the transformation of a given Lotka-Volterra system to the so-called replicator form. By this transformation, the dynamic system trajectory is, in fact, projected on the simplex $\sum x_i = 1$.

We first give all necessary information about some details of the replicator equation formalism. We introduce into the Lotka-Volterra equations

$$Fx_i = \sum_j G_{ij} x_j \,, \qquad x_i \geqq 0 \,,$$

new variables $y_i$ in form of barycentric coordinates,

$$y_i = x_i / \sum_j x_j = x_i y_n \quad \text{with} \quad y_n = 1/\sum x_j \,.$$

Then we have obviously

$$\sum y_j = 1 \,.$$

The new variables $y_i$ can be interpreted as probabilities, as the relative weight of the species $x_i$ in the whole ecosystem.

---

[1]) The studies in connection with the replicator equation were stimulated by many problem discussions with Prof. Dr. P. Schuster, Prof. Dr. K. Sigmund and Dr. J. Hofbauer, Universität Wien. We thank Dr. M. Voigt, Forschungsbereich Mathematik und Informatik der Akademie der Wissenschaften der DDR, for providing us with his approach of using equilibrium points of the replicator equations as fuzzy decomposition of the competition graph.

With this substitution, we now transform the given Lotka-Volterra system and get

$$Fy_i = Fx_i + Fy_n = \sum_j G_{ij}x_j + Fy_n ,$$

$$Fy_n = -y_n\Phi \quad \text{with} \quad \Phi = \sum_{k,j} x_k G_{kj} x_j = y_n^{-2} \sum_{k,j} y_k G_{kj} y_j .$$

Here $\Phi = \sum_{k,j} x_k G_{kj} x_j$ is a flow function, which should be interpreted as the actual necessary supply of the system as a whole. Consequently, we get

$$Fy_i = \frac{1}{y_n}(\sum G_{ij}y_j - \Psi) \quad \text{with} \quad \Psi = \sum y_k G_{kj} y_j .$$

By changing the time scale, meaning the velocity of the systems motion along its trajectory, we can omit the factor $1/y_n$. This has no influence on the geometry of the trajectories, meaning on the qualitative properties of the system. Thus, we get the replicator equation [11]

$$Fy_i = \sum G_{ij}y_j - \Psi .$$

This is again a Lotka-Volterra equation with a common system-dependent resource $\Psi$ as a flow for the supply of the whole ecosystem. This equation has a very interesting invariance property.

We can alter all columns $G_{.j}$ about a certain constant $\Delta_j$ additively without any influence on the form of the replicator equation, because of

$$Fy_i = \left(\sum_j (G_{ij} + \Delta_j)\, y_j - \sum_{k,j} y_k(G_{kj} + \Delta_j)\, y_j\right)$$

$$= \left(\sum_j G_{ij}y_j - \sum_{k,j} y_k G_{kj} y_j\right) .$$

If we modified the original given Lotka-Volterra system in the same way as

$$Fx_i = \sum (G_{ij} + \Delta_j)\, x_j$$

we would get quite different Lotka-Volterra systems. They all, however, have the same geometry of projected trajectories on the simplex, and might differ only in the velocity of running along their trajectories and in the drift motion to infinity. We could thus eliminate all autocatalytic terms by putting

$$\Delta_j = -G_{jj} .$$

Besides the already mentioned differences in the behaviour, the numerical properties of all these systems can be quite different.

We now give, using the Riccati representation of the replicator equation, another representation which can be very useful, especially for the purposes of computations. For simplicity, we assume that all $G_{ii} \neq 0$ (the result we are going to derive does not depend on this assumption). We will now transform the replicator equation

$$Fy_i = \sum_j G_{ij}y_j - \sum_{k,j} G_{kj}y_k y_j$$

into the Riccati representation.

For this purpose, we introduce the additional term variables $y_{kj}$ with

$$y_{kj} = y_k y_j ,$$

$$Fy_{kj} = \sum_r (G_{kr} + G_{jr})\, y_r - 2 \sum_{r,s} G_{rs} y_{rs} .$$

From this Lotka-Volterra description of the replicator form, we get the following Riccati representation

$$du_i/dt = -\prod_r u_r^{-G_{ir}/G_{rr}} \prod_{r,s} u_{rs}^{-1/2} G_{rr} ,$$

$$du_{kj}/dt = 2 \prod_r u_r^{-(G_{kr}+G_{jr})/G_{rr}} \prod_{r,s} u_{rs}^{-1} G_{kj} .$$

We have

$$y_i = w_i/w$$

with the notations

$$w_i = \prod_r u_r^{-G_{ir}/G_{rr}} , \qquad w = \prod_{r,s} u_{rs}^{1/2} .$$

Obviously we have

$$\sum_i w_i = w \quad \text{because of} \quad \sum y_i = 1 .$$

We will now transform the replicator form into the state-variable description by the variables $w_i$. We get

$$Fw_i = -\sum_r \frac{G_{ir}}{G_{rr}} Fu_r \quad \text{and, with} \quad Fu_r = -G_{rr}x_r ,$$

$$Fw_i = \sum_j G_{ij}w_j/w \quad \text{with} \quad w = \sum w_i .$$

For $w$ itself, which plays the character of a drift variable, we get the equation

$$Fw = \Psi .$$

The advantage of this representation lies in thesimple possibility of rescaling. Always when one of the variables $w_i$ exceeds a certain value $K$, we can divide all $w_j$ and $w$ by the threshold value $K$ and continue the simulation without any change in the variables $y_j = w_i/w$.

We now introduce the cluster notion into the Lotka-Volterra dynamics. At first, we are interested in all equilibrium points $y^*$ of the replicator equation. Obviously, equilibrium points can be roughly characterized as follows: There is an index set $J \subseteq (1, 2, \ldots, n)$ with $w_j^* = 0$ for $j \in J$. For the other components, the following hôlds:

$$w_j > 0 , \qquad j \in (1, 2, \ldots, n) - J , \qquad \sum_{j\in(1,\ldots,n)-J} G_{ij}w_j^* = 0 , \qquad i = 1, 2, \ldots, n .$$

We interpret the species $j \in (1, 2, \ldots, n) - J$ as a cluster (biotope) and the corresponding $w_j^* > 0$ as relative weights of the different species in the cluster.

We are especially interested in stable clusters, in the sense that small fluctuations of the values of the state-variables or of the parameters do not change the cluster according to the above definition. Moreover, in the course of time, this situation should not change autonomously. It is sufficient for this property that there exists a stable equilibrium point or a stable limit cycle fulfilling the cluster conditions. Of special interest are such equilibrium points which are stable attractors. Many of the stable attractors are evolutionarily stable strategies (Maynard-Smith [12]) with the following definition:

$y^*$ is an ESS, if for all $y \in S_n$ (simplex in the state-space) holds,
$(y, Gy^*) < (y^*, Gy^*)$ (equilibrium condition) or if
$(y, Gy^*) = (y^*, Gy^*)$ then follows $(y, Gy) < (y^*, Gy^*)$ (stability condition).

This means that an ESS is at least a local maximum of the flow function $\Psi$. This definition ensures that if $y^*$ is evolutionarily stable, then any deviation from $y^*$ will be less fit than $y^*$ according to the consumption of the available flow $\Psi$. Such a deviation will be promptly wiped out. If we consider all stable equilibrium points $y^*$ simultaneously in the interpretation as clusters, we can assume the coexistence of these clusters if the corresponding area offers enough niches for settlements. In this sense, we can consider the set of all such equilibrium points as a cluster decomposition of the set of all species. Because some species can exist at the same time in different clusters, we are dealing obviously with a fuzzy cluster decomposition in which the corresponding weights $y_i^* > 0$ can be interpreted as memberships $U(i, k)$ in the sense discussed earlier, where $i$ denotes the considered species and $k$ the cluster. We will now introduce control influences on this kind of graph decomposition. Applying the invariance property of the replicator equation we first eliminate all original reproduction rates $G_{ii}$ of the species $i$. Then we introduce control variables $u_i$, $i = 1, 2, \dots, n$, which act on the reproduction rates of the species $i$. Then the competition matrix of the replicator equation changes in the following way

$$\tilde{G} = G + \operatorname{diag}(u_i) .$$

The corresponding (modified) replicator equation can be written in the following form

$$Fy_i = (\sum G_{ij}y_j - \sum y_k G_{kj} y_j) + (u_i y_i - \sum u_k y_k^2) .$$

The stationary solutions of these equations have the property that, for $u_i \to -\infty$, the evolutionarily stable strategy $y_i^* = 1/n$ will be attained, i.e. the barycentre of the simplex $S_n$ is the only stable attractor (we must mention that all species are saved because of the simplex normalisation condition; in reality, a global trend toward extinction is superimposed). In this condition, all species lose their individuality, and we get something like one cluster of a dying world.

For $u_i \to \infty$ we observe the reverse flow on the simplex $S_n$; more and more, the species $i$ orient themselves to their individual benefits, meaning that the $u_i$ show here a similar influence as the diagonal discussed for the fixed-point iteration in Section 7.1. All corners of the simplex are now stable attractors and the whole population is decomposed into clusters, each one corresponding to only one of the species. This limit case is something like overfeeding from outside, meaning that the species must not cooperate for their gain out of the competition structure.

For $u_i \to -\infty$, we understand the resulting global state as the state of uniform cooperation, and, for $u_i \to \infty$, as the state of uniform competition. This implies a scaling of the global state to the controlled replicator equation under the influence of the control variables $u_i$ according to the following scheme

$$u_i \to -\infty \;\underset{\xleftarrow{\text{increasing cooperation}}}{\xrightarrow{\text{increasing competition}}}\; u_i \to \infty$$

In the following we will discuss in more detail the simplified case with $u_i = u$ for all $i = 1, 2, \dots, n$.

The modified competition matrix is given by

$$\tilde{G} = G + uE \quad \text{with the unity matrix } E.$$

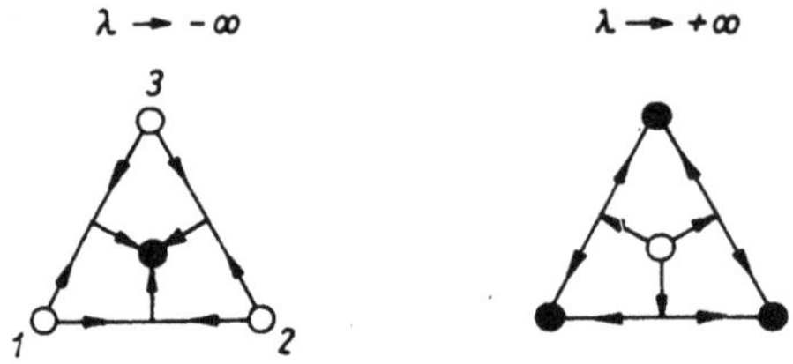


**Fig. 89** The influence of λ upon source and sink behaviour

An equilibrium point in the interior of the simplex $S_n$ can be computed by the equation

$$y^*(u) = \tilde{G}(u)^{-1} f / (f, \tilde{G}(u)^{-1} f)$$

with $f_i = 1$ for $i = 1, 2, \dots, n$.

The inverse matrix $\tilde{G}(u)^{-1}$ can be computed by a modified method of Leverrier

$$\tilde{G}(u)^{-1} = \sum_{i=0}^{n-1} B_i u^{n-i-1} / \sum_{j=0}^{n} d_j u^{n-j}$$
$$= \operatorname{adj} \tilde{G}(u)/d(u) \quad \text{with} \quad d_0 = 1 .$$

Therefore, the corresponding equilibrium point is given by

$$y^*(u) = \operatorname{adj} \tilde{G}(u)/(f, \operatorname{adj} \tilde{G}(u)\, f) .$$

Equilibrium points on the boundary of the simplex $S_n$ can be determined by eliminating corresponding columns and rows from the controlled competition matrix $\tilde{G}$ and then applying the same procedure as above described.

Now we illustrate this method by some simple examples. We interpret the increase of $u$ as growing of the selection pressure.

1. We consider the following replication equation given by the competition matrix

$$G = \begin{pmatrix} 0 & 1 & 2 \\ 1 & 0 & 4 \\ 2 & 4 & 0 \end{pmatrix}.$$

Fig. 90 shows the cluster configurations depending on the selection pressure.

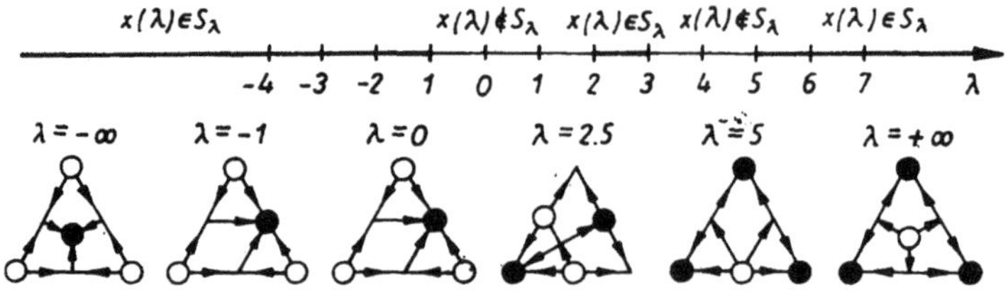


**Fig. 90** Qualitative results dependent upon the selection pressure λ

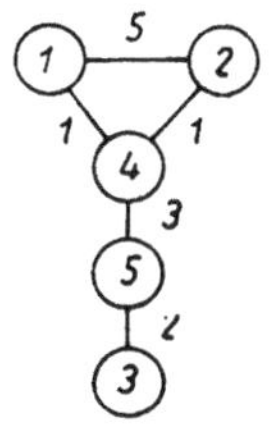


**Fig. 91** Examplé of a simple compound structure

2. Clustering of a connection graph (Partition problem). We consider a connection graph according to Fig. 91 and interpret the number of connections between the elements as components of the competition matrix. With increasing selection pressure, we get the cluster decompositions as shown in Fig. 92.

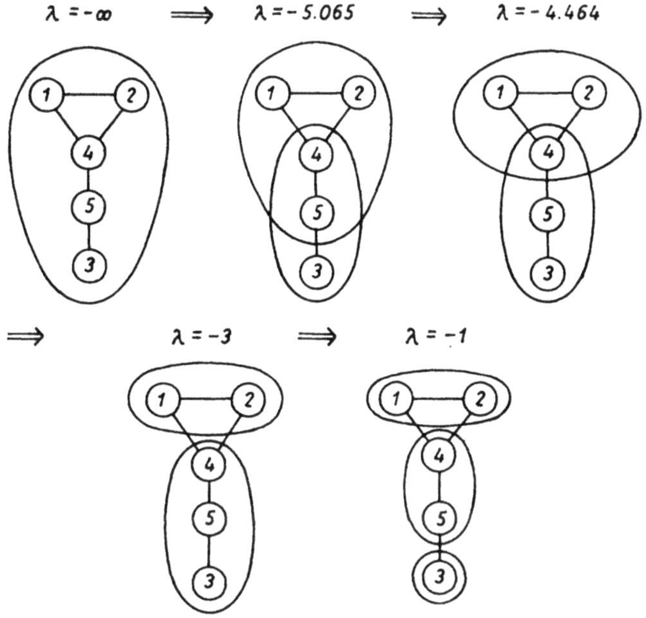


**Fig. 92** Structures obtained in dependency upon the selection pressure $\lambda$ for the compound structure in fig. 91

# 8. Some Qualitative Properties of Nonlinear Systems in Lotka-Volterra Representation, Riccati Representation or Replicator Form

In this chapter, we discuss some methodological studies of qualitative properties of nonlinear systems, making use of the representation forms introduced in this monograph.

## 8.1. *Growth Tendencies of the Biomass of Species and the Relation to the Demanded Flow*

We rely for the following considerations on the fact that many of the stable equilibrium points of the replicator equation are *evolutionarily stable strategies* (ESS) [12]:

$y^*$ is an ESS, if, for all $y \in S_n$ the following holds
$y^*Gy^* > yGy^*$ or if it is the case that
$y^*Gy^* = yGy^*$ then it holds that $y^*Gy^* > yGy$.

We interpret this as a local maximum of the flow function $\Psi$ in the replicator equation.

Equivalent to this definition of an ESS is the demand:

$$y^*Gy > yGy \quad \text{for all } y \in S_n$$

and belonging to a certain neighbourhood of $y$ (proof for this is contained in [16]).

Under the conditions $\sum y_i = 1$ and $\sum p_i = 1$, the term function

$$P(y) = y_r^{p_r}$$

has a unique maximum for $y_r = p_r$ (see, for example, Bongard [73]). From these findings we get the important statement that, from the dynamics of the replicator equation yields, all ESS are stable equilibrium points, meaning that they are stable attractors for the dynamic behaviour of the Lotka-Volterra equations. Unfortunately, the reverse statement does not hold. There are other stable attractors which are not necessarily ESS.

This property is based on the fact that $P(y)$ with $p = y^*$ and $y^*$ an ESS is a Ljapunov function. From the local characterisation of an ESS mentioned above it follows that $dP/dt > 0$ and $y$ approaches from a neighbourhood necessarily the ESS, then $y = y^*$. We return now to the Lotka-Volterra representation and introduce some ecological interpretations for quantities connected with the dynamic behaviour

$$Fx_i = \sum G_{ij}x_j \quad \text{with} \quad x_i \geqq 0 .$$

If we interpret $x_i$ as the individual biomass of species $i$ in the considered area, then the whole biomass of the ecological system is given by

$$B = \sum x_i .$$

It is interesting to watch how the biomass $B$ changes under the dynamics of the given Lotka-Volterra system.

Obviously, we get for the change of $B$ in time the relation

$$\mathrm{d}B/\mathrm{d}t = \sum x_k G_{kj} x_j = \Phi \,.$$

Therefore, $\Phi$ is the flow necessary for the system to follow its autonomous dynamics. The system organizes this flow itself out of the competition and cooperation given by the Lotka-Volterra equation. For an asymmetric Lotka-Volterra system

$$G_{ij} = -G_{ji}$$

apparently the flow $\Phi = 0$, meaning that in this case the whole biomass $B$ is a constant of motion (conservative system).

When $\Phi > 0$, meaning that $G$ defines a positive definite quadratic form in some part of the state-space, the biomass $B$ can only increase, and the system as a whole shows the behaviour of a predator.

For $\Phi < 0$, meaning that $G$ defines a negative definite quadratic form in some part of the state-space, the biomass $B$ can only decrease, and the system as a whole shows then the behaviour of a prey.

If we pass to the replicator equation, we redefine the driving force only as a relative quantity in comparison with the flow $\Phi$. This leads to something like a trend elimination and to a differentiation between the species independent of the flow. That means that we can formally perform this transfer only by subtracting the flow term $\Phi$ from the driving forces on the right hand sides of the Lotka-Volterra equation. We then get

$$Fx_i = \sum G_{ij} x_j - \Phi \,.$$

Then $B = \sum x_i$ will automatically be a first integral.

It is useful to interpret from this point of view the role of an ESS. We already mentioned that a disturbed ecosystem exposes an adaptive behaviour, in the sense that very often its strategy $y$ approaches an ESS $y^*$, which usually yields a maximisation of the flow $\Psi$ in the replicator equation. We will now demonstrate the consequence of this fact in terms of the original state-variables $x_i$ of the Lotka-Volterra system.

Between the flow $\Phi$ of the Lotka-Volterra system and the corresponding flow $\Psi$ of the replicator equation, we observed the relation

$$\Psi = \Phi / B^2 \,.$$

If we could guarantee a constant value $\Psi_0$ of the flow of the replicator equation, then the consequence would be a quadratic law between biomass $B$ and corresponding flow $\Phi$ in the Lotka-Volterra equation

$$\Phi = \Psi_0 B^2 \,.$$

We will show that such a quadratic law indeed holds, at least in the neighbourhood of an ESS. We have immediately

$$F\Phi = F\Psi - 2FB \,,$$

and in the neighbourhood of an ESS we know that $\Psi$ increases. This leads to the following two important consequences:

1. Let $\Phi > 0$. That means that the ecosystem is, on the whole, a predator. Then immediately we get from $\mathrm{d}\Psi/\mathrm{d}t > 0$ two consequences

$$(\Phi/\Phi_0) > (B/B_0)^2$$

and

$$B(t) > 1/(1/B_0 - \Phi_0 t/B_0^2) \,.$$

This is astonishing, because, for us, the assumption is likely that the effort of flow $\Phi$ should be proportional to the biomass $B$ which we must supply. This quadratic growth law is very important because it shows that the effort for the supply of large systems is very high, and leads necessarily to hyperbolic growth of the biomass $B$. The reason for this phenomenon seems to be the growth oriented organisational effort to secure a high cooperativity within the whole system by building up growth oriented structures. (In some sense the statement ist true that "growth is waste".) Therefore, in a late stage of the system (saturation, maturation) the structuring tendency to decompose the whole system into smaller units and, therefore, the partition of the whole supply flow into smaller flows supplying the units can provide some advantages.

Therefore, we must usually expect a certain equilibrium between the tendency to grow bottom-up with the aim of infinite increase of the biomass, and the decomposition top-down because of the advantages of decomposing large systems into smaller ones, which can be more easily supplied. This may be the reason for the known tendency in natural systems to decay if restrictions of the supply flow arise. This simultaneously provides a second advantage, because the smaller systems are more suitable for construction of new, better fitted structures by recombination.

2. $\Phi < 0$, meaning the ecosystem is, on the whole, a prey. We get from $\mathrm{d}\Psi/\mathrm{d}t > 0$ the consequence

$$(\Phi/\Phi_0) < (B/B_0)^2 \,.$$

It is necessary to mention that the biomass is now decreasing, meaning that the system orients itself upon very small values of $(\Phi/\Phi_0)$. The system tries to make the parasitic flow which is exhausting the whole system very small, thus braking the extinction process which is threatening the system's future. The consequence for the time trajectory of the biomass is then

$$B(t) > 1/(1/b_0 - \Phi_0 t/B_0^2) \,.$$

We must take into account that now $\Phi_0 < 0$ holds. Therefore, the degradation of the biomass of the system goes more slowly than does the corresponding hyperbolic extinction process. The system succeeds in doing this obviously because it has an organisational reserve. Structure from this point of view is something like a stored reserve, which has the task in the extensive phase of the system to lead to increasing growth rates, and in the degradation phase to secure a very slow extinction process. This finding is not unusual at all in the case where there exist autocatalytic driving forces for the species in the Lotka-Volterra system; already for the simplest system we have

$\mathrm{d}x/\mathrm{d}t = Kx^2$ and with the interpretation $x = B$ we have

$$\mathrm{d}B/\mathrm{d}t = \Phi = \mathrm{d}x/\mathrm{d}t = KB^2 \,.$$

However, it is important to stress that these phenomena are not necessarily bound at the existence of primary autocatalytic driving forces; moreover, we can assume that all $G_{ii} = 0$. Nevertheless, the quadratic supply law considered above can be observed, if the system has an ESS. Therefore, hyperbolic growth can be generated by coupling effects within a complex system.

Volterra himself [1] excluded the possibility of autocatalytic reproduction rates $G_{ii} > 0$ in the Lotka-Volterra equation, perhaps mainly for reasons of numerical stability, and admitted only for $G_{ii} < 0$. Our analysis shows that we should not exclude explosive phenomena like hyperbolic growth, but that $G_{ii} < 0$ is by no means a guarantee of avoiding hyperbolic growth. We propose to restrict the consideration of the quadratic law between biomass $B$ and flow $\Phi$ not only to the neighbourhood of an ESS.

For any simulation study in connection with Lotka-Volterra equations, we should control some quantities with global significance, placing useful information for the interactive influence on the simulated object in the computer. Therefore, we propose to observe the following quantities:

1. biomass $B = \sum x_i$,
2. flow $\Phi = \sum x_k G_{kj} x_j$,
3. quadratic supply law gain $\Psi = \Phi/B^2$,
4. complexity measure $\varkappa = \Phi/B$.

We call $\varkappa$ a complexity measure because the earlier the explosion in the hyperbolic growth law takes place, the larger the value of $\varkappa_0 = \Phi_0/B_0$ is. It can be very useful to follow the time trend of because this gives us the possibility of showing that, during the extensive growth phase, the complexity monotonously increases, originated by an aim-oriented structuring process in the background of the considered system.

## 8.2. *A Necessary Condition For Recurring Motions in Lotka-Volterra Representations*

For the general Lotka-Volterra equation

$$Fx_i = \sum G_{ij} x_j$$

we already have the result

$$\mathrm{d}B/\mathrm{d}t = \Phi \quad \text{and} \quad B = \sum x_i \quad \text{and} \quad \Phi = \sum x_k G_{kj} x_j \,.$$

From this we can conclude at once that, if $\Phi > 0$ holds everywhere, meaning that if the quadratic form $\Phi$ is positive definite, then $B$ is a global Ljapunov function and the point $\infty$ is a stable attractor. In this case, $B$ is monotonously increasing and a saturation in a finite area is impossible. Such a system especially cannot possess limit cycles.

If $\Phi < 0$ holds everywhere, then the quadratic form $\Phi$ is negative definite and $B$ again is a global Ljapunov function, but now the point 0 is a stable attractor. In this case, $B$ is monotonously decreasing. Also in this case, a saturation in an area not certaining 0 is not possible, especially also in this case no limit cycle can exist.

The cases of special interest are those where the sign of $\Phi$ changes in the state-space, meaning where positivity and negativity of $\Phi$ are locally restricted. To gain some insight into this case, we extract from the whole system's behaviour in the $n$-dimensional state-space the projection on an arbitrary two-dimensional plane given by $x_i = x$ and $x_j = y$. We can obtain this projected behaviour easily if we put the state-variables of all other species to zero. Therefore, we get a second-order Lotka-Volterra system

$$Fy = G_{11}y + G_{12}x ,$$
$$Fx = G_{21}y + G_{22}x .$$

Now we have $B = x + y$ and $\Phi = G_{11}y^2 + (G_{21} + G_{12})\,xy + G_{22}x^2$. We can gain new insights only in the case where $\Phi$ is semidefinite, meaning that the equation $\Phi = 0$ has real roots in the positive cone. Without a restriction of generality we can assume $G_{22} \neq 0$. Then we get

$$(x/y)_0 = -\frac{(G_{21} + G_{12})}{2G_{22}} \pm D$$

with

$$D = \frac{(G_{12} + G_{21})^2}{4G_{22}^2} - \frac{G_{11}}{G_{22}} .$$

In the $(x, y)$ plane, we get, depending on the position of these straight lines, cones with changing signs of the flow $\Phi$.

**Fig. 93** Basic constellations in the phase plane

For our considerations only the three cases shown in Fig. 93 are of interest. The assignment of + or − signs defines the flow direction. For +, the system is a predator and is pumped, and for −, the system is a prey and will be exploited. In a sector designated with +, $B$ is an increasing, and in a sector signed with −, $B$ is a decreasing local Ljapunov function. This does not mean, however, that a system which starts in one of these sectors remains in that sector all the time. On the contrary, in a finite area, the system can pass the frontier from one sector to an adjoining sector, can change its character; being perhaps first a predator, it can later become a prey and vice versa.

We will now study the phenomena which might occur if the system passes the frontier between adjoining sectors. For this purpose we consider the quantity

$$K(t) = K(x, y) = x/y$$

in the course of time $t$.

Obviously we have

$$FK(t) = Fx - Fy = (G_{21} - G_{11})\,y + (G_{22} - G_{12})\,x .$$

For $(G_{21} - G_{11}) > 0$ and simultaneously $(G_{22} - G_{12}) > 0$, the function $K(t)$ can only increase monotonously, and the motion of the trajectory shows, according to this statement, that the straight line corresponding to $K(t)$ exposes a rotation and a drift to the left.

If the signs are opposite, that means $(G_{21} - G_{11}) < 0$ and $(G_{22} - G_{12}) < 0$ then we get a right rotation instead of a left rotation of $K(t)$, with the consequence that now the trajectory necessarily shows a right drift. In both cases, the frontiers of adjoining sectors are necessarily passed exactly once; afterwards, the system stays in the corresponding sector continuously.

The case where $(G_{21} - G_{11})(G_{22} - G_{12}) < 0$ is also of interest. Now $K(t)$ does not necessarily show a monotonous behaviour, but it can approach a constant $K_0$, namely on a straight line

$$y/x = -(G_{22} - G_{12})/(G_{21} - G_{11}) .$$

In such a case, the system stays in that sector to which this straight line belongs.

The special qualitative findings we have just considered, which are of importance for all two-dimensional planes in the higher-dimensional state-space of a Lotka-Volterra system, can be generalised by making use of the formerly obtained results in connection with first integrals in term form. For this purpose, we consider in the positive cone $x_i \geqq 0$ arbitrary terms of the form

$$K_k(t) = K_k(x_1, x_2, \dots, x_n) = \prod_r x_r^{k_r} .$$

The index $k$ should be understood as a shortation of the exponent vector $k = (k_r)$. Such a function $K_k(t)$ can be built up for every vector $k$ with real components. We are interested in the time behaviour of these functions $K_k(t)$ along the system's trajectories.

We consider the logarithmic derivative $FK_k(t)$ and get

$$FK_k(t) = k^T G x .$$

From this expression, it follows that if $k^T G \geqq 0$ holds for a vector $k$, then $K_k(t)$ can only increase; if, conversely, $k^T G \leqq 0$ holds, then $K_k(t)$ can only decrease. This means that, in such a case, $K_k(t)$ is, in both cases, a global Ljapunov function. This has important consequences for the trajectories in the state-space, because

$$K_k(t) = \prod x_r^{k_r} = \text{const}$$

is a family of hyperfaces in the state-space with a trend enforced over time by $K_k(t)$. For $k > 0$ or $k < 0$, these are hyperboloid-like hyperfaces which, depending on the increase or decrease of $K_k(t)$, either approach $\infty$ or 0.

If the signs of the components of $k$ change, then we can rewrite the hyperfaces in the following way

$$K_k(t) = \prod_{k_r > 0} x_r^{k_r} / \prod_{k_s < 0} x_s^{|k_s|} = \text{const}.$$

These are *paraboloid-like* hyperfaces which, according to the growth of $K_k(t)$, turn into the direction of coordinate hyperfaces as boundaries of the positive cone. That means that, if $K_k(t)$ is in fact such a global Ljapunov function in the positive cone, then all the trajectories are forced to follow a walking direction to $\infty$ or to 0, or into the direction to a coordinate hyperface. This has the consequence that if at least for one vector $k$ the corresponding $K_k(t)$ is a global Ljapunov function then an iterative reversion of the direction of motion is impossible, and, no limit cycle can occur in the state-space. From these considerations, we get a very restrictive necessary condition for the existence of limit cycles (or, moreover, of returning motions such as spirals).[1]) *A limit cycle in the Lotka-Volterra system*

$$Fx_i = \sum G_{ij} x_j$$

*can only exist if the following necessary condition holds. In every vector of the form*

$$k^T G$$

*a change of the sign of the components must occur.* It is interesting to mention that this condition can be extended immediately to the multinomial differential equation in the generalised form, namely: Let

$$K_k(t) = \Phi^{-1}\left(\sum k_r \Phi(x_r)\right) .$$

These should be positive *term-expressions* in the positive cone. The corresponding multinomial system as generalisation of the Lotka-Volterra equations is then

$$\mathrm{d}\Phi(x)/\mathrm{d}t = A\Phi^{-1}\left(a\Phi(x_r)\right) .$$

From this we immediately get

$$\mathrm{d}\left(K_k(t)\right)/\mathrm{d}t = k^T A \Phi^{-1}\left(a\Phi(x_r)\right) .$$

If $\Phi(u)$ is a strict monotonous function, we can draw conclusions at once on the strict monotony of $K_k(t)$, if all components of $k^T A$ have the same sign. All the conclusions hold for the multinomial differential equations with $\Phi(u) = \ln u$. Therefore, the question of the existence of limit cycles can be checked immediately by the multinomial differential equations without a complete transformation into the Lotka-Volterra equations.

As a side effect, we again get, for the first integrals of multinomial differential equations, the result which we already obtained in connection with the normal form theory.

*If the vector k is a leftannulator of A, that means*

$$k^T A = 0$$

*then*

$$K_k(t) = \Phi^{-1}\left(k^T \Phi(x)\right) = \text{const.}$$

*is a first integral of the generalized multinomial differential equation.* We conclude this section with some simple examples demonstrating the applicability of the derived condition.

1. Returning motions in the usual two-dimensional Lotka-Volterra system

$$Fx_1 = e_1 x_3 - g_{12} x_2 ,$$

$$Fx_2 = -e_2 x_3 + g_{21} x_1 ,$$

$$Fx_3 = 0 .$$

---

[1]) We thank Dr. Jansen, ZKI, for deriving the necessary condition for returning motions, and, moreover, for the construction of the shift-cones if this necessary condition is not fulfilled.

Here we have

$$k^T A = (k_2 g_{21}, -k_1 g_{12}, k_1 e_1 - k_2 e_2) .$$

In the usual Lotka-Volterra system, all parameters contained in the model are positive.

From Fig. 94 we see that there is a change of the signs of the components everywhere in the plane $(k_1, k_2)$. Therefore, returning motions are possible here. In fact, we now know that this system has cyclic motions as characteristic trajectories.

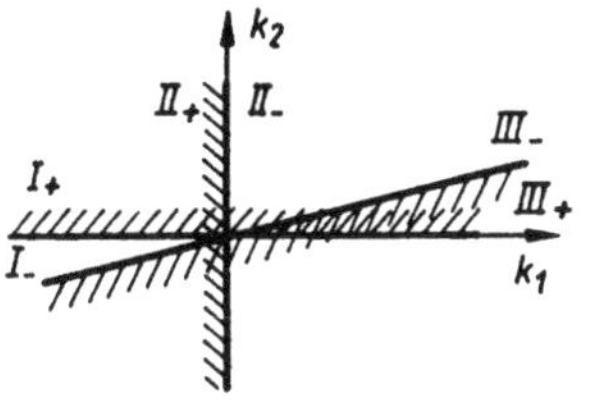


**Fig. 94** Returning motions problem for the usual Lotka-Volterra predator prey system of the second order

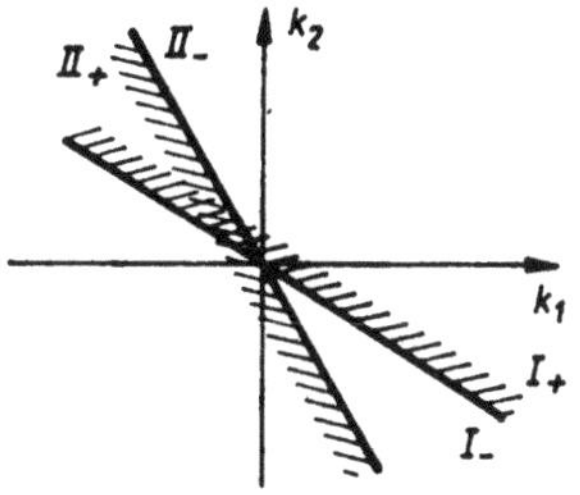


**Fig. 95** Nonexistence of a returning motion for the autocatalytic Lotka-Volterra system of the second order

2. Returning motions in the second-order autocatalytic Lotka-Volterra system without external resources.

$$Fx_1 = e_1 x_1 - g_{12} x_2 ,$$

$$Fx_2 = -e_2 x_2 + g_{21} x_1 .$$

Now we have

$$k^T A = (k_1 e_1 + k_2 g_{21}, -k_1 g_{12} - k_2 e_2) .$$

Fig. 95 shows that the two intersecting straight lines in the $(k_1, k_2)$ plane always define a sector $x$ in which both components of $k^T A$ have the same sign. Therefore, no returning motions, and especially no limit cycles, are possible.

3. Returning motions in Lotka-Volterra system with autocatalytic interactions of higher order.

$$Fx_1 = e_1 x_1^{a_1} x_2^{\lambda_2} - g_{12} x_1^{\mu_1} x_2^{b_2} ,$$

$$Fx_2 = -e_2 x_2^{a_2} x_1^{\lambda_1} + g_{21} x_2^{\mu_2} x_1^{b_1} .$$

It is important that we now have four generally independent algebraic terms. We immediately apply our condition to this multinomial differential equation, and get

$$k^T A = (k_1 e_1, -k_1 g_{12}, -k_2 e_2, k_2 g_{21}) .$$

If we assume that all parameters have only positive values, then obviously every such vector shows different signs for its components. Therefore, the necessary condition for returning motions (and, therefore, perhaps also for limit cycles) is fulfilled.

## 8.3. *Shift-Cone Construction for Lotka-Volterra Equations*

We will first discuss the more general case of multinomial differential equations, and afterwards draw the corresponding conclusions for the Lotka-Volterra equations. We study the case where the system of linear inequalities

$$k^T A \geqq 0$$

has a solution $k \neq 0$, meaning that the necessary condition for the nonexistence of returning motions is fulfilled. The set of all solutions $k$ of this system of linear inequalities generates a convex cone, which we designate as $\mathcal{K}^*$. Obviously, the columns $A_{.j}$ of $A$ themselves define a convex cone, which we designate with $\mathcal{K}$, and which is contained in the dual cone of $\mathcal{K}^*$, defined by the set of inequalities

$$k^T \xi \geqq 0 \quad \text{for all} \quad k \in \mathcal{K}^* .$$

From the multinomial differential equation

$$\mathrm{d}\Phi(x)/\mathrm{d}t = A\Phi^{-1}(a\Phi(x))$$

with $\Phi^{-1}(a\Phi(x)) \geqq 0$ we get

$$k^T \,\mathrm{d}\Phi(x)/\mathrm{d}t = k^T A\Phi^{-1}(a\Phi(x))$$

and, therefore, it holds

$$\text{for} \quad k \in \mathcal{K}^* , \qquad k^T \,\mathrm{d}\Phi(x)\,\mathrm{d}t \geqq 0$$

and

$$\mathrm{d}\Phi(x)/\mathrm{d}t \in \mathcal{K} .$$

By integration, we get

$$\Phi(x(t)) - \Phi(x(t_0)) \in \mathcal{K} \quad \text{for} \quad t \geqq t_0 ,$$

$$\Phi(x(t)) - \Phi(x(t_0)) \in -\mathcal{K} \quad \text{for} \quad t \leqq t_0 .$$

This is very important, because it shows that, if there exists a shift-cone (as we will call this construction further on), then the system's trajectory $\xi(t) = \Phi(x(t))$ in the $\xi$ space obtained by nonlinear scaling by $\Phi$ in every reference point $\xi_0 = \Phi(x(t_0))$ necessarily enters into the cone $\mathcal{K}$ and arrives in $\xi_0$, leaving the corresponding cone $-\mathcal{K}$. Fig. 96 shows this construction in the $\xi$ space schematically.

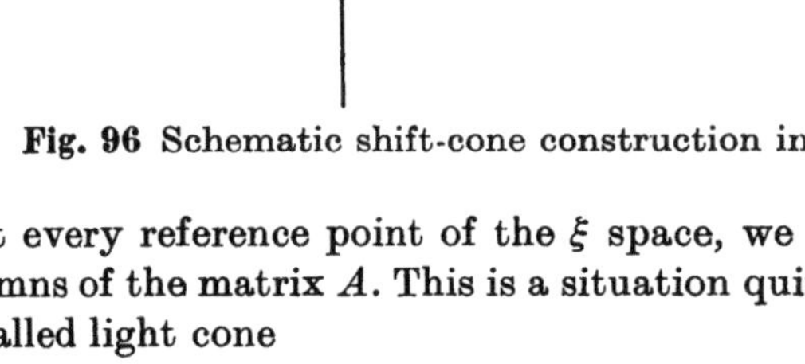


**Fig. 96** Schematic shift-cone construction in the $\xi_1$, $\xi_2$-plane

At every reference point of the $\xi$ space, we have the same cone spanned by the columns of the matrix $A$. This is a situation quite similar to what we know about the so-called light cone

$$x^2 + y^2 + z^2 - c^2t^2 \leqq 0$$

in the special theory of relativity. Real motion, or world trajectories, must be always contained within the light cone because the highest possible velocity is that of light velocity $c$. From this construction, it clearly follows that, under the assumed condition, $\mathcal{K}^* \neq (0)$, no returning motions are possible.

We will now clarify a little further the cone construction introduced above. Let rank $(A) = s \leqq n$. We choose from the columns $A_{.J}$ of $A$ one possible set of $s$ linearly independent columns, and bring them by commutation onto the first places $J = 1, 2, \ldots, s$. Then we choose $A_{.J} = e_J$ as partial vector base in the $n$-dimensional space and adjoin this partial base by other $n - s$ linearly independent vectors $e_j$ to a complete base in the $n$-dimensional space.

The corresponding dual base $f^j$ is uniquely determined by the biorthogonality relations

$$(f^j, e_i) = \delta_i^j .$$

With the help of the dual base $f^j$, we will now characterize the cone $\mathcal{K}^*$. We represent any $k \in \mathcal{K}^*$ as a linear combination in the dual base

$$k = \sum \varkappa_i f^i .$$

From our definition, $k^T A_{.J} \geqq 0$ holds for all columns of $A$. From this, we get $\varkappa_i \geqq 0$ for $i = 1, 2, \ldots, s$, $\varkappa_i = 0$ for $i = s + 1, \ldots, n$. Obviously, the cone $\mathcal{K}^*$ is cylindrical in the coordinates $\varkappa_i$ for $i = s + 1, \ldots, n$.

Now we will draw the corresponding consequences for Lotka-Volterra equations

$$F x_i = \sum G_{ij} x_j .$$

Here all results hold, because it is the special case of $A = G$. If rank $(G) = n$, then the columns $G_{.j}$ always span a convex cone $\mathcal{K}$. The corresponding dual cone $\mathcal{K}^*$ is then spanned by the dual base $f^j$, corresponding to a vector base $e_j = G_{.j}$. This means that, for a Lotka-Volterra system with full rank a shift-cone always exists, and, therefore, no returning motions are possible. The consequence is that the more interesting dynamic phenomena of Lotka-Volterra systems occur only if we have a system with reduced rank.

Here we meet again the problem of robust rank. A rank reduction can take place, depending upon random effects connected with the concrete quantitative values of the matrix elements (the competition coefficients), thus leading to interesting but nonrobust dynamic phenomena. However, a rank reduction can be enforced for the structural reason that there are no interactions at all between certain pairs of species $i$, $j$. In such a case, robust interesting dynamic phenomena, for example robust limit cycles, might occur.

## *8.4. Interpretation of the Shift-Cone as an Autonomous Preference Structure of the System's Behaviour in Comparison with Pareto-Optimality in Polyoptimization*

In polyoptimization, problems are studied with several objectives $Q_i$, $i = 1, 2, \ldots, n$, instead of one global objective $Q$, and the values of the objectives $Q_i$, dependent upon a set of common control variables $a_j$, $j = 1, 2, \ldots, n$, are compared with each other with the aim to find most promising variants.

Following the convenience in real-life problems, a variant $Q_i'$ is considered preferable (dominant) to another variant $Q_i$ (if we are interested in maximizing all criteria) if the following inequalities hold

$$Q_i' \geqq Q_i$$

and for at least one $i_0$, we have the strong inequality

$$Q_{i_0}' > Q_{i_0} .$$

Thus, we intend to avoid all nonpromising variants, which are dominated in this Pareto preference sense. For the exclusion of nonpromisable variants we, therefore, make use of a preference cone which is the positive orthant. With a homogeneous preference relation not dependent upon the position in the decision space, we use this Pareto cone everywhere, for the comparison. All remaining variants in a set of possible variants, after having performed all these dominance comparisons, are called *efficient solutions* of the corresponding polyoptimization task. Fig. 97 shows the Pareto preference construction, together with an example for the efficient solution set.

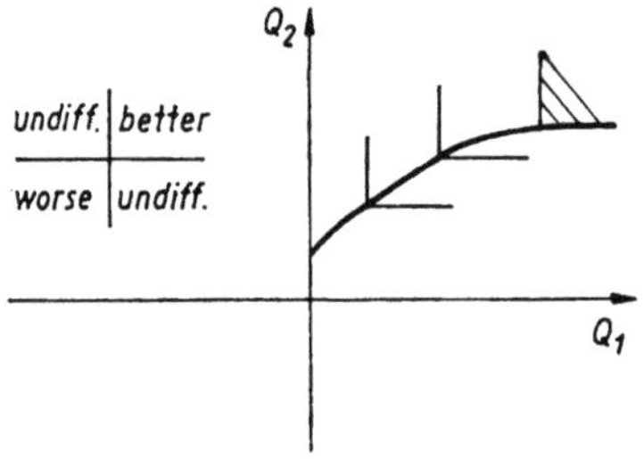


**Fig. 97** Pareto preference cone construction

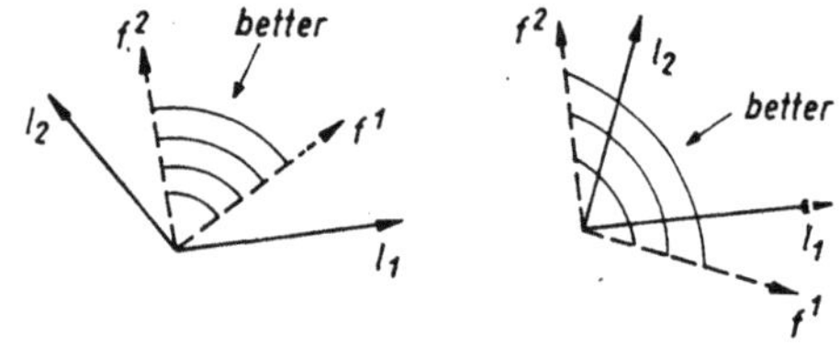


**Fig. 98** The Pareto preference relation interpreted in the control-space

It is sometimes useful to interpret the Pareto preference relation immediately in the control space $a_1, a_2, \ldots, a_m$, because we must there find a good seeking procedure for efficient controls leading us to efficient variants in the objective space. Such an interpretation of the Pareto cone in the control-space can be given easily, if we linearize the given generally nonlinear functions

$$Q_i = Q_i(a_1, a_2, \ldots, a_m)$$

by considering

$$\Delta Q_i = (\operatorname{grad} Q_i, \Delta a) = (e_i, \Delta a); \qquad e_i = \operatorname{grad} Q_i .$$

For simplicity, we assume $m = n$ and Rank $(\partial Q_i/\partial a_j) = n$. Then, obviously, the convex cone spanned by all the gradient vectors grad $Q_i$ corresponds uniquely to the local Pareto cone $\Delta Q_i \geqq Q$. Therefore, expressed in the language of the control-space the role of the Pareto-preference is taken over by a cone preference, given by the cone and spanned by the grad $Q_i$. This interpretation can immediately be used for the design of good seeking procedures for efficient controls [27].

Under the condition that the cone $\mathcal{K}^*$ is not empty, for multinomial differential equations in the transformed space $\xi = \Phi(x)$, a shift-cone $\mathcal{K}$ exists, which can be interpreted as an autonomous preference cone of the dynamic system. For this purpose, we interpret $\xi_i = \Phi(x_i) = Q_i$ as objectives of a polyoptimization task. $Q_i$, in this case, are then measures of the biomass of the individual species $x_i$. It seems feasible to assume that all species should try to increase their biomass $\xi_i$ as much as possible, meaning that one would expect a Pareto cone as the shift-cone in the objectives space. However, against these individual wishes and aspirations, the restricted resources and interaction between the different species for the exploitation of the available resources are acting. These effects together lead to a deformation of the Pareto cone as preference cone into the shift-cone $\mathcal{K}$.

Of special interest is the case where we measure the biomass in a logarithmic scale meaning that $\xi_i = \ln x_i$. If the shift-cone $\mathcal{K}$ is completely contained in the Pareto cone $\mathcal{K}$ in the $\xi$ space, then obviously the interactive structure is compatible with the individual aspirations of unrestricted growth of the individual biomasses. This, however, is, in general, not the case. That means that the common autonomous preference of the system as a whole can be beneficial for some species and detrimental to others, to the point of threatening them with extinction.

We will interpret this situation for the Lotka-Volterra system for the hyperlogistic EVOLON model.

For this model, we derived the following equations in Section 4.1

$$Fx_0 = x_1 ,$$

$$Fx_1 = (k - 1)\, x_1 - wlx_2 ,$$

$$Fx_2 = (k + w - 1)\, x_1 - w(l - 1)\, x_2$$

(non-normalised, but regular case).
$x = x_0$ obeys the hyperlogistic differential equation

$$\mathrm{d}x/\mathrm{d}t = Kx^k(B - x^w)^l$$

leading to an $s$-formed transition curve with a saturation. This means that the dynamic trajectories of solutions $(x_1, x_2)$ of the last two equations should, for all feasible parameter constellations approach a state with $x_1 = 0$.

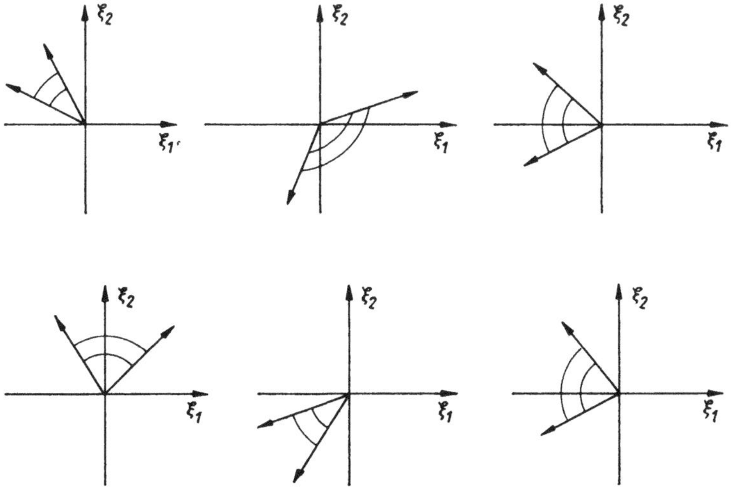


**Fig. 99** All possible shift-cone constellations for the hyperlogistic model

We will check to see whether the shift-cone constructions, in this case, are compatible with this demand.

Let us assume that the system matrix is regular.

$$G = \begin{pmatrix} (k-1) & -wl \\ (k+w-1) & -w(l-1) \end{pmatrix}$$

Fig. 99 shows the shift-cones for all possible constellations of signs of the matrix elements. The only shift-cone constructions which do not imply the reachability of large negative values of $\xi_1 = \ln x_1$ would obviously be sign constellations

$$\begin{matrix} + & + \\ + & - \end{matrix} \quad \text{or} \quad \begin{matrix} + & + \\ + & + \end{matrix} \quad \text{or} \quad \begin{matrix} + & + \\ - & - \end{matrix}$$

which is not admitted for positive values of $w$ and $l$.

## 8.5. *The Shift-Cone Problem for General Lotka-Volterra Equations: The Interactive Structure in Lotka-Volterra Equations Interpreted as a Differential Game*

We now consider in detail possible cone constructions in connection with Lotka-Volterra equations, or, more generally, with multinomial differential equations.

Because of the dynamic equations

$$\mathrm{d}\Phi(x)/\mathrm{d}t = A\Phi^{-1}(a\Phi(x))$$

the velocity vector $\mathrm{d}\xi/\mathrm{d}t$ of the trajectory $\xi = \Phi(x(t))$ is directed into the convex hull of all the columns vectors $A_{.J}$ of the matrix $A$. We consider as a basic geometric construction the convex hull

$$\mathcal{H} = H(A_{.J})$$

of all columns of the matrix $A$ given by all vectors of the form

$$\sum \eta_J A_{.J} \quad \text{with} \quad \sum \eta_J = 1 \quad \text{and} \quad \eta_J \geqq 0 \,.$$

We denote with Int $\mathcal{H}$ the set of all inner points of $\mathcal{H}$. We must distinguish two cases.

1. $0 \in \operatorname{Int} \mathcal{H}$

This means that 0 is a linear combination of the columns $A_{.J}$ with non-negative coefficients

$$0 = \sum \eta_J A_{.J} \,.$$

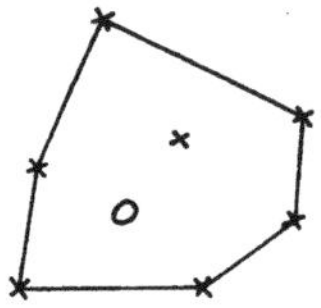


**Fig. 100** The situation when 0 belongs to the convex hull of the columns of a given matrix

This means in the Lotka-Volterra case, that there exists a stationary solution $\eta_J$ of Lotka-Volterra equations.

In the multinomial case, we must then solve the following system of linear equations

$$\Phi(\eta_J) = \sum a_{Jr}\Phi(x_r) .$$

Only then, if this system of linear equations has non-negative solutions, can we conclude also that the multinomial differential equation has a stationary solution. If, conversely, there exists a stationary solution with

$$0 = \sum \eta_J A_{.J}$$

then we can conclude that $0 \in \operatorname{Int} \mathscr{H}$.

We now construct the convex cone $\mathscr{K}$ generated from $\mathscr{H}$ with the tip in 0. We state that this convex cone $\mathscr{K}$ coincides with the whole linear space $\mathscr{A}$ spanned by the columns $A_{.J}$. This can be proved in the following way: We choose from all $A_{.J}$ a set of $s$ linearly independent vectors. They generate together with 0, a convex cone contained in $\mathscr{K}$ as a subcone. Because 0 belongs to Int $\mathscr{H}$ with a certain $s$-dimensional neighbourhood $\mathscr{N}$, $-\mathscr{K} \cap \mathscr{N}$ also belongs to this neighbourhood and is also $s$-dimensional. However, a cone $\mathscr{K}$ together with $-\mathscr{K}$, both being $s$-dimensional, span with their vectors in linear combinations with non-negative coefficients, the whole $s$-dimensional space. In this case, obviously, no shift-cone exists. We shall continue to study the details of this case later.

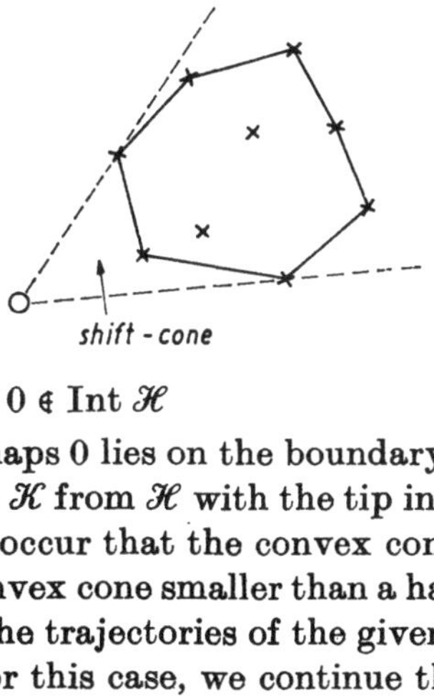


**Fig. 101** The shift-cone construction for the case where 0 does not belong to the convex hull of the columns of a given matrix

2. $0 \notin \operatorname{Int} \mathscr{H}$

Perhaps 0 lies on the boundary of $\mathscr{H}$ or outside of $\mathscr{H}$. We again construct the convex cone $\mathscr{K}$ from $\mathscr{H}$ with the tip in 0. If 0 belongs to the boundary of $\mathscr{H}$, it can, but must not, occur that the convex cone $\mathscr{K}$ is a whole half-space. If $0 \notin \mathscr{H}$, then $\mathscr{K}$ is always a convex cone smaller than a half-space. In both cases, $\mathscr{K}$ plays the role of a shift-cone for the trajectories of the given multinomial differential equation system.

For this case, we continue the interpretation for Lotka-Volterra equations

$$Fx_i = \sum G_{ij}x_j .$$

If there is a shift-cone $\mathscr{K}$, the trajectory velocity $\mathrm{d}\xi/\mathrm{d}t$ with $\xi = \ln x$ is always directed into this shift-cone.

In this case, we consider the flow function

$$\Phi = \sum x_k G_{kj} x_j .$$

Unfortunately, it cannot be stated that $\Phi$ is always either positive definite or negative definite. That does not necessarily mean that the existence of a shift-cone leads to the consequence that either all species $x_i$ are predators or all species $x_i$ are preys depending upon the sign of the flow $\Phi$. The character of the system as a whole can change by passing the hyperphase $\Phi = 0$.

We now continue the consideration of case 1 with $0 \in \operatorname{Int} \mathcal{H}$. Relying on the interpretation just given, we raise the question that we want to decompose the set of all columns $A_{.J}$ in such a way that we form, from all $A_{.J}$ with $J \in I$ (a chosen index set), together with 0, a convex cone $\mathcal{K}$, and demand that all $A_{.J}$ with $J \in \bar{I}$ (the rest of the columns) are contained in the cone $-\mathcal{K}$. We will check to see under which conditions such a decomposition is possible. We can state that such a decomposition is always possible, but, in general, we meet many different solutions, and it must be also admitted that the cone $\mathcal{K}$ can be a whole half-space.

In the following, we give the idea of the proof for this statement. We choose any subset of the columns $A_{.J}$, the convex hull $\mathcal{H}$ of which does not contain 0, and construct the convex cone $\mathcal{K}$ from this convex hull $\mathcal{H}$ together with 0. Now we skip all columns $A_{.J}$ contained in $-\mathcal{K}$. We now pick up any column which belongs neither to $\mathcal{K}$ nor to $-\mathcal{K}$. Then, we extend the cone $\mathcal{K}$ in such a way that the considered column belongs to the boundary of the extended convex cone. We denote the new cone again as $\mathcal{K}$, and continue this construction as described above.

After a finite number of steps, we get the cone decomposition we wanted[1]), meaning that all columns $A_{.J}$ now either belong to the convex cone $\mathcal{K}$ or to the convex cone $-\mathcal{K}$. If we apply this construction to the competition matrix $G$ of the Lotka-Volterra equations, we get the following property: We can decompose the Lotka-Volterra equations as follows

$$Fx_i = \sum_{j \in I_1} G_{ij} x_j^1 + \sum_{j \in I_2} G_{ij} x_j^2 .$$

Herein

$$G_{.j} \text{ for } j \in I_1 \text{ belongs to } \mathcal{K} ,$$

$$G_{.j} \text{ for } j \in I_2 \text{ belongs to } -\mathcal{K} ,$$

$$I_1 \vee I_2 = (1, 2, \ldots, n) .$$

$\mathcal{K}$ as well as $-\mathcal{K}$ are now shift-cones. This can be seen easily, for example, for $\mathcal{K}$, if we put all $x_j^2 = 0$ (for $-\mathcal{K}$, if we put all $x_j^1 = 0$).

We build the bilinear forms

$$\Phi_1 = \sum x_k G_{kj} x_j^1 \quad \text{and} \quad \Phi_2 = \sum x_k G_{kj} x_j^2$$

which are the interaction energies of the species $x_j^1$ with all species, respective to the species $x_j^2$ with all other species. We would like to interpret the species corresponding to one cone, for example to $\mathcal{K}$, as permanent predators, and the species corresponding to the other cone $-\mathcal{K}$ as permanent prey. This interpretation, though is, in general, not possible because of changing signs of $\Phi_1$ or $\Phi_2$. But it generally holds that, for $j \in I_1$, the cone $\mathcal{K}$ is a shift-cone if the corresponding species only exist while the species $j \in I_2$ have been extincted. In the same way for $j \in I_2$ the cone $-\mathcal{K}$ can be considered as a shift-cone.

In the sense of the decomposition of the competition structure of a given Lotka-Volterra system into shift-cones, we can consider the Lotka-Volterra equation as a differential game with two aggregated players, namely, the two communities correspond-

---

[1]) All algorithmic ideas for the cone decomposition of an arbitrary competition graph were elaborated in good cooperation with aspirant Dipl. Math. Nguyen Hai Bao, Centre of Sciences, Hanoi, SRV.

ing to the two shift-cones. The two players would be permanent predators vs. permanent prey if, in the dynamics of the differential game, the predators always get a positive gain and the prey always lose something, meaning if the following conditions hold:

$$\Phi_1 = \sum x_k G_{kj} x_j^1 \geqq 0 ,$$

$$\Phi_2 = \sum x_k G_{kj} x_j^2 \leqq 0 .$$

However, these conditions imply that $G_{ij} \geqq 0$ for $j \in I_1$ respective to $G_{ij} \leqq 0$ for $j \in I_2$. This is usually not the case. To better understand the sign;ficance of the shift-cone decomposition for the dynamics of Lotka-Volterra systems, it seems to be necessary to better understand the roles that the different species play in the case where a unique shift-cone exists.

It is obvious that all $G_{.j} \geqq 0$, for they themselves define a shift-cone which corresponds to a positive definite flow $\Phi$, and the corresponding community consists then of permanent predators.

In the same way, $G_{.j} \leqq 0$ leads to a shift-cone corresponding to a community of permanent prey.

How, though, is the dynamic interaction between possible permanent predators (resp. permanent prey) within the community of all species determining a shift-cone?

How does the sign-changing of the flow $\Phi$ which takes place in the interaction process between permanent predators (or prey) and the rest of the community define a common shift-cone?

## 8.6. *Analysis of Volterra Systems with Fuzzy Methods*[1])

We will try to obtain by *fuzzification*, of Volterra systems models, with the help of which we will be able to roughly study the behaviour of the original given system. This can be important from the point of view also of the robustness of some types of motion. We study systems of the form

$$Fx_i = \sum G_{ij} x_j , \qquad i = 1, \ldots , n, n + 1 ,$$

with the constant resource

$$x_{n+1} = 1 .$$

A possibility which is frequently used to roughly study the behaviour of a system given by differential equation is to study discretely the values of all time functions. We shall study the values of $x_i$, discretely, as well as those of the given competition coefficients $G_{ij}$.

We decompose the value area of $x_i$ into ordered classes (intervals) $A, B, C, \ldots, G, H$, and describe these classes by one-dimensional membership functions in the sense of the fuzzy system theory [44]. $A$ then corresponds to the fuzzy set *very small* and $H$ to

[1]) The results of the section 8.6 are provided by Dr. Straube, ZKI. This study was initiated by collective problem discussions in the seminar 'Polyoptimization and Fuzzy Model-Building', Akademie der Wissenschaften der DDR.

the fuzzy set *very large*. The fuzzy sets $B \dots G$ lie then between the fuzzy sets *very small* and *very large*.

To the value area of all competition coefficients $G_{ij}$, we introduce the following fuzzy sets

| NL NM NS | NU | PS PM PL |
|---|---|---|
| negative $G_{ij}$ | 0 | positive $G_{ij}$ |

N = negative, P = positive, L = large, M = medium, S = small.

In the following studies, NU is alway a non-fuzzy one-element set with the membership value of 1 only where $G_{ij} = 0$, and the membership value 0 for all $G_{ij} \neq 0$. We agree that this will preserve the non-coupling property in the original Lotka-Volterra system for the rough model, too.

We denote the partial relative growth rate for the influence of $x_j$ on $x_i$ via $G_{ij}$ by

$$\frac{\dot{x}_{ij}}{x_i} = G_{ij}x_j\,.$$

To the value area of these partial relative growth rates, we introduce the unordered fuzzy relative growth rates by

| CN4 CN3 CN2 | CN1 | CNU CP1 CP2 CP3 CP4 |
|---|---|---|
| negative $\frac{\dot{x}_{ij}}{x_i}$ | 0 | positive $\frac{\dot{x}_{ij}}{x_i}$ |

as we had previously introduced fuzzy sets for $G_{ij}$. The connection between the fuzzy classes of $x_j$, the fuzzy values of $G_{ij}$, and the fuzzy partial relative growth rates will be established by a fuzzy algorithm (GX). This algorithm is based on the operation table shown in Table 10.

**Table 10.** Operation table for the fuzzy algorithm GX (The notation corresponds to the fuzzy simulation programm FUZSIM)

| | PB | PM | PS | NU | NS | NM | NB |
|---|---|---|---|---|---|---|---|
| A | CP1 | CP1 | CP1 | CNU | CN1 | CN1 | CN1 |
| B | CP1 | CP1 | CP1 | CNU | CN1 | CN1 | CN1 |
| C | CP2 | CP1 | CP1 | CNU | CN1 | CN1 | CN2 |
| D | CP2 | CP1 | CP1 | CNU | CN1 | CN1 | CN2 |
| E | CP3 | CP2 | CP1 | CNU | CN1 | CN2 | CN3 |
| F | CP3 | CP2 | CP1 | CNU | CN1 | CN2 | CN3 |
| G | CP4 | CP3 | CP2 | CNU | CN2 | CN3 | CN4 |
| H | CP4 | CP3 | CP2 | CNU | CN2 | CN3 | CN4 |

In the next step, all partial relative growth rates $\dot{x}_{ij}/x_i$ for all $j$ and fixed $i$ must be summed up to the relative growth rate $\dot{x}_i/x_i$ for $x_i$. To avoid the addition of fuzzy sets, we organize this aggregation procedure in a special way which we will now explain.

We substitute the derivative $\dot{x}_i$ by the Euler forward first differences and get

$$x_i(k+1) = x_i(k)\left(1 + h\left[\frac{\dot{x}_i}{x_e}\right](k)\right).$$

On the basis of the partial relative growth rates $[\dot{x}_{ij}/x_i]$ $(k)$, we get by Euler forward first differences in connection with $x_i(k)$ the partial response $x_{ij}(k)$. The corresponding fuzzy sets are then linked, using with the fuzzy algorithm RATE. The corresponding operation table is shown in Table 11. To avoid unwanted saturation phenomena for $x_i$, the smallest fuzzy class A and the largest fuzzy class H are decomposed here into A0, A1, A2, A3, A4 resp. H0, H1, H2, H3 and H4. In the fuzzy algorithm GX, they all correspond to the class A resp. the class H.

**Table 11.** Operation table for the fuzzy algorithm RATE

| | CP4 | CP3 | CP2 | CP1 | CNU | CN1 | CN2 | CN3 | CN4 |
|---|---|---|---|---|---|---|---|---|---|
| A4 | A0 | A1 | A2 | A3 | A4 | A4 | A4 | A4 | A4 |
| A3 | B | A0 | A1 | A2 | A3 | A4 | A4 | A4 | A4 |
| A2 | C | B | A0 | A1 | A2 | A3 | A4 | A4 | A4 |
| A1 | D | C | B | A0 | A1 | A2 | A3 | A4 | A4 |
| A0 | E | D | C | B | A0 | A1 | A2 | A3 | A4 |
| B | F | E | D | C | B | A0 | A1 | A2 | A3 |
| C | G | F | E | D | C | B | A0 | A1 | A2 |
| D | H0 | G | F | E | D | C | B | A0 | A1 |
| E | H1 | H0 | G | F | E | D | C | B | A0 |
| F | H2 | H1 | H0 | G | F | E | D | C | B |
| G | H3 | H2 | H1 | H0 | G | F | E | D | C |
| H0 | H4 | H3 | H2 | H1 | H0 | G | F | E | D |
| H1 | H4 | H4 | H3 | H2 | H1 | H0 | G | F | E |
| H2 | H4 | H4 | H4 | H3 | H2 | H1 | H0 | G | F |
| H3 | H4 | H4 | H4 | H4 | H3 | H2 | H1 | H0 | G |
| H4 | H4 | H4 | H4 | H4 | H4 | H3 | H2 | H1 | H0 |

We avoid an addition of the fuzzy partial relative growth rates by computing a sequence of single responses according to the following algorithm:

Step 0: Choose row $i$ and put $j = 0$ ,

Step 1: Put $x_{i0}(k + 1) = x_i(k)$ ,

Step 2: $j := j + 1$ ,

Step 3: Determine from $\big(x_j(k), G_{ij}\big)$, by application of GX, the fuzzy relative growth rate $\dfrac{\dot{x}_{ij}}{x_i}(k)$ ,

Step 4: Determine from $\left(\left[\dfrac{\dot{x}_{ij}}{x_i}\right](k),\ x_{ij-1}(k + 1)\right)$, by application of RATE, the partial response $x_{ij}(k + 1)$,

Step 5: If $j < n + 1$, then go to Step 2, else continue with

Step 6: We now put

$$x_i(k + 1) := x_{ij}(k + 1)$$

Remark: Because we apply the fuzzy algorithms GX and RATE, the values of all $x_i$ are the fuzzy sets A, B, C, ... .

By application of the simulation system FUZSIM [123] we simulated the following characteristic examples of Lotka-Volterra system $x_5 = 1$. The equations and the trajectories of the solutions are shown in the Figs. 102 ... 105, the indices designate the corresponding components of the state-vector.

Example 1 (Fig. 102): The original Volterra system has a fixed point at (1, 1, 1, 1, 1) and a stable limit cycle around this fixed point.

$$G = \begin{pmatrix} -0.4 & 1.0 & -0.6 & -1.0 & 1.0 \\ -1.0 & -0.4 & 1.0 & -0.6 & 1.0 \\ -0.6 & -1.0 & -0.4 & 1.0 & 1.0 \\ 1.0 & -0.6 & -1.0 & -0.4 & 1.0 \end{pmatrix}$$

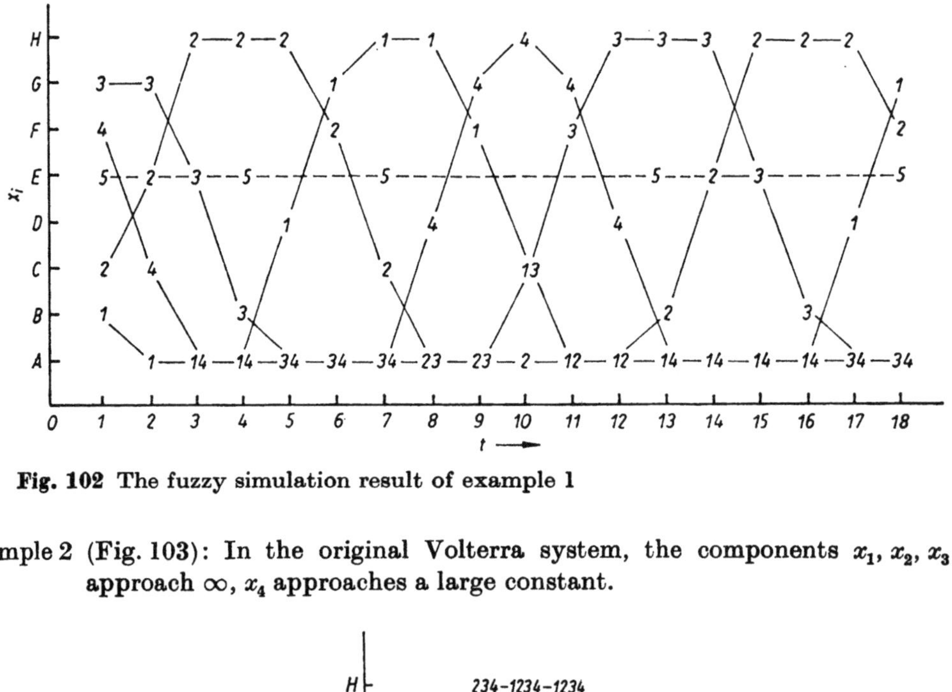


**Fig. 102** The fuzzy simulation result of example 1

Example 2 (Fig. 103): In the original Volterra system, the components $x_1$, $x_2$, $x_3$ approach $\infty$, $x_4$ approaches a large constant.

$$G = \begin{pmatrix} -1 & -1 & 2 & -1 & 0 \\ 2 & 1 & 1 & 0 & 1 \\ 1 & 0 & 2 & 1 & 1 \\ 3 & 2 & 0 & -1 & 1 \end{pmatrix}$$

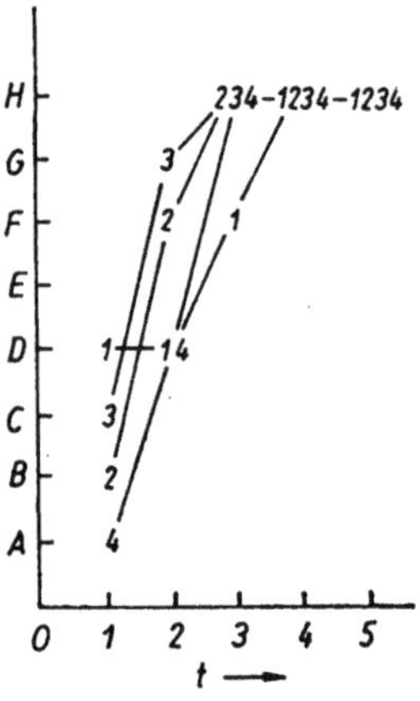


**Fig. 103** The fuzzy simulation result for example 2.

Example 3 (Fig. 104): In the original Volterra system, the components $x_1$, $x_2$, $x_3$ approach 0, $x_4$ approaches a small constant.

$$G = \begin{bmatrix} 1 & 1 & -2 & -1 & 0 \\ -2 & -1 & -1 & 0 & -1 \\ -1 & 0 & -2 & -1 & -1 \\ -3 & -2 & 0 & 1 & -1 \end{bmatrix}$$

**Fig. 104** The fuzzy simulation result for example 3

$$G = \begin{bmatrix} -3 & 1 & 1 & 0 & 1 \\ 1 & 0 & -3 & 1 & 1 \\ 4 & -1 & -4 & 1 & 0 \\ -5 & 2 & 5 & -1 & -1 \end{bmatrix}$$

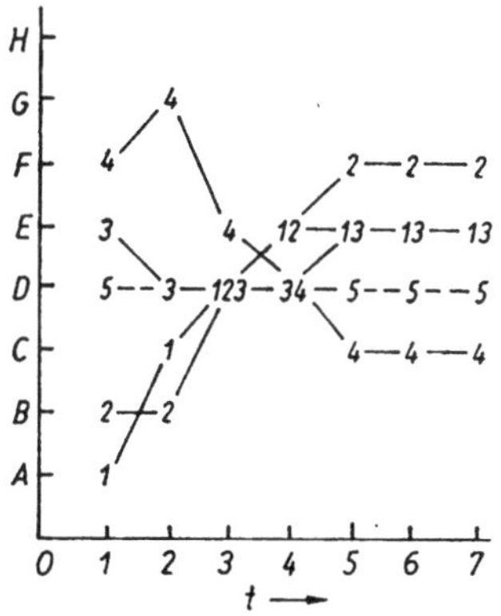


**Fig. 105** The fuzzy simulation result for example 4

Example 4 (Fig. 105): Here the original Volterra system has fixed points lying on a straight line through (1, 1, 1, 1, 1). The trajectories approach one of these fixed points.

Example 5 (Fig. 106): The original Volterra system is the so-called hypercycle. It has a stable limit cycle around the fixed point (1, 1, 1, 1, 1).

Examples 1, 2, 3 and 4 model in a satisfactory way the original trajectories (in example 1, we recognize the cyclic trajectory especially in the component $x_2$). In example 5, the cyclic behaviour is expressed especially by the component $x_4$, and by the incomplete oscillations of the other components. The periods are, however, too long (12 s in example 1 instead of 5.1 s of the original system, 17 s instead of 11.7 s, as

$$G = \begin{bmatrix} -K & K & \cdot & \cdot & \cdot \\ -K & \cdot & K & \cdot & \cdot \\ -K & \cdot & \cdot & K & \cdot \\ -K & \cdot & \cdot & \cdot & K \end{bmatrix}$$

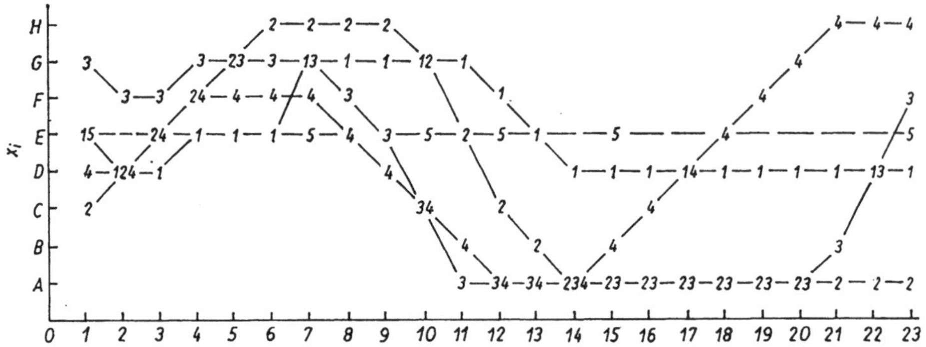


**Fig. 106** The fuzzy simulation result for example 5

in example 5). This might be caused by the large step width of 1 s, leading to the consequence that the simulation cannot follow very quickly the dynamics of the original system. It is remarkable that, despite this influence, the qualitative behaviour of the original system is reflected quite well by this rough fuzzy model. Therefore, we believe that the rough fuzzy simulation may be a suitable tool to obtain information about the qualitative behaviour and the robustness of the system against changes of the system's parameters in any case leading to discussable hypotheses, which can be checked afterwards by traditional methods.

## 8.7. *Generation of Growth Functions by Chains of Higher Aggregated Modules*

As base module (elementary functional principle of growth) we use now

$$\mathrm{d}x/\mathrm{d}t = Kx^k y \quad \text{instead of} \quad \mathrm{d}x/\mathrm{d}t = Kxy\,.$$

In such a higher aggregated description of modules the "atoms" show autocatalytical growth already (hyperbolic, parabolic instead of only exponential growth). Based on these higher aggregated modules we get the following chain concept:

$$\mathrm{d}x_i/\mathrm{d}t = K_i x_i^{k_i} x_{i+1}\,, \qquad i = 0, 1, 2, \ldots\,.$$

We call these more general chains in comparison with exponential chains: Riccati chains.

May $x_0 = x(t)$ be a given growth process, for example, with an $s$-formed transition curve. Then we are interested to expand the signal $x_0(t)$ into a Riccati chain. For this purpose we have to identify the corresponding coefficients $K_i$ and $k_i$ of the chain expansion

$$\mathrm{d}x_i/\mathrm{d}t = K_i x_i^{k_i} x_{i+1}\,, \qquad i = 0, 1, \ldots, n\,, \qquad x_{n+1}(t) \equiv 1\,.$$

Obviously for a given sequence of exponents $k_i$ the corresponding parameters $K_i$ can be determined in the usual way by

$$K_i = \mathrm{d}x_i/\mathrm{d}t|_{t=0}$$

if we base the expansion on the normality conditions $x_i(0) = 1$ for $i = 0, 1, \ldots$. In this case, with given values of the exponents $k_i$, the iterative functions $x_{i+1}$ are computed by

$$x_{i+1}(t) = \frac{\mathrm{d}x_i/\mathrm{d}t}{K_i x_i^{k_i}}.$$

It is of interest that we can propose an approach for "optimum" estimation of the cooperativity exponents $k_i$, if they are not apriori given. This approach cannot be applied always, but it leads to success in a lot of cases.

"Identification" of the cooperativity exponents $k_i$: We build the following help expressions

$$\varkappa_i(t) = \frac{\ddot{x}_i}{\dot{x}_i^2} x_i \quad \text{and} \quad \lambda_i(t) = \frac{\dot{x}_{i+1}}{x_{i+1}} \frac{x_i}{\dot{x}_i}.$$

From the Riccati chain equations we get

$$\varkappa_i(t) = k_i + \lambda_i(t) .$$

It is important to state that the functions $\lambda_i(t)$ have fixed signs independent of time $t$.

Therefore, we get obviously the following identification rule:

If sign $(\lambda_i(t)) > 0$ and there exists a $t_0$ with $\lambda_i(t_0) = 0$, then $k_i = \mathrm{Min}\, \varkappa_i(t)$.

If sign $(\lambda_i(t)) < 0$ and there exists a $t_0$ with $\lambda_i(t_0) = 0$, then $k_i = \mathrm{Max}\, \varkappa_i(t)$.

We will now apply this identification approach for the cooperativity exponents $k_i$ for some different growth functions concepts.

*Identification of the EVOLON as a Riccati chain*

For the hyperlogistic EVOLON model $\mathrm{d}x/\mathrm{d}t = Kx^k(B - x^w)^l$ we get

$$x_0 = x ,$$

$$\mathrm{d}x_0/\mathrm{d}t = Kx_0^k x_1 \quad \text{with} \quad x_1 = (B - x_0^w)^l ,$$

$$\mathrm{d}x_1/\mathrm{d}t = K_1 x_1^{(2l-1)/l} x_2 \quad \text{with} \quad K_1 = -wlK , \qquad x_2 = x_0^{w+k-1} ,$$

$$\mathrm{d}x_2/\mathrm{d}t = K_2 x_2^{(w+2k-2)/(w+k-1)} x_1 \quad \text{with} \quad K_2 = K(w + k - 1) .$$

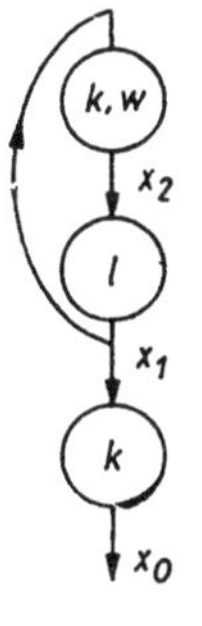


**Fig. 107** Realization of the hyperlogistic system by a hypercycle of order $r = 2$ with higher aggregated modules

This is obviously in the Riccati chain representation in essential a hypercycle of order $r = 2$, as Fig. 107 schematically shows.

*Riccati chain models for some probability distributions*

For $dx/dt = K\,e^{\varkappa(t-a)^k}$ with integer $k$, we get

$$dx/dt = K\,e^{\varkappa(-a)^k}\,x_1\,,$$

$$dx_1/dt = \varkappa k(-a)^{k-1}\,x_1x_2\,,$$

$$dx_2/dt = \frac{k-1}{(-a)}\,x_2^{(k-2)/(k-1)}\,.$$

Especially for the Gaussian normal distribution with

$$K = 1/(\sqrt{2\pi}\,\sigma)\,,\qquad \varkappa = -1/(2\sigma^2)\quad\text{and}\quad k = 2$$

we get the following representation being characteristic for normal distributions

$$dx/dt = \frac{1}{\sqrt{2\pi}\,\sigma}\,e^{-a^2/(2\sigma^2)}\,x_1\,,$$

$$dx_1/dt = \frac{a}{\sigma^2}\,x_1x_2\,,$$

$$dx_2/dt = -1/a\,.$$

For the $\Gamma$-distribution $dx/dt = Kt^k\,e^{-at}$ we get

$$dx/dt = Kx_1\,,$$

$$dx_1/dt = -ax_1x_2\,,$$

$$dx_2/dt = \frac{k}{a}\,x_2^2x_3\,,$$

$$dx_3/dt = -2x_2^{3/2}\,.$$

For the Cauchy distribution (arctan growth function)

$$dx/dt = \frac{\lambda}{\pi\left(\lambda^2 + (t-\mu)^2\right)}$$

we get

$$dx/dt = \frac{\lambda}{\pi(\lambda^2 + \mu^2)}\,x_1\,,$$

$$dx_1/dt = \frac{2\mu}{(\lambda^2 + \mu^2)}\,x_1^{3/2}x_2\,,$$

$$dx_2/dt = \frac{\lambda^2}{(\lambda^2 + \mu^2)}\,x_2x_1\,.$$

*Despite of the obvious similarity this is no special case of the EVOLON*

For a modified half-circle distribution

$$x(t) = \left(1 - (1-t)^w\right)^k \quad\text{we get}$$

$$dx/dt = kwx^{1-1/kw}x_1\,,$$

$$dx_1/dt = -(w-1)\, x_1^{(w-2)/(w-1)} x_2 \,,$$

$$dx_2/dt = -\frac{(w+1)}{kw}\, x_2^{1+kw/(w+1)} \,.$$

*Some other examples*

We consider the generating function of the Bessel functions

$$x(t) = e^{3/2(t-1/t)}$$

and get for it the following Riccati chain

$$dx/dt = \frac{z}{2}\, x x_1 \,,$$

$$dx_1/dt = -2x_1^{3/2} x_2 \,,$$

$$dx_2/dt = -3x_2^{4/3} x_1^{-1/2} \,.$$

For the same function $x(t)$ we can get at the same time the following network solution as an alternative to the Riccati chain solution

$$x(t) = y_1 y_2 \,,$$

$$dy_1/dt = \frac{z}{2}\, y_1 \,,$$

$$dy_2/dt = \frac{z}{2}\, y_2 y_3 \,,$$

$$dy_3/dt = -2y_3^{3/2} \,.$$

We consider the generating function of the Laguerre functions

$$x(t) = \frac{e^{-zt/(1-t)}}{(1-t)}$$

and get the following network for it

$$x(t) = y_1 y_2 \,,$$

$$dy_1/dt = y_1^2 \,,$$

$$dy_2/dt = -z y_2 y_1^2 \,.$$

We consider the generating function of Hermite' polynoms

$$x(t) = e^{-t^2+2tz}$$

and get the following solution

$$dx/dt = 2x x_1 \,,$$

$$dx_1/dt = -1 \,.$$

We look for a network solution similar to a Riccati chain for Pearson's differential equation

$$dx/dt = x_1 \,,$$

$$dx_1/dt = x_1(t-a)/(b_0 + b_1 t + b_2 t^2)$$

and get the following solution

$$\begin{aligned} dx/dt &= x_1 , \\ dx_1/dt &= x_1 x_2 x_3 , \\ dx_2/dt &= 1 , \\ dx_3/dt &= -x_3^{3/2} x_4 , \\ dx_4/dt &= \frac{(4b_0 b_2 - b_1^2)}{2} x_3^{3/2} . \end{aligned}$$

We can extend the structure design principle formerly based on the logarithmic derivative for the case of use of Riccati modules as base modules.

This is always possible and we get the following results:

— The field of application of the structure design principle based on modules of the form

$$dx/dt = Kx^k y$$

coincides with that for the base module logarithmic derivative

$$dx/dt = Kxy .$$

— We can get by the extended structure design principle a standard description of admissable ordinary differential equations of the form

$$dz_i/dt = K_i z_i^{k_i} \prod_{r \in I_i} z_r .$$

This is a Riccati representation with Cobb-Douglas production functions where all elasticities besides the autocatalytic ones can be only 0 or 1.

## 8.8. *Qualitative Properties of Exponential Chains*

With the base module $dx/dt = Kxy$ exponential chains

$$dx_i/dt = K_i x_i x_{i+1} , \qquad i = 0, 1, \dots , N ,$$

are constructed with

$$x_{N+1}(t) = y(t) \geqq 0$$

and the normalisation condition $x_i(0) = 1$. For a given function $x_0(t) = x(t)$ we seek an approximation by a suitable exponential chain expansion.

As we already know the coefficients $K_i$ can be determined as follows

$$K_i = F^{i+1} x_0(t)|_{t=0} \quad \text{with} \quad F = d \ln/dt .$$

If we integrate the exponential chain beginning with $x_{N+1}(t) = 1$ down to the bottom we get the corresponding proxy $x_{0N}(t)$ for the given signal $x_0 = x(t)$. The function $x_{0N}(t)$ we call the exponential tangent of order $N + 1$ to $x_0(t)$ in $t = 0$.

Instead of $t = 0$ we can use any other reference time-point $t_0 \geqq 0$ and determine their the coefficients $K_i$ and the corresponding chain approximation (exponential tangent) in this point.

For studying the qualitative behaviour of arbitrary exponential chains we can in principle restrict on

$$K_i = \begin{cases} +1 \,, \\ -1 \,. \end{cases}$$

For the corresponding signal $x_{0N}(t)$ at the bottom of the chain we introduce the following notation

$$x_{0N}(t) = \mathrm{e}(s_0, s_1, \dots, s_N) \quad \text{with} \quad s_i = \operatorname{sign} K_i = \begin{cases} + \\ - \end{cases}.$$

We call these functions coordinate functions of the exponential chain of order $N + 1$. Obviously for the first and second order coordinate functions we can get the following expressions by integration

$$\mathrm{e}(s_0) = \mathrm{e}^{s_0 t} \,,$$

$$\mathrm{e}(s_0, s_1) = \mathrm{e}^{(s_0/s_1)\,\mathrm{e}^{(s_1 t - 1)}} \,.$$

If we take into account, that $\mathrm{e}^{K\int_0^t}$ is for $K > 0$ a positive monotonous and for $K < 0$ a positive antitonous operator, we get the following monotony relationships

$$\mathrm{e}(+) > 1 > \mathrm{e}(-) \,,$$

$$\mathrm{e}(+,+) > \mathrm{e}(+) > \mathrm{e}(+,-) > 1 > \mathrm{e}(-,-) > \mathrm{e}(-) > \mathrm{e}(-,+);$$

$$\mathrm{e}(+,+,+) > \mathrm{e}(+,+) > \mathrm{e}(+,+,-) > \mathrm{e}(+) > \mathrm{e}(+,-,-) > \mathrm{e}(+,-) >$$
$$> \mathrm{e}(+,-,+) > 1 > \mathrm{e}(-,-,+) > \mathrm{e}(-,-) > \mathrm{e}(-,-,-) >$$
$$> \mathrm{e}(-) > \mathrm{e}(-,+,-) > \mathrm{e}(-,+) > \mathrm{e}(-,+,+);$$

All following results of this Section are based on this half-order relationships of the binary vectors up to the order $N + 1$

$$(s_0, s_1, \dots, s_N) \,.$$

This splitting of the lower order coordinate functions into coordinate functions of the next higher order can be transparently represented as shown in Fig. 108.

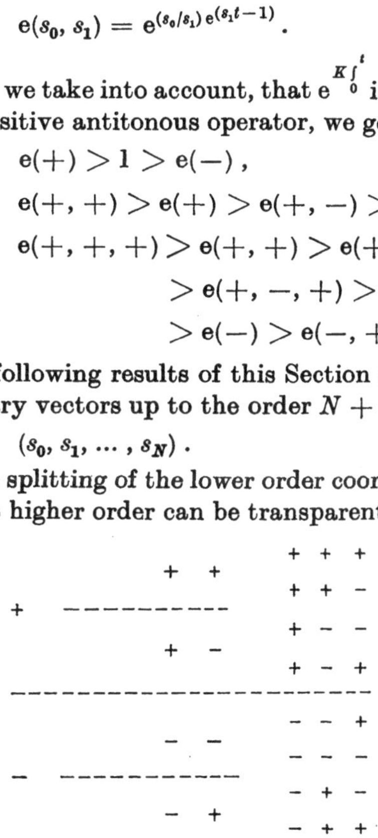


**Fig. 108** The half-order of binary vectors corresponding to the monotony of coordinate functions for $l = 1$, 2, and 3

The corresponding linearly ordered set of binary vectors of length $l$ we designate with or $(W_l)$, the corresponding inverse order we designate with $\overline{\text{or}}\,(W_l)$. Obviously the generation rule for the construction of or $(W_{l+1})$, iteratively from or $(W_l)$ is given by

$$\text{or}\,(W_{l+1}) = \begin{matrix} +\ \text{or}\,(W_l) \,, \\ -\ \overline{\text{or}}\,(W_l) \,. \end{matrix}$$

This is apparently a tree-like branching process.

Let us now characterize the followers of length $k$ to a given binary word $W_l$ of length $l$ in the half-ordered description. May be $\Pi W_l = \Pi s_i$ with $W_l = (s_0, s_1, \dots, s_{l-1})$. Then we get the following result. If $\Pi W_l = +1$ then for given $W_l$ and an arbitrary $W_k$ we have

$$(W_l, W_k) = \begin{matrix} W_l + \text{or}\,(W_{k-1}) \\ W_l - \overline{\text{or}}\,(W_{k-1}) \,. \end{matrix}$$

If $\Pi W_l = -1$ then for given $W_l$ and an arbitrary $W_k$ we have

$$(W_l, W_k) = \begin{matrix} W_l - \text{or}\,(W_{k-1}) \\ W_l + \overline{\text{or}}\,(W_{k-1}) \,. \end{matrix}$$

We introduce now the following important notations (monotony classes)

$$<W_l = \begin{matrix} W_l + \text{or}\,(W) & \text{if} & \Pi W_l = +1 \,, \\ W_l - \text{or}\,(W) & \text{if} & \Pi W_l = -1 \end{matrix}$$

and

$$<W_l = \begin{matrix} W_l - \overline{\text{or}}\,(W) & \text{if} & \Pi W_l = +1 \,, \\ W_l + \overline{\text{or}}\,(W) & \text{if} & \Pi W_l = -1 \end{matrix}$$

where $W$ is an arbitrary binary word.

For given $l$ ($l = 0, 1, 2, \dots$) we introduce in correspondence to the coordinate functions $\mathrm{e}(s_0, s_1, \dots, s_{l-1})$ the so-called *monotony classes for trajectories* in the following way. We consider families of curves with

$$>\mathrm{e}(s_0, s_1, \dots, s_{l-1}) = \begin{matrix} \mathrm{e}(s_0, s_1, \dots, s_{l-1}, +K_l) & \text{for} & \Pi s_i = +1, \\ \mathrm{e}(s_0, s_1, \dots, s_{l-1}, -K_l) & \text{for} & \Pi s_i = -1; \end{matrix}$$

$$<\mathrm{e}(s_0, s_1, \dots, s_{l-1}) = \begin{matrix} \mathrm{e}(s_0, s_1, \dots, s_{l-1}, -K_l) & \text{for} & \Pi s_i = +1 \,, \\ \mathrm{e}(s_0, s_1, \dots, s_{l-1}, +K_l) & \text{for} & \Pi s_i = -1; \end{matrix}$$

$>\mathrm{e}(s_0, s_1, \dots, s_{l-1})$ is called upper and $<\mathrm{e}(s_0, s_1, \dots, s_{l-1})$ lower monotony class of $\mathrm{e}(s_0, s_1, \dots, s_{l-1})$.

Obviously these bands are completely covered by the corresponding curve families parametrized by the parameter $K_l < 0$. It can be seen easily that all these monotony classes together are a decomposition of the whole positive quadrant ($t \geqq 0$, $x_0 \geqq 0$) into disjoint bands.

All the curves $\mathrm{e}(s_0, s_1, \dots, s_{l-1}, s_l K_l)$ together can be considered as generalised straight lines in the positive quadrant determining a generalised convexity notion in this part of the plane, if the reference point is allowed to shift into an arbitrary point $(t_0, x(t_0))$ [120]. Remark: Instead of restricting on chains with coefficients $+1$ and $-1$ we can introduce in the same way for an arbitrary exponential chain similar coordinate functions

$$\mathrm{e}(s_0 K_0, s_1 K_1, \dots, s_{l-1} K_{l-1})$$

assuming $K_i > 0$.

Then we get corresponding monotony classes which we designate as follows

$$>\mathrm{e}(s_0 K_0, s_1 K_1, \dots, s_{l-1} K_{l-1}) \quad \text{and} \quad <\mathrm{e}(s_0 K_0, s_1 K_1, \dots, s_{l-1} K_{l-1}) \,.$$

The monotony classes can be considered as classes in the sense of classification theory. At the same time we are going to consider the monotony classes $>\mathrm{e}(s_0 K_0,$

$s_1K_1, \ldots, s_{l-1}K_{l-1}$), $<e(s_0K_0, s_1K_1, \ldots, s_{l-1}K_{l-1})$ as the states of a finite automaton and we will study the state-transitions of this automaton, if we either adjoin another base module of the exponential chain with the parameters $s_0K_0, s_1K_1, \ldots, s_{l-1}K_{l-1}$ at the top or at the bottom.

Let us at first adjoin another base module at the top of the exponential chain, meaning that we consider now the chain with the parameters

$$(s_0K_0, s_1K_1, \ldots, s_{l-1}K_{l-1}, sK) .$$

We know from the considerations presented above that the corresponding coordinate function

$$e(s_0K_0, s_1K_1, \ldots, s_{l-1}K_{l-1}, sK)$$

either belongs to the state

$$>e(s_0K_0, s_1K_1, \ldots, s_{l-1}K_{l-1}) \quad \text{or} \quad <e(s_0K_0, s_1K_1, \ldots, s_{l-1}K_{l-1})$$

depending on the sign of $s$:

If $\Pi s_i = +1$ then for $s = +$ $e(s_0K_0, \ldots, sK)$ belongs to $>e(s_0K_0, \ldots, s_{l-1}K_{l-1})$ and for $s = -$ to $<e(s_0K_0, \ldots, s_{l-1}K_{l-1})$.

If $\Pi s_i = -1$ then for $s = +$ $e(s_0K_0, \ldots, sK)$ belongs to $<e(s_0K_0, \ldots, s_{l-1}K_{l-1})$ and for $s = -$ to $>e(s_0K_0, \ldots, s_{l-1}K_{l-1})$.

Thus in any case a unique state of our automaton is specified. It is important to mention that this state will be invariant if we adjoin more and more modules at the top level of the exponential chain. From this result an important property follows: If we classify the possible outputs $x_0(t)$ of a chain which begins with the parameter sequence $s_0K_0, s_1K_1, \ldots, s_{l-1}K_{l-1}$ with resolution force $l$ (that means if we use $2^l$ monotony classes), the behavioural type of all longer chains is uniquely determined by the initial chain

$$s_0K_0, s_1K_1, \ldots, s_{l-1}K_{l-1}, sK .$$

Now we adjoin another base module at the bottom level of the exponential chain with the parameters $s_0K_0, s_1K_1, \ldots, s_{l-1}K_{l-1}$, that means we consider the following chain specified by the parameter sequence

$$sK, s_0K_0, \ldots, s_{l-1}K_{l-1} .$$

The generated bottom signal $x_0(t)$ of this chain can be interpreted as if we had adjoined the base module with the parameter $s_{l-1}K_{l-1}$ at the top-level of the chain specified by

$$sK, s_0K_0, \ldots, s_{l-2}K_{l-2} .$$

We meet here the difficulty that by this manipulation all coordinate functions are quantitatively changed.

But it is important to mention that qualitatively the corresponding state transition depends only on the signs $s_i$.

Therefore, we look first for the corresponding state transition for the case where all $K_i = 1$ for all $i$. In this case the following transition rule can be easily derived:

$$s \gtrless e(s_0, s_1, \ldots, s_{l-1}) = \begin{matrix} >e(s, s_0, \ldots, s_{l-2}) & \text{if} \quad \Pi s_i \cdot s = 1 , \\ <e(s, s_0, \ldots, s_{l-2}) & \text{if} \quad \Pi s_i \cdot s = -1 . \end{matrix}$$

Herein $s$ designates the operator corresponding to the chain extension by adjoining a base module with the operator $s$ at the bottom. This transition rule is proven by the following consideration: We have to consider the chain with the parameters $s, s_0, \dots, s_{l-2}$ together with the input $x_{l-1}(s_{l-1}K_{l-1})$ at the top. Obviously for $s_{l-1} = +1$ we have $x_{l-1} > 1$. If the chain $s, s_0, \dots, s_{l-2}$ contains an even number of antitonous operators ($s_i = -1$) we get obviously $>$ and otherwise $<$. For $s_{l-1} = -1$ the conclusion can be drawn similarly.

Most important is the fact that we really get a state-transition and that none of the states is splitted up. The result can be illustrated by rather interesting graphs corresponding to our classifying automata. Fig. 109 shows the graphs for the resolution forces $l = 0, 1, 2$ and 3. From these graphs we recognize immediately the remarkable

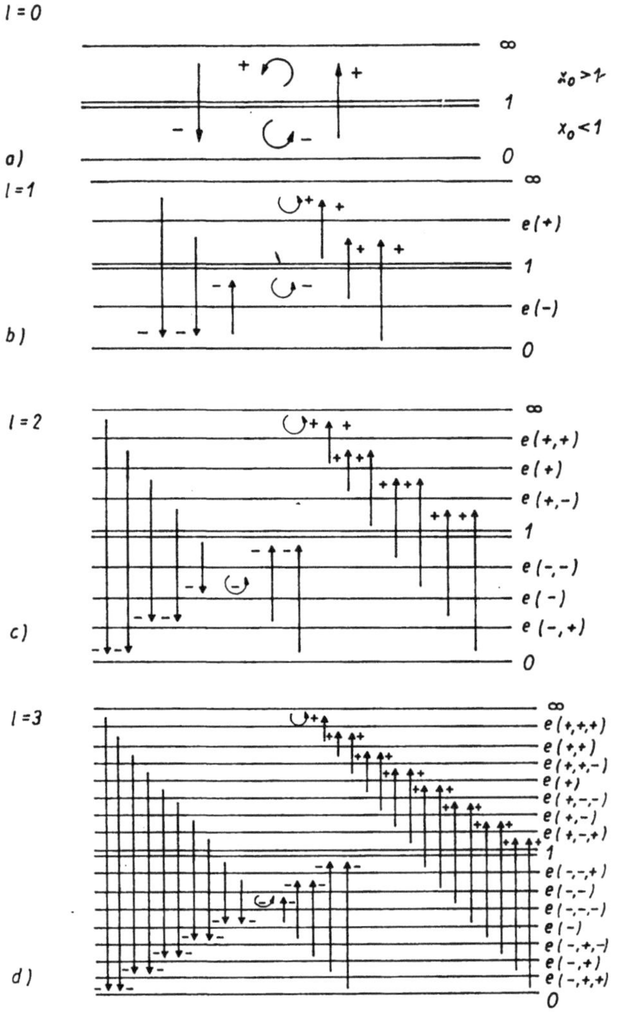


**Fig. 109** The state-transition graphs for the automata corresponding to the monotony classes of resolution forces $l$
a) $l = 0$; b) $l = 1$; c) $l = 2$; d) $l = 3$

role which is played by hyperbolic growth respective hyperbolic decay for the qualitative behaviour of exponential chains.

It is remarkable, that hyperbolic growth is a monotonous class attractor which can be approached if we adjoin at the bottom of the chain more and more driving base modules (+), whereas the hyperbolic decay is a focus-like class attractor which can be approached, if we adjoin at the bottom level of the chain more and more damping modules (−).

The state-transition is a little more complicated if we allow for $K_i \neq 1$. Then we have obviously

$$sK \gtrless e(s_0K_0, s_1K_1, \dots, s_{l-1}K_{l-1}) = \begin{matrix} > e(sK, s_0K_0, \dots, s_{l-2}K_{l-2}), & s\Pi s_i = 1, \\ < e(sK, s_0K_0, \dots, s_{l-2}K_{l-2}), & s\Pi s_i = -1. \end{matrix}$$

This result we have to interpret in the following way: We should have at our hands for every set of specified coefficients $(K_0, K_1, \dots, K_{l-1})$, $K_i > 0$ a page exposing the concrete quantitative picture with the coordinate functions

$$e(s_0K_0, s_1K_1, \dots, s_{l-1}K_{l-1})$$

in a corresponding diagram showing their time-dependence.

On all these pages the monotony between these coordinate functions and the automata graphs are invariant, but the concrete form of the monotony classes (bands) depends on the concrete parameter sequence $(K_0, K_1, \dots, K_{l-1})$.

Let us now describe the corresponding state-transition: We start in a given state on the graph belonging to the page $(K_0, K_1, \dots, K_{l-1})$. Now we adjoin a base module with the parameter $sK$ at the bottom of the considered chain. Then we have to pass first onto the page with the parameters $(K, K_0, \dots, K_{l-2})$, determine there the corresponding state specified by $(s_0, s_1, \dots, s_{l-1})$ and $>$ or $<$ respectively and perform in this graph the state transition to $(s, s_0, \dots, s_{l-2})$ and $>$ or $<$ according to the above described transition rule.

Similarly we can discuss the robustness of the qualitative behaviour if the parameters $K_i$ are independently disturbed by amplitude-bounded random noise not changing the signs of the parameters. We shall now use the monotony and classification properties which we have described above to answer the question which functions of the type

$$e(s_0K_0, s_1K_1, \dots)$$

describe saturation processes, meaning that they approach a non-zero constant. Without notating the $K_i$ explicitly (because the existence of a bounded solution is a generic question) we characterize explicitly all bounded coordinate functions $e(s_0, s_1, \dots)$. On the level $l = 1$ we meet only functions of the type

$$e(+) \sim \exp(Kt), \qquad K > 0, \text{ approaching } \infty$$

and

$$e(-) \sim \exp(-Kt) \quad \text{approaching } 0.$$

That means in this case no saturation function can exist. On the level $l = 2$ we meet the following two saturation functions,

$$e(+, -) \sim \exp\left(-\frac{K_0}{K_1}(\exp(-K_1t) - 1)\right), \quad K_0 > 0, \quad K_1 > 0,$$

approaching $e^{K_0/K_1}$,

and

$$e(-,-) \sim \exp\left(\frac{K_0}{K_1}\left(\exp\left(-K_1 t\right) - 1\right)\right)$$

approaching $e^{-K_0/K_1}$.

It follows that

$$<e(+,-) = e\left(+,-,+,\overline{\text{or}}\,(W)\right)$$

and

$$>e(-,-) = e\left(-,-,+,\text{or}\,(W)\right)$$

where $W$ designates an arbitrary binary word $W$, consist only of saturation functions. Now the important question arises, whether in

$$e(+,-) = e\left(+,-,-,\text{or}\,(W)\right)$$

or in

$$e(-,-) = e\left(-,-,-,\overline{\text{or}}\,(W)\right)$$

are contained more saturation functions. The answer is positive. We shall now describe the corresponding construction for the determination of all these additional saturation functions. Obviously $e(+,-,-)$ is no saturation function, it shows a similar behaviour as $e(+)$, meaning it will grow infinitely. Therefore we split into

$$>e(+,-,-) = e\left(+,-,-,-,+,\text{or}\,(W)\right) \quad \text{with no chance for saturation}$$

and

$$<e(+,-,-) = e\left(+,-,-,-,\overline{\text{or}}\,(W)\right).$$

We see, that

$$e(+,-,-,-)$$

is a saturation function with the consequence that also all functions in

$$<e(+,-,-,-) = e\left(+,-,-,-,+,\overline{\text{or}}\,(W)\right)$$

are saturation functions.

We have now to check the band

$$>e(+,-,-,-) = e\left(+,-,-,-,-,\text{or}\,(W)\right).$$

We see similarly as above that

$$e(+,-,-,-,-)$$

is no saturation function. Therefore, we have to consider

$$e\left(+,-,-,-,-,-,\overline{\text{or}}\,(W)\right)$$

and get the result that

$$e(+,-,-,-,-,-)$$

and

$$<e(+,-,-,-,-,-) = e\left(+,-,-,-,-,-,+,\overline{\text{or}}\,(W)\right)$$

are all saturation functions. Similarly we can find all the saturation functions for the other cases

$$e(-,-).$$

Therefore, we have derived the following complete results: All saturation functions in exponential chains are of the types

$$e(+, -) \qquad e(+, -, \text{even number of } -),$$

$$e(+, -, +, \overline{\text{or}}\,(W)) \qquad e(+, -, \text{even number of } -, +, \overline{\text{or}}\,(W))$$

respectively

$$e(-, -) \qquad e(-, -, \text{even number of } -),$$

$$e(-, -, +, \text{or}\,(W)) \qquad e(-, -, \text{even number of } -, +, \text{or}\,(W)).$$

We will interpret this general and rather transparent result by coding every sign-sequence into the interval [0, 1]. This can be done, for example, by assigning the following digits

$$- \sim 0,$$

$$\text{empty} \sim 1,$$

$$+ \sim 2$$

and representing the position $i = 0, 1, 2, \ldots$ by $3^{-(i+1)}$.

Obviously we get for the corresponding codes the following values

$$e(+, -, \text{even number } 2k \text{ of } -) \sim 2/3 + 1/18 \times 1/9^k,$$

$$e(+, -, \text{even number } 2k \text{ of } -, +, \overline{\text{or}}\,(W)) \sim 2/3 + 2/27, 3/27 \times 1/9^k$$

respectively

$$e(-, -, \text{even number } 2k \text{ of } -) \sim 1/18 \times 1/9^k,$$

$$e(-, -, \text{even number } 2k \text{ of } -, +, \text{or}\,(W)) \sim 2/27, 3/27 \times 1/9^k.$$

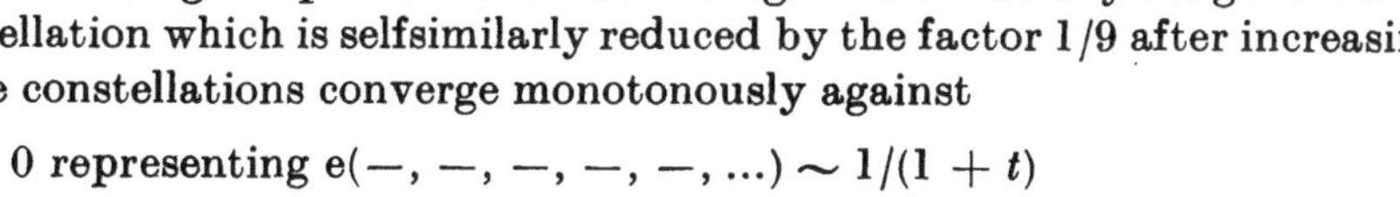


**Fig. 110** Code representation of all saturation types which can be realized by an exponential chain

Therefore we get a picture as shown in Fig. 110. Obviously we get for $k = 0$ a basic constellation which is selfsimilarly reduced by the factor 1/9 after increasing $k$ by one. These constellations converge monotonously against

$$0 \text{ representing } e(-, -, -, -, -, \ldots) \sim 1/(1 + t)$$

from above, and

$$2/3 \text{ representing } e(+, -, -, -, \ldots) \sim 1 + t$$

from below.

If we reconsider again the process for the construction of all saturation types we can interpret the adjoining of corresponding top modules as a bifurcation process. We should remark that the chosen code is not convenient because it does not preserve the monotony property between the coordinate functions. We shall see that this disadvantage can be corrected easily.

Let us first compute for all bounded coordinate functions the corresponding saturation levels. Obviously we get for

$$\mathrm{e}(s, -, \text{even number } 2k \text{ of } -, +, \text{or } (W))$$

in general the following saturation level

$$\mathrm{e}^{s(K_0/K_1)\mathrm{e}^{(K_2/K_3)\mathrm{e}^{(K_4/K_5)\cdots\mathrm{e}^{(K_{2f}/(K_{2f+1})K)}}}}$$

where $K$ (with $0 < K < 1$) is the saturation level of $\mathrm{e}(-, -, +, \text{or } (W))$.

From this result we can derive easily monotony preserving coding approaches for the coordinate functions $\mathrm{e}(s_0K_0, s_1K_1, \ldots)$. For an arbitrary "base" $a > 1$ we define the code for the coordinate functions as follows

$$\mathrm{e}(s_0K_0, s_1K_1, \ldots) = a^{s_0K_0a^{s_1K_1a^{s_2K_2\cdots}}} = z\,.$$

This code can be transformed monotonously onto the interval [0, 1], for example, by

$$w = z/(1 + z)\,.$$

For example we get with $a = 1.5$ for the coordinate functions $\mathrm{e}(s_0, s_1, s_2)$ the following codes

| | | | | | | | |
|---|---|---|---|---|---|---|---|
| $+++$ | $++-$ | $+--$ | $+-+$ | $--+$ | $---$ | $-+-$ | $-++$ |
| 0.68 | 0.63 | 0.58 | 0.56 | 0.44 | 0.42 | 0.37 | 0.32 |

# 9. Some Large-Scale Applications of the Lotka-Volterra Concept

## *9.1. Lotka-Volterra Approach and Control Systems*[1]

We will first try to apply the Lotka-Volterra approach to multivariable linear systems. At first glance, one would think that it makes no sense to simulate the behaviour of a linear system

$$\mathrm{d}z_i/\mathrm{d}t = \sum A_{ij} z_j$$

with the Lotka-Volterra approach. Why should we leave the linear superposition principle for the price of a higher-dimensional nonlinear representation? Linear systems, though, have some properties which lead us very simply to a regular Lotka-Volterra representation.

By shifting the origin of the coordinate system, we can achieve, in most cases, the aim that the trajectories of a linear system stay always in the positive cone $z_i \geqq 0$. This shift is, for linear systems, already sufficient for regularisation, because multiple differentiation is not necessary for the transformation of a given multivariable linear system into a Lotka-Volterra representation.

By dividing the velocities $\mathrm{d}z_i/\mathrm{d}t$ by $z_i$, we get

$$Fz_i = \sum A_{ij} z_j / z_i \, .$$

This is a special case for a system of the form

$$Fz_i = \sum A_i^{rs} z_r / z_s$$

which we call a *Volterra ratio system.*

These equations can, after one additional step, be transformed into the Lotka-Volterra equation by the introduction of the additional state-variables

$$z_{rs} = z_r / z_s \, .$$

We then get the following Lotka-Volterra representation

$$Fz_i = \sum A_i^{rs} z_{rs} \, ,$$

$$Fz_{rs} = Fz_r - Fz_s = \sum \left(A_r^{mn} - A_s^{mn}\right) z^{mn} \, .$$

For the special case of a multivariable linear system, we get

$$Fz_i = \sum A_{ij} z_{ji} \, ,$$

$$Fz_{ji} = \sum \left(A_{jr} z_{rj} - A_{ir} z_{ri}\right) .$$

---

[1]) We thank Dr. Breitenecker, Institut für Technische Mathematik, Technische Universität Wien, for his useful ideas and proposals to apply the Lotka-Volterra approach for linear control systems.

For simulation purposes, the sign of the autocatalytic term is of importance, meaning the sign of

$A_{ii}$ in the linear equations,

$-A_{ij}$ in the nonlinear equations.

We assume that it is usually good for a reliable situation to get rid of positive autocatalytic summands.

Because the Lotka-Volterra description is highly redundant, we propose to skip all equations with positive autocatalytic summands. If the remaining equations are more than or equal to $n$, we should reduce the number of equations to $n$ and simulate the corresponding multinomial system on the computer. It is not certain, that this measure helps much in general, but it can provide some help for those equations where the diagonal terms are dominant over the other terms. This heuristic approach seems to be useful for the simulation of linear systems, by transforming them into nonlinear equations under reduction of this system, as we described above.

Using the same heuristic approach, we will now analyse which possibilities we have for the simulation of the simplest systems in linear control theory. We begin with a second-order linear lumped system with constant parameters

$$\mathrm{d}^2x/\mathrm{d}t^2 + \frac{\vartheta}{m}\mathrm{d}x/\mathrm{d}t + \frac{cx}{m} = \frac{y}{m}.$$

We introduce state-variables, as usual, by

$$x_0 = x\,,$$

$$x_1 = \mathrm{d}x/\mathrm{d}t$$

and get the well known state-variable description in the form

$$\mathrm{d}x_0/\mathrm{d}t = x_1\,,$$

$$\mathrm{d}x_1/\mathrm{d}t = y/m - \frac{\vartheta}{m}x_1 - \frac{c}{m}x_0\,.$$

Now we pass to the Lotka-Volterra description

$$Fx_0 = x_{10}\,, \qquad x_{10} = x_1/x_0\,,$$

$$Fx_1 = y_{01} - \frac{\vartheta}{m} - \frac{c}{m}x_{01} \quad \text{with} \quad y_{01} = \frac{y}{mx_1}\,, x_{01} = x_0/x_1\,,$$

$$Fx_{10} = y_{01} - \frac{\vartheta}{m} - \frac{c}{m}x_{01} - x_{10}\,,$$

$$Fx_{01} = -y_{01} + \frac{\vartheta}{m} + \frac{c}{m}x_{01} + x_{10}\,,$$

$$Fy_{01} = Fy - y_{01} + \frac{\vartheta}{m} + \frac{c}{m}x_{01}\,.$$

Taking into account the redundancy of this description we can reduce it to a two-dimensional analytical representation. We choose for this aim, the following set of equations

$$Fx_0 = x_{10} ,$$

$$\mathrm{d}x_{10}/\mathrm{d}t = \frac{y}{mx_0} - \left(x_{10}^2 + \frac{\vartheta}{m} x_{10} + \frac{c}{m}\right).$$

For the further discussion the roots of the characteristic polynomial are important

$$a_{1/2} = -\frac{\vartheta}{2m} \pm \sqrt{\frac{\vartheta^2}{4m^2} - \frac{c}{m}} = -a \pm r .$$

We rewrite the equations into an easily well interpretable form

$$Fx_0 = w - a \quad \text{putting} \quad x_{10} + a = w ,$$

$$\mathrm{d}w/\mathrm{d}t = \frac{y}{mx_0} - (w^2 - r^2) .$$

Now we should distinguish two cases:

$$r \quad \text{real for} \quad \frac{\vartheta^2}{4m^2} - \frac{c}{m} > 0 ,$$

$$r = \mathrm{j}w_0 \quad \text{for} \quad \frac{\vartheta^2}{4m^2} - \frac{c}{m} < 0 .$$

For $r$ real, it is useful to rewrite the equation in the following form: With

$$r - w = v ,$$

$$Fx_0 = r - a - v ,$$

$$\mathrm{d}v/\mathrm{d}t = -\frac{y}{mx_0} - v(2r - v) .$$

This, obviously, is a superposition of a logistic driving force with the force originated by the input signal $y$.

For $r$ imaginary, there are no logistics at work because of the oscillations with the circular frequency $w_0$, and we should, in this case, write the equations in the following way

$$Fx_0 = w - a ,$$

$$\mathrm{d}w/\mathrm{d}t = \frac{y}{mx_0} - (w^2 + w_0^2) .$$

It is remarkable that, for $y = 0$, in both cases these Riccati-like equations possess closed analytical solutions. These formulas are of great interest because they demonstrate that the behaviour of linear systems is not as very different from the behaviour of rate-coupled systems as one might think.

We will now show how we can use this approach for the general linear system with constant parameters. We consider the $n$-th order linear system with constant parameters

$$\left(\sum_{i=0}^{n} \mathrm{d}_i p^{n-i}\right) x(t) = Ky(t) , \qquad p = \mathrm{d}/\mathrm{d}t , \qquad a_0 = 1 .$$

In the field of real numbers, the characteristic polynomial can be decomposed into linear factors corresponding to the real roots, and into quadratic polynomials corresponding to a pair of conjugate complex roots. All roots can occur in a certain multiplicity, but for the application of our approach, this fact plays no important role. We can then write the characteristic polynomial in the following form

$$\sum_{i=0}^{n} \mathrm{d}_i p^{n-i} = \prod_{l} ((p + a_l)^2 - r_l^2) \prod_{j} (p + l_j) .$$

For a factor of the form

$$(p + a_l)^2 - r_l^2$$

we already gave a standard representation with our approach. We now give also a standard representation for a linear factor

$$(p + l_i) .$$

That means that we must consider the following differential equation

$$\mathrm{d}x/\mathrm{d}t + l_i x = Ky , \qquad x = x_0$$

with the consequence

$$Fx_0 = K \frac{y}{x_0} - l_i .$$

In this case the intermediary state-variable $w$ is not necessary, we have only the damping term superimposed immediately upon the term originated by the input.

According the principle of series programming in analogue computing, we can represent the general linear system by a chain of these models of first- and second-order, corresponding to the roots of the characteristic equation. This is possible because we can solve any equation of the form

$$P_1(p)\, P_2(p)\, x(t) = Ky$$

by the following chain

$$P_2(p)\, x(t) = x_1(t) ,$$

$$P_1(p)\, x_1(t) = Ky(t) .$$

We will now demonstrate this general chain approach for an interesting example from hydrodynamics.

Giving a relevant description of the transition phenomenon from laminar to turbulent flow in dependence on the Reynolds number $R$ has frequently been tried. There is a linear model which seems to reflect the corresponding phenomena correctly, described by the Orra-Sommerfeld equation

$$(u - c)\,(f^{(2)} - \alpha^2 f) = -\frac{\mathrm{j}}{\alpha R}\,(f^{(4)} - 2\alpha^2 f^{(2)} + \alpha^4 f) .$$

The corresponding characteristic equation has the following root distribution

$$a_{1/2} = \pm\alpha ,$$

$$a_{3/4} = \pm \sqrt{\alpha^2 + \mathrm{j}(u - c)\,\alpha R} = \pm r .$$

Thus the problem can be represented as follows

$$((p - r)(p + r))\, x_1 = 0\,,$$
$$((p - \alpha)(p + \alpha))\, x_0 = x_1\,.$$

Our standard approach gives us the following chain of two modules

$$Fx_1 = w\,,$$
$$\mathrm{d}w/\mathrm{d}t = -(w - r)(w + r) \quad \text{logistic (complex)},$$
$$Fx_0 = v\,,$$
$$\mathrm{d}v/\mathrm{d}t = \frac{x_1}{x_0} - (v - \alpha)(v + \alpha) \quad \text{logistic (real)}\,.$$

We will discuss briefly the character of the solution $x_1$ of the first module. We first represent $r$ as a complex number. We introduce

$$\alpha^2 = \varrho \cos \varphi \quad \text{and} \quad (u - c)\,\alpha R = \varrho \sin \varphi\,.$$

Then we have

$$\varrho = \sqrt{\alpha^4 + \alpha^2(u - c)^2 R^2}\,,$$
$$\tan \varphi = (u - c)\, R/\alpha\,.$$

For $r$ we then get the solution

$$r = \varrho^{1/2}(\cos \varphi/2 + \mathrm{j} \sin \varphi/2)\,.$$

We rewrite the first module into the logistic form

$$w = r - u\,,$$
$$Fx_1 = r - u\,,$$
$$\mathrm{d}u/\mathrm{d}t = -u(2r - u)\,.$$

Now we solve the logistic equation with complex *saturation value* $B$

$$\mathrm{d}u/\mathrm{d}t = Ku(B - u)$$

which yields the result

$$u(t) = \frac{BC \exp (BKt)}{1 + C \exp (BKt)}\,.$$

We recognise the following complex logistic behaviour

$$u(t) \begin{matrix} 0 & \text{for real part } (BK) > 0\,, \\ B & \text{for real part } (BK) < 0\,. \end{matrix}$$

In our case we see that $u(t) \to 0$, that means

$$x_1 \to \mathrm{e}^{rt} = \mathrm{e}^{\varrho^{1/2} \cos \varphi/2t}\, \mathrm{e}^{j\psi t}$$

with

$$\Psi = \varrho^{1/2} \sin \varphi/2\,.$$

The second module has in respect to the influence from $R$, no specific consequences, because we then get only a modified exponent

$$\varrho^{1/2} \cos \varphi/2 \pm \alpha\,.$$

Therefore, the exponent of the amplitude increases proportionally to $R^{1/2}$, whereas, with increasing $R$, the angle $\varphi/2$ approaches $\pi/4$. At the same time, the frequency of the oscillation increases too proportional to $R^{1/2}$. Therefore, the process is, with increasing $R$, mainly determined asymptotically by an unstable oscillation with an exponentially growing amplitude proportional to $R^{1/2}$, and a growing frequency, also proportional to $R^{1/2}$. This exposes a possible way to chaos, as is already known from other sources, for example [121].

Now we apply the Lotka-Volterra approach on second-order time-invariant linear systems. This is a modification of our approach for linear systems with constant parameters.

We introduce for the corresponding differential equation

$$\mathrm{d}^2x/\mathrm{d}t^2 + f(t)\,\mathrm{d}x/\mathrm{d}t + g(t)\,x = y(t)$$

with time-dependent damping function $f(t)$ and the time-dependent spring parameter $g(t)$, the state-variables, according to the linear system theory,

$$x_0 = x\,,$$
$$x_1 = \mathrm{d}x/\mathrm{d}t$$

and get then the linear state-space description

$$\mathrm{d}x_0/\mathrm{d}t = x_1\,,$$
$$\mathrm{d}x_1/\mathrm{d}t = -f(t)\,x_1 - g(t)\,x_0 + y\,.$$

Now we pass, similar to what we have done previously to the Lotka-Volterra description,

$$Fx_0 = x_{10}\,, \qquad Fx_1 = -f(t) - g(t)\,x_{01} + y_{01}\,,$$
$$Fx_{10} = -f(t) - g(t)\,x_{01} + y_{01} - x_{10}\,,$$
$$Fx_{01} = -Fx_{10}, \qquad Fy_{01} = Fy - Fx_1\,.$$

As a sufficient reduced description, we choose

$$Fx_0 = x_{10}\,,$$
$$\mathrm{d}x_{10}/\mathrm{d}t = \frac{y}{x_0} - \left(x_{10}^2 + f(t)\,x_{10} + g(t)\right)\,.$$

Now we have on the right-hand side, something like a characteristic polynomial with time-variant coefficients, and, therefore, we are able to determine the roots and split the characteristic polynomial into linear factors. This is remarkable because we would not do that if we wanted to apply Fourier or Laplace transformation.

For the corresponding (time-variant) roots we obviously get

$$a_{1/2} = -\frac{f(t)}{2} \pm \sqrt{\frac{f^2}{4} - g}\,,$$
$$a_{1/2} = a(t) \pm r(t)\,.$$

We introduce instead of $x_{10}$ a modified state-variable

$$w = x_{10} + \frac{f(t)}{2}$$

and get the following system

$$Fx_0 = w - f(t)/2\,,$$
$$\mathrm{d}w/\mathrm{d}t = \frac{y}{x_0} + \frac{\mathrm{d}f/\mathrm{d}t}{2} - (w^2 - r^2)\,.$$

In this case, that part which was formerly logistic is now represented in the form

$$\mathrm{d}w/\mathrm{d}t = H(t) - w^2$$

with

$$H(t) = \frac{\mathrm{d}f/\mathrm{d}t}{2} + \frac{f^2}{4} - g\,.$$

We already know that, for the case $H(t) \rightleftharpoons$ const, a closed analytical solution exists. Let us now clarify somewhat the properties of the Riccati equation

$$\mathrm{d}w/\mathrm{d}t = H(t) - w^2\,.$$

The solution to this equation can be got by application of the Riccati representation. We get first the following Lotka-Volterra equations

$$Fw = w_1 - w \quad \text{with} \quad w_1 = \frac{H(t)}{w}\,,$$
$$Fw_1 = FH - w_1 + w$$

which leads immediately to the following Riccati representation

$$\mathrm{d}u/\mathrm{d}t = u_1\,, \qquad w = u^{-1}u_1\,,$$
$$\mathrm{d}u_1/\mathrm{d}t = Hu\,, \qquad w_1 = Hu_1^{-1}u$$

or, equivalent to this, we get the following representation

$$\mathrm{d}^2u/\mathrm{d}t^2 - H(t)\,u = 0 \quad \text{with} \quad w = \frac{\mathrm{d}u/\mathrm{d}t}{u}\,.$$

This representation is well-known; but it was not studied in connection with a feedback module of the type presented above. This presentation is called in the literature the *normal form* of a Riccati equation. For the solution of this normal form, an interesting recursive procedure was proposed. We will here present this procedure, and, afterwards try to apply it to get the solution of our second-order feedback module. The iteration procedure acts immediately on the equation

$$\mathrm{d}w/\mathrm{d}t + w^2 = H(t)\,.$$

For it, we seek a solution in the form

$$w = w_0 + w_1 + w_2 + \dots$$

with some degrees of freedom in partitioning, which we use in the following special way

$$w_0^2 = H(t)\,,$$
$$2w_0w_1 + \mathrm{d}w_0/\mathrm{d}t = 0\,,$$
$$2w_0w_2 + w_1^2 + \mathrm{d}w_1/\mathrm{d}t = 0\,,$$
$$2w_0w_3 + 2w_1w_2 + \mathrm{d}w_2/\mathrm{d}t = 0\,,$$
$$\dots\dots\dots\dots\dots\dots$$
$$2w_0w_i + 2w_1w_{i-1} + \dots + \mathrm{d}w_{i-1}/\mathrm{d}t = 0\,.$$

This is a chain, from which we can recursively determine all summands $w_i$ of $w$.

We shall now see how we can apply this recursive procedure for our feedback module. Our problem has the following structure:

$$Fx_0 = w - G(t) ,$$

$$\mathrm{d}w/\mathrm{d}t + w^2 = H(t) = h(t) + \frac{y}{x_0} .$$

There are several possibilities, for example

1. We assume that we have already constructed

$$w^* = \sum_{i=0}^{n} w_i .$$

Then, we compute a new function $x_0(t)$ from the first equation, and determine all $w_i$ up to $i = n$ anew, in corrected form. We obtain an additional summand $w_{n+1}$ and reiterate this step.

2. We assume that we have already constructed

$$w^* = \sum_{i=0}^{n} w_i .$$

Then, we compute a new $x_0$ from the first equation and iterate now, for a fixed $n$, both equations up to stabilisation of the solutions. We then proceed to the next summand $w_{n+1}$.

We will try now to generalise our approach on the case of $n$-order time-variant linear differential equations. In this case, we do not have the possibility of splitting the characteristic equation into factors as we did for an $n$-order linear system with constant parameters.

First, we introduce state-variables in the same manner as it is usually done in linear system theory.

$$\left(\sum_{i=0}^{n} a_i(t)\, p^{n-i} x(t)\right) = y(t) ; \qquad a_0(t) \equiv 1 ,$$

$$p^i x = x_i , \qquad i = 0, 1, \dots , n-1 ,$$

with the consequence

$$\mathrm{d}x_i/\mathrm{d}t = x_{i+1} \quad \text{for} \quad i = 0, 1, \dots , n-2 ,$$

$$\mathrm{d}x_{n-1}/\mathrm{d}t = -\sum_{i=1}^{n} a_i(t)\, x_{n-i} + y(t) .$$

Now we pass to the Lotka-Volterra representation

$$Fx_i = x_{i+1,i} , \qquad i = 0, 1, \dots , n-2 , \qquad x_{i+1,i} = x_{i+1}/x_i ,$$

$$Fx_{n-1} = -\sum_{i=1}^{n} a_i(t)\, x_{n-i,n-1} + y/x_{n-1} ,$$

$$Fx_{i+1,i} = x_{i+2,i+1} - x_{i+1,i} , \qquad i = 0, 1, \dots , n-3 ,$$

$$Fx_{n-1,n-2} = -\sum_{i=1}^{n} a_i(t)\, x_{n-i,n-1} + y/x_{n-1} - x_{n-1,n-2} ,$$

$$x_{n-i,n-1} = x_{n-i,n-i+1} \cdot x_{n-i+1,n-i+2} \cdots x_{n-2,n-1} ,$$

$$Fx_{n-i,n-1} = -Fx_{n-1,n-2} - Fx_{n-2,n-3} - \dots - Fx_{n-i+1,n-i} .$$

In general, there is no simplification to be seen. Therefore, we study this more complicated situation for the cases $n = 3$ and $n = 4$.

$$x^{(3)} + f(t)\, x^{(2)} + g(t)\, x^{(1)} + h(t)\, x = y\,.$$

With $x_0 = x$, $x_1 = x^{(1)}$ and $x_2 = x_0^{(2)}$, we get the following state-variable description

$$dx_0/dt = x_1\,,$$

$$dx_1/dt = x_2\,,$$

$$dx_2/dt = -f(t)\, x_2 - g(t)\, x_1 - h(t)\, x_0 + y\,.$$

For the Volterra representation, we get

$$Fx_0 = x_{10}\,, \qquad Fx_1 = x_{21}\,,$$

$$Fx_2 = -f - gx_{12} - hx_{02} + y/x_2\,, \qquad y_{02} = y/x_2\,,$$

$$Fx_{10} = x_{21} - x_{10}\,, \qquad Fx_{21} = -f - gx_{12} - hx_{02} + y/x_2 - x_{21}\,,$$

$$Fx_{12} = -Fx_{21}\,, \qquad Fx_{02} = x_{10} + f + gx_{12} + hx_{02} - y/x_2\,.$$

We choose from this system the following reduced representation

$$Fx_0 = x_{10}\,, \qquad Fx_1 = x_{21}\,,$$

$$dx_{21}/dt = y/x_1 - h/x_{10} - (x_{21}^2 + fx_{21} + g)\,,$$

$$dx_{10}/dt = x_{10}x_{21} - x_{10}^2\,.$$

The last two equations of this are coupled Riccati equations.

If we assume $x_0$ and $x_1$ as given and also the summand $y/x_1$, the iterative procedure described above can be applied. Then, new values of $x_0$ and $x_1$ can be determined in a way similar to what has been said for the second-order system.

Let us now proceed to the fourth-order system

$$x^{(4)} + fx^{(3)} + gx^{(2)} + hx^{(1)} + mx = y\,.$$

For this case the same considerations can be repeated, and, at the end, we get the following equations, which can also be solved by at least two iteration levels, one for the state-variables $x_{10}$, $x_{21}$ and $x_{32}$ and the other for the state-variables $x_0$, $x_1$ and $x_2$

$$Fx_0 = x_{10}\,, \qquad Fx_1 = x_{21}\,, \qquad Fx_2 = x_{32}\,,$$

$$Fx_{10} = x_{21} - x_{10}\,,$$

$$Fx_{21} = x_{32} - x_{21}\,,$$

$$dx_{32}/dt = -(x_{32}^2 + fx_{32} + g) + y/x - h/x_{21} - m/(x_{10}x_{21})\,.$$

We can proceed in the same manner also for higher dimensions, but we have to expect that the Riccati iteration for the determination of approximated solutions will become more and more complicated.

As a possible application in control theory we design now a dynamic inverter on the base of an adaptive PD-controller. We consider a time-variant linear second-order differential equation

$$d^2x/dt^2 + f(t)\, dx/dt + g(t)\, x = y$$

arising from a feedback loop of a control line with constant parameters

$$d^2x/dt^2 + a\, dx/dt + bx = y$$

with an adaptive PD-controller with

$$y_c = \tilde{f}(t)\, dx/dt + \tilde{g}(t)\, x$$

so that the closed loop circuit has the equation

$$d^2x/dt^2 + a\, dx/dt + bx = y - \left(\tilde{f}(t)\, dx/dt + \tilde{g}(t)\, x\right) .$$

Therefore, we have

$$f(t) = a + \tilde{f}(t) \quad \text{and} \quad g(t) = b + \tilde{g}(t) .$$

Our aim is the design of the adaptive controller, meaning the determination of effective expressions for the controller functions $\tilde{f}(t)$ and $\tilde{g}(t)$. For this purpose, we first transform the considered differential equation into a state-variable description as it is usually done

$$x_0 = x , \qquad dx_0/dt = x_1 , \qquad dx_1/dt = -f(t)\, x_1 - g(t)\, x_0 + y .$$

Now we pass to nonlinear equations using the ideas from the Lotka-Volterra approach

$$Fx_0 = x_{10} ,$$

$$dx_{10}/dt = \frac{y}{x_0} - \left(v^2 + f(t)\, v + g(t)\right) .$$

Using the roots of the characteristic equations, we introduce a modified state-variable in place of $x_{10}$

$$w = x_{10} + f(t)/2 .$$

Now we get the following clear description

$$Fx_0 = w - f(t)/2 ,$$

$$\begin{aligned} dw/dt &= \frac{df/dt}{2} + \frac{y}{x_0} + \frac{f(t)^2}{4} - g(t) - w^2 , \\ &= H(t) - w^2 \end{aligned}$$

with

$$H(t) = \frac{df/dt}{2} + \frac{y}{x_0} + \frac{f(t)^2}{4} - g(t) .$$

The second equation is a Riccati equation with the important attraction property that if $H(t) > 0$, then $\sqrt{H(t)}$ is a stable attractor for $w$, which is moving in time.

From the state $w(t)$ the distance to the stable sink $\sqrt{H}$ will be $\sqrt{H} - w$, and $dw/dt$ is then the velocity of approaching this sink. An estimation for the time distance to the sink is then obviously given by

$$(\sqrt{H} - w)/(dw/dt) = 1/(\sqrt{H} + w) .$$

This means that, for approaching the stable sink very quickly, it is useful to have large positive values of $H(t)$. We propose to achieve high values of $H(t)$ by choosing a large but *constant* value for $f(t)$

$$f(t) = a + K_2 .$$

We hope that, by adjusting $g(t)$ in the right way, we can very quickly follow a given signal $y(t)$ with the output signal $x(t)$. Under these assumptions, we have

$$H(t) \sim 1 + \frac{f(t)^2}{4} - g(t)\,, \qquad w \sim \sqrt{H(t)}\,,$$

and, most importantly we should have

$$\frac{\mathrm{d}y/\mathrm{d}t}{y} \sim \sqrt{H(t)} - f(t)/2\,.$$

From these considerations, we get the following condition for the adaptation of $g(t)$:

$$g(t) \sim 1 - \frac{(\mathrm{d}y/\mathrm{d}t)^2}{y} - f\,\frac{(\mathrm{d}y/\mathrm{d}t)}{y}$$

and

$$H(t) \sim \frac{f^2}{4} + \frac{(\mathrm{d}y/\mathrm{d}t)^2}{y} + f\,\frac{(\mathrm{d}y/\mathrm{d}t)}{y}\,.$$

To avoid compensation phenomena and a reduction of the necessary value of $H(t)$, we demand

$$f > \left|\frac{\mathrm{d}y/\mathrm{d}t}{y}\right|.$$

The proportional part of the controller must then be changed according to the function

$$g(t) = 1 - b - (Fy)^2 - fFy\,.$$

After substitution of this result into the equation of the closed circuit, we get

$$\mathrm{d}^2x/\mathrm{d}t^2 + f\,\mathrm{d}x/\mathrm{d}t + (1 - Fy^2 - fFy)\,x = y$$

as a solution which provides a feed-forward influence of $y$ on the parameter of the proportional part of the controller.[1] Obviously we get the following error signal estimation for the closed circuit

$$\Delta = \frac{\mathrm{d}^2y/\mathrm{d}t^2}{y} - Fy^2 = \mathrm{d}^2 \ln y/\mathrm{d}t^2$$

(relative error in percent of the input signal $y(t)$). If we admit only small relative errors, we must demand that this logarithmic curvature of the input signal $y(t)$ lies within the following bounds

$$-\Delta < \mathrm{d}^2 \ln y/\mathrm{d}t^2 < \Delta\,.$$

This is a local condition in every reference point $(t_0, y(t_0))$, and it defines curved bounded local cones for the derivative $\mathrm{d}y/\mathrm{d}t$. The form of these cones can easily be determined by integration of these inequalities in the neighbourhood of any reference point. It is worth mentioning, that, for all exponentials $y(t) = K \exp(at)$, no error of the closed circuit occurs. One could ask why we propose this adaptation and why we do not simply use the following expression

$$\mathrm{d}^2x/\mathrm{d}t^2 + f\,\mathrm{d}x/\mathrm{d}t + \left(1 - \frac{\mathrm{d}^2y/\mathrm{d}t^2}{y} - fFy\right)x = y\,.$$

---

[1]) We thank Dr. Lunze and Dr. Billerbeck, ZKI, for checking the proposal for a follower of a control system by theoretical studies and computer simulations.

This is apparently an equation which is identically fulfilled for $y = x$. However, we are convinced that this adaptive procedure could not work very well because we have, in this case, no such convenient attraction property as we obtained before by passing to the corresponding Riccati equation. We think that this last approach would fail because of numerical instability. Nevertheless, this approach should also be checked in comparison with the Volterra approach proposed above in concrete simulation studies.

We will now complete the considerations for our approach by giving some information about the stability behaviour.

— Stability analysis for input signals $y = K\,e^{\lambda t}$ with $\lambda$ real. The equation for the closed loop system

$$d^2x/dt^2 + f\,dx/dt + (1 - Fy^2 - fFy)\,x = y = K\,e^{\lambda t}$$

then has the form

$$d^2x/dt^2 + f\,dx/dt + (1 - \lambda^2 - f\lambda)\,x = K\,e^{\lambda t}\,.$$

This is a linear time-invariant system with the following stability-property: The system is stable if and only if

$$-\frac{f}{2} - \sqrt{\frac{f^2}{4} + 1} \leqq \lambda \leqq -\frac{f}{2} + \sqrt{\frac{f^2}{4} + 1}\,.$$

Therefore, the class of exponentials for which the closed circuit is stable is restricted; the upper level of $\lambda$ is smaller than 1. The lower level is the less restricted the larger the value of $f$ is.

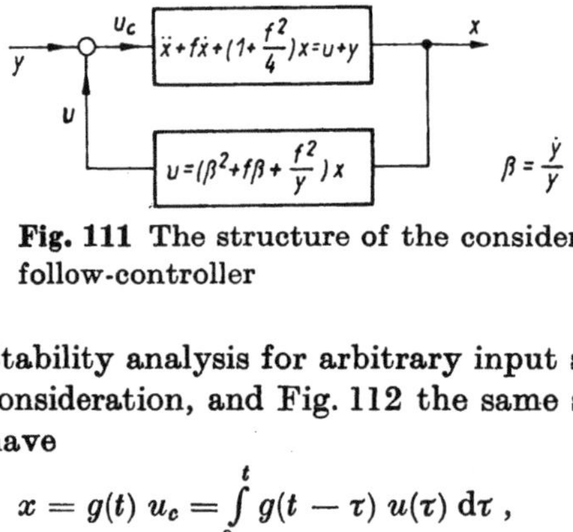


**Fig. 111** The structure of the considered follow-controller

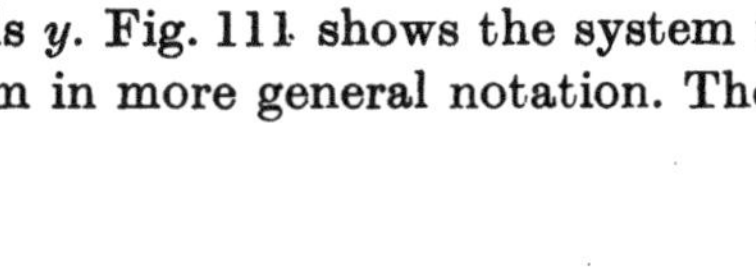


**Fig. 112** The structure of the follow-controller in more general notation

— Stability analysis for arbitrary input signals $y$. Fig. 111 shows the system under consideration, and Fig. 112 the same system in more general notation. Then we have

$$x = g(t)\,u_c = \int_0^t g(t - \tau)\,u(\tau)\,d\tau\,,$$

$$g(t) = 0 \quad \text{for} \quad t < 0\,,$$

$$u(t) = 0\,,$$

$$u = k(t) \cdot x\,,$$

$$u_c = u + y\,.$$

By upper estimations of the subsystems we are looking now for a majorant of the closed circuit.

Then we have

$$|x(t)| \leqq \bar{g}(t) * |y(t)|$$

with

$$\bar{g}(t) = |g(t)| + |g(t)| * \bar{k} * \bar{g}(t)$$

with $\bar{k} \geqq |k(t)|$ for all $t$. For the majorant system, we can check the stability by the following criterion

$\bar{g}(t)$ represents a stable system if the following holds

$$\int_0^\infty |g(t)| \, dt \cdot \bar{k} < 1$$

because of $|x(t)| \leqq \bar{g} * |u|$, then the original system is also stable in the input-output behaviour.

We are now able to apply this result to our problem. For the forward plant, we have

$$G(s) = x(s)/u_a(s) = 1 \Big/ \left(s^2 + fs + \frac{f^2}{4} + 1\right) = 1/(1 + (s + f/2)^2)$$

and

$$g(t) = e^{-f/2} \sin t \, .$$

The consequence is

$$\int_0^\infty |g(t)| \, dt \leqq \int_0^\infty e^{-f/2} \, dt = 2/f \, .$$

For the backward plant, we get

$$k(t) = \beta^2(t) + f\beta(t) + \frac{f^2}{4} \quad \text{with} \quad \beta = Fy \, .$$

The consequence is

$$|k(t)| \leqq \bar{k} = \left(\bar{\beta} + \frac{f}{2}\right)^2 \quad \text{with} \quad |\beta(t)| \leqq \bar{\beta} \, .$$

The stability condition by upper estimation leads to the condition

$$\int_0^\infty |g(t)| \, dt \, \bar{k} \leqq \frac{2}{f} \left(\bar{\beta} + \frac{f}{2}\right)^2 < 1$$

or

$$\bar{\beta}^2 + f\bar{\beta} + \frac{f^2}{4} - \frac{f}{2} < 0$$

with the consequence

$$|Fy| \leqq -\frac{f}{2} + \sqrt{\frac{f}{2}} \, .$$

The boundary for $|Fy|$ must be always smaller than 1/4.

— The behaviour of the closed system as a follow plant. We demand for the closed system

$$d^2x/dt^2 + f \, dx/dt + (1 - Fy^2 - fFy) \, x = y$$

that holds

$$x(t) \sim y(t) \, .$$

We put

$$x(t) = y(t) + e(t)$$

with the follow error $e(t)$ obeying the differential equation

$$\mathrm{d}^2e/\mathrm{d}t^2 + f\,\mathrm{d}e/\mathrm{d}t + (1 - Fy^2 - fFy)\,e = -y\,\mathrm{d}Fy/\mathrm{d}t\,.$$

The follow-error approaches 0 for $y(t) = K\,\mathrm{e}^{\lambda t}$ with

$$1 - \lambda^2 - f\lambda > 0\,.$$

## 9.2. *A System Study For an Adaptive Follow-Controller*

We connect this approach for a *hyperlogistic controller* with the idea from Section 9.1 to make use of the good follow-property of the Riccati equation

$$\mathrm{d}w/\mathrm{d}t = H(t) - w^2 \quad \text{for} \quad H(t) > 0\,.$$

Our purpose is to elaborate a hardware variant for a nonlinear follow-controller as a generalisation of the wellknown PID-concept from control theory.

We introduce some new ideas for the design of a hyperlogistic controller as basis for a future hardware realisation.

### *Some Ideas for the Design of the Hyperlogistic Controller*

At first we generalise the follower-property of the equation

$$\mathrm{d}w/\mathrm{d}t = H(t) - w^2$$

by passing to the hyperlogistic differential equation

$$\mathrm{d}z/\mathrm{d}t = Kz^k(H(t) - z^w)^l\,.$$

As we know from Section 1.11 this is a suitable model for the EVOLON transition, meaning that for positive $H(t)$ the state-variable $z(t)$ is approaching monotonously $(H(t))^{1/w}$ in an $s$-formed transition and offers to us all the richness of dynamic properties contained within the EVOLON model. We apply this hyperlogistic differential equation as integrator-like plant in the controller using $z$ as the actual control influence on the control system. Now we are going to implement the aim of the control into the moving saturation level $H(t)$. This can be done in several different ways depending on the concrete control aim. We give here only some possible examples for the realisation of this aim. $y$ may be the input signal and $x$ the corresponding output signal of a one-dimensional control system, that means

$$x = Gy$$

with $G$ being the transfer operator of the control system. For a follow-controller we put

$$H(t) = (y + z - x)^w\,.$$

We demand that very quickly $z$ approaches $y + z - x$ or reversely, meaning that follows

$$x \sim y\,.$$

For an integral-like controller we put

$$H(t) = (x_s + z - x)^w .$$

Herein $x_s(t)$ is the demanded trajectory of the control-variable $x$. We demand that very quickly $z$ approaches $x_s + z - x$ or reversely, meaning that follows

$$x \sim x_s .$$

For a proportional-like controller we put

$$H(t) = (\mathrm{d}x_s/\mathrm{d}t + z - \mathrm{d}x/\mathrm{d}t)^w .$$

We demand that very quickly $z$ approaches $\mathrm{d}x_s/\mathrm{d}t + z - \mathrm{d}x/\mathrm{d}t$ or reversely, meaning that follows

$$\mathrm{d}x/\mathrm{d}t \sim \mathrm{d}x_s/\mathrm{d}t .$$

It is necessary to take into account, that the hyperlogistic plant itself shows the behaviour of an integrator. It is not automatically guaranteed that $z$ follows with high quality after the signal $\left(H(t)\right)^{1/w}$ in any case. We must introduce special measures for achieving this property for a suitable hardware realisation. Firstly we must admit that not $z$ follows $\left(H(t)\right)^{1/w}$ but reversely it may occur that $\left(H(t)\right)^{1/w}$ follows after $z$ depending on the signs of these expressions. Further on, we cannot exclude that the arguments of the power functions change the sign. This leads usually to serious problems in the numerical evaluation of the power function.

To get a realistic variant of the hyperlogistic controller we introduce a modified power function. For this purpose we follow the considerations:

- $z^k$ should be used only to strengthen the autocatalytic effect especially to influence the velocity of change of the growth rate $\mathrm{d}z/\mathrm{d}t : z$.

Therefore, $z^k$ must not change its sign. Here we introduce the following modified power function

$$\overline{z^k} = \begin{cases} z^k & \text{for} \quad z > \mathrm{EPS} , \\ 0 & \text{for} \quad -\mathrm{EPS} \leqq z \leqq \mathrm{EPS} , \\ (-z)^k & \text{for} \quad z < -\mathrm{EPS} . \end{cases}$$

The parameter EPS determines a certain interval of insensitivity of the controller. On one hand it gives the controller some robustness, on the other hand, from this insensitivity interval some serious problems may arise.

- On the other hand the controller increment should be able to change its sign. Therefore we demand the sign-changing property from the expression within the brackets of the hyperlogistic controller.

For the realisation of the change of the sign we introduce the following modified power function

$$\overline{\overline{z^p}} = \begin{cases} z^p & \text{for} \quad z > \mathrm{EPS} , \\ 0 & \text{for} \quad -\mathrm{EPS} \leqq z \leqq \mathrm{EPS} , \\ -(-z)^p & \text{for} \quad z < -\mathrm{EPS} . \end{cases}$$

This concept we have to apply three times within the bracket expression.

— From these considerations we get the following controller concept

$$\mathrm{d}z/\mathrm{d}t = \overline{Kz^k}\,\overline{\overline{(\overline{(\Delta + z)^w} - \overline{\overline{z^w}})}}^{\,l}$$

$\Delta$ is the corresponding difference signal depending on the concrete control aim.

— If $z^k$ or a term in the brackets run into the insensitivity interval$(-\mathrm{EPS}, \mathrm{EPS})$ the corresponding term will vanish.
If for example $z$ itself will be within the insensitivity interval, it can never leave this interval, meaning that the controller will be out of function. In such a case we need a measure to restart the controller.

This can be done by chosing a random increment, for example,

LET $z = \sigma \cdot (2\,\mathrm{RND} - 1)$

where RND is a uniformly distributed random variable in $0, +1$.

Now we discuss how the hyperlogistic controller can substitute the classical PID-regulator. The classical PID-regulator consists of three parallel parts from the functional point of view

— proportional part $z_1 = K_P\Delta$,
— integral part $\mathrm{d}z_2/\mathrm{d}t = K_I\Delta$,
— differential part $z_3 = K_D\,\mathrm{d}\Delta/\mathrm{d}t$.

The hyperlogistic controller offers only integrally acting plants. A hyperlogistic part with $H(t) = (\mathrm{d}\Delta/\mathrm{d}t + z)^w$ generates something like a proportional part after integrating. Because of the integral constant it is possible that $z$ stabilizes on a level which is able to compensate a large rest-$\Delta$ without forcing $\Delta$ to vanish. Therefore, in any case we need a generalisation of the pure proportional part. If we are interesting also in this case to apply the hyperlogistic principle we should introduce the corresponding proportional part in the following way

$$z_1 = K_P\overline{z^k}\,\overline{\overline{(\overline{(\Delta + z)^w} - \overline{\overline{z^w}})}}^{\,l} \quad \text{with} \quad \Delta = x_s - x\,.$$

This is obviously a one-dimensional nonlinear mapping with a unimodal right hand side. But we know that these mappings can produce a rich dynamics. The most simple case

$$z_1 = K_P z(B - z)$$

already leads by iterative application to a bifurcation tree and the famous Feigenbaum sequence. Therefore, our proposed proportional part must be studied in detail with the aim either to avoid the bifurcation phenomena or to make use of them.

Having thus agreed about the proportional part of the controller we can now construct a hyperlogistic analogy to the classical PID-regulator in the following way.

- proportional part $z_1' = K_P\overline{z_1^k}\,\overline{\overline{(\overline{(\Delta_1 + z_1)^w} - \overline{\overline{z_1^w}})}}^{\,l}, \quad \Delta_1 = x_s - x$
- integral part $\mathrm{d}z_2/\mathrm{d}t = K_I\overline{z_2^k}\,\overline{\overline{(\overline{(\Delta_2 + z_2)^w} - \overline{\overline{z_2^w}})}}^{\,l}, \quad \Delta_2 = x_s - x$
- differential part $\mathrm{d}z_3/\mathrm{d}t = K_D\overline{z_3^k}\,\overline{\overline{(\overline{(\Delta_3 + z_3)^w} - \overline{\overline{z_3^w}})}}^{\,l}, \quad \Delta_3 = \ddot{x}_s - \ddot{x}.$

In the different plants several different sets of parameters might be used.

In future research the most suitable parameter configurations must be identified and rules must be derived for an adaptive feedback from the control loop variables to the choice of good parameter configurations of the controller. Fig. 113 shows how the hyperlogistic controller cooperates with the control plant in a closed loop regime.

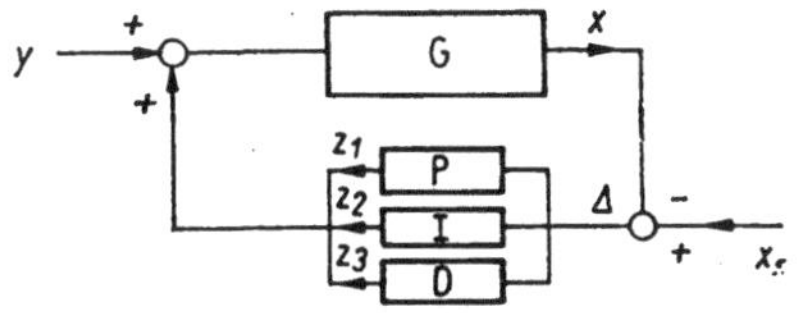


**Fig. 113** Scheme of a control circuit with a hyperlogistic controller in the feedback loop

By applying the hyperlogistic controller the control plant now gets the modified input signal $y + z_1' + \varkappa z_2 + \lambda z_3$ which generates the control variable $x$ according to the relationship

$$x = G(y + z_1' + \varkappa z + \lambda z_3) .$$

Here $\varkappa$ and $\lambda$ are another pair of adjustable control parameters.

## 9.3. *Control of Robots Using Lotka-Volterra Models*[1])

In the mathematical analysis of industrial robots, we meet some types of nonlinear dynamic systems. These are, for example

— the equations of motions of the mechanical systems
— the coordinate transformations
— the control algorithms

We need, for the description of mechanical systems with exogenous driving forces, non-homogenous Lotka-Volterra equations of the form

$$Fx_i = \sum G_{ij}x_j + B_i + C_i(y) , \qquad i = 1, 2, \dots , n .$$

$x_i$ are the state-variables of the system, $y$ is the vector of the input signals, $G_{ij}$ are the competition coefficients, $B_i$ are resource constants and $C_i(y)$ are certain functional relations between the components of the input vector $y$.

Because of the influence of our structure design principle, we get for $C_i$ expressions of the following form, in the case of mechanical systems

$$C_i(y) = \sum (C_{ij}^1 y_j + C_{ij}^2 F y_j + C_{ij}^3 F \, \mathrm{d}y_j/\mathrm{d}t) .$$

Usually, the matrices $C_{ij}^r$, $r = 1, 2, 3,$ and the vector $B$ are sparse matrices. Now we shall analyse the character of the equations of motions of mechanical systems. For a robot with $n$ degrees of freedom the equations of motions, expressed in generalised coordinates are

$$M(q) \, \mathrm{d}^2q/\mathrm{d}t^2 + N(q, \mathrm{d}q/\mathrm{d}t) = Q .$$

---

[1]) We thank Dr. Jansen, ZKI, for providing us with his study in applying the Lotka-Volterra approach on joint-robots.

Here is $Q$ the vector of all applied forces respectively of the rotation moments. $M(q)$ is a state-dependent invertable matrix. The components of $N(q, dq/dt)$ are quadratic forms in the $q_i$ with coefficients depending on the $dq_i/dt$.

With the usual state-variable approach, we get, from the given system of the equations of motions, the following state-space description

$$dp/dt = M^{-1}(q)\,\bigl(Q - N(q, dq/dt)\bigr)\,, \qquad dq/dt = p\,.$$

To this system of equations, we can now apply the structure design principle.

We now consider the equations of motion for a joint robot with two degrees of freedom.

The corresponding mechanical equations are

$$A_{11}\,d^2q_1/dt^2 \qquad\qquad + B_{121}\,dq_1/dt\,dq_2/dt + B_{122}(dq_2/dt)^2 = Q_1\,,$$
$$A_{21}\,d^2q_1/dt^2 + A_{22}\,d^2q_2/dt^2 + B_{211}(dq_1/dt)^2 = Q_2 + m_2 g l_{11}\sin q_2$$

with the following expressions for the coefficient functions

$$\begin{aligned}
A_{11} &= \theta_1^{11} + \theta_2^{11} + (\theta_2^{22} - \theta_2^{11})\sin^2 q_2 + m_2(l_{23}^2 + l_{21}^2\sin^2 q_2) + m_1 l_{13}^2\,,\\
A_{21} &= -m_2 l_{21} l_{23}\cos q_2\,,\\
A_{22} &= \theta_2^{33} + m_2 l_{21}^2\,,\\
B_{121} &= (\theta_2^{22} - \theta_2^{11} + m_2 l_{21}^{21})\sin q_2\cos q_2\,,\\
B_{122} &= m_2 l_{21} l_{23}\sin q_2\,,\\
B_{211} &= -B_{121}\,.
\end{aligned}$$

The quantities $m$, $l$, and $\theta$ designate geometrical and mechanical constants of the robot. $g$ is the earth acceleration. We introduce the following notations

$$\begin{aligned}
s_1 &= \theta_1^{11} + \theta_2^{11} + m_2 l_{23}^2 + m_1 l_{13}^2\,,\\
s_2 &= \theta_2^{22} - \theta_2^{11} + m_2 l_{21}^2\,,\\
s_3 &= m_2 g l_{21}\,,\\
s_6 &= m_2 l_{21} l_{23}\,,\\
s_{12} &= \theta_2^{33} + m_2 l_{21}^2\,,\\
s_4 &= s_6^2/(s_2\cdot s_{12})\,.
\end{aligned}$$

Now we can use our structure design principle to get Lotka-Volterra equations for this robot system. First we get expressions for the state-variables $y_i$ which must be introduced in addition

$$y_1 = Fq_1 = p_1/q_1\,, \qquad y_2 = Fq_2 = p_2/q_2\,,$$

$$Fp_1 = -\frac{B_{121}}{A_{11}}p_2 - \frac{B_{122}}{A_{11}}\frac{p_2^2}{p_1} + \frac{Q_1}{A_{11}}\frac{1}{p_1} = y_3 + y_4 + y_5\,,$$

$$\begin{aligned}
Fp_2 &= \frac{A_{21}}{A_{22}}\frac{B_{121}}{A_{11}}p_1 + \frac{A_{21}}{A_{22}}\frac{B_{122}}{A_{11}}p_2 - \frac{A_{21}}{A_{22}}\frac{Q_1}{A_{11}}p_2 -\\
&\quad - \frac{B_{121}}{A_{22}}\frac{p_1^2}{p_2} + \frac{s_3}{A_{22}}\frac{\sin q_2}{p_2} + \frac{Q_2}{A_{22}}\frac{1}{p_2}\\
&= y_6 + s_4 y_3 + y_7 + y_8 + y_9 + y_{10}\,,
\end{aligned}$$

$$y_{11} = p_2 \frac{\cos q_2}{\sin q_2}, \quad y_{12} = p_2 \frac{\sin q_2}{\cos q_2}, \quad y_{13} = \frac{2s_2 \sin q_2 \cos q_2}{s_1 + s_2 \sin^2 q_2} = FA_1 .$$

With this notation, we get the following Lotka-Volterra equations:

$$\begin{aligned}
Fq_1 &= y_1 ,\\
Fq_2 &= y_2 ,\\
Fp_1 &= y_3 + y_4 + y_5 ,\\
Fp_2 &= s_4 y_3 + y_6 + y_7 + y_8 + y_9 + y_{10} ,\\
Fy_1 &= -y_1 + y_3 + y_4 + y_5 ,\\
Fy_2 &= - y_2 + s_4 y_3 + y_6 + y_7 + y_8 + y_9 + y_{10} ,\\
Fy_3 &= s_4 y_3 + y_6 + y_7 + y_8 + y_9 + y_{10}\\
&\quad - y_{12} - y_{13} ,\\
Fy_4 &= (2s_4 - 1)\, y_3 - y_4 - y_5 + 2y_6 + 2y_7 + 2y_8 + 2y_9 + 2y_{10}\\
&\quad + y_{11} - y_{13} ,\\
Fy_5 &= - y_3 - y_4 - y_5 - y_{13} + FQ_1 ,\\
Fy_6 &= y_3 + y_4 + y_5 + 2y_{11} - y_{12} - y_{13} ,\\
Fy_7 &= - s_4 y_3 - y_6 - y_7 - y_8 - y_9 - y_{10}\\
&\quad - y_{12} - y_{13} + FQ_1 ,\\
Fy_8 &= (2 - s_4)\, y_3 + 2y_4 + 2y_5 - y_6 - y_7 - y_8 - y_9 - y_{10} + y_{11}\\
&\quad - y_{12} ,\\
Fy_9 &= - s_4 y_3 - y_6 - y_7 - y_8 - y_9 - y_{10} + y_{11} ,\\
Fy_{10} &= - s_4 y_3 - y_6 - y_7 - y_8 - y_9 - y_{10} + FQ_2 ,\\
Fy_{11} &= s_4 y_3 + y_6 + y_7 + y_8 + y_9 + y_{10} - y_{11}\\
&\quad - y_{12} ,\\
Fy_{12} &= s_4 y_3 + y_6 + y_7 + y_8 + y_9 + y_{10} + y_{11}\\
&\quad + y_{12} ,\\
Fy_{13} &= s_4 y_3 + y_6 + y_7 + y_8 + y_9 + y_{10} + y_{11}\\
&\quad - y_{12} - y_{13} .
\end{aligned}$$

We could introduce an additional state-variable $y_{14} = s_4 y_3$ with $Fy_{14} = Fy_3$, so that we would meet, on the right hand sides of the Lotka-Volterra equations, only integer coefficients.

We did not achieve a regular Lotka-Volterra system in the structure design. Therefore, if $q_1$, $q_2$, $p_1$, $p_2$, $\sin q_1$, $\cos q_2$, $Q_1$, $Q_2$ pass zero, the corresponding state-variables $y_1$, $y_2$, $y_4$, $y_5$, $y_8$, $y_9$, $y_{10}$, $y_{11}$ and $y_{12}$ would expose parasitic singularities which greatly restrict in simulation experiments. Because of $A_{11} > 0$ and $A_{22} > 0$ the additional state-variables $y_3$, $y_6$, $y_7$, $y_{13}$ never get such singularities. The mechanical model can be simplified without a great influence on the accuracy if we put all coefficients $B$ in the equations equal to zero. Then, the additional state-variables $y_3$, $y_4$, $y_6$ and $y_7$ can be omitted. For this system simplified in such a way, we will now derive regular Lotka-Volterra equations.

We have now

$$A_{11}\,\mathrm{d}p_1/\mathrm{d}t = Q_1\,,$$

$$A_{21}\,\mathrm{d}p_1/\mathrm{d}t + A_{22}\,\mathrm{d}p_2/\mathrm{d}t = Q_2 + s_3 \sin q_2\,.$$

For the regularisation approach, we will assume that all variables $q_1$, $q_2$, $p_1$, $p_2$, $Q_1$ and $Q_2$ for technological reasons, are bounded by resp. $\bar{q}_1$, $\bar{q}_2$, ... .

Further on, we introduce $\bar{s}$, $\bar{c} > 1$ and put $\bar{A}_{21} = \bar{c}s_6$. Then, we pass, in the mechanical equations, to shifted variables according to the substitutions

$$\tilde{q}_1 = q_1 + \bar{q}\,, \quad \tilde{q}_2 = q_2 + \bar{q}\,, \quad \tilde{p}_1 = p_1 + \bar{p}\,, \quad \tilde{p}_2 = p_2 + \bar{p}_2\,,$$

$$\tilde{Q}_1 = Q_1 + \bar{Q}_1\,, \quad \tilde{Q}_2 = Q_2 + \bar{Q}_2\,, \quad \tilde{A}_{21} = A_{21} + \bar{A}_{21} \quad \text{(with new notations)},$$

$$\sin q_2 + \bar{s} \quad \text{and} \quad \cos q_2 + \bar{c} \qquad \text{(without new notations)}.$$

Then we get the following shifted equations of motion

$$A_{11}\,\mathrm{d}\tilde{p}_1/\mathrm{d}t = \tilde{Q}_1 - \bar{Q}_1\,,$$

$$\tilde{A}_{21}\,\mathrm{d}\tilde{p}_1/\mathrm{d}t - \bar{A}_{21}\,\mathrm{d}\tilde{p}_1/\mathrm{d}t + A_{22}\,\mathrm{d}\tilde{p}_2/\mathrm{d}t = \tilde{Q}_2 - \bar{Q}_2 + s_3(\sin q_2 + \bar{s}) - s_3\bar{s}\,.$$

We now introduce additional state-variables, which must have all non-negative values.

$$Fq_1 = \frac{\tilde{p}_1 - \bar{p}_1}{\tilde{q}_1} = y_1 - y_2 \quad \text{with} \quad y_1 = \frac{\tilde{p}_1}{\tilde{q}_1}\,, \qquad y_2 = \frac{\bar{p}_1}{\tilde{q}_1}\,,$$

$$Fq_2 = \frac{\tilde{p}_2 - \bar{p}_2}{\tilde{q}_2} = y_3 - y_4 \quad \text{with} \quad y_3 = \frac{\tilde{p}_2}{\tilde{q}_2}\,, \qquad y_4 = \frac{\bar{p}_2}{\tilde{q}_2}\,,$$

$$Fp_1 = y_5 - y_6 \quad \text{with} \quad y_5 = \frac{\tilde{Q}_1}{\tilde{p}_1 A_{11}}\,, \qquad y = \frac{\bar{Q}_1}{\tilde{p}_1 A_{11}}\,,$$

$$Fp_2 = y_7 + y_8 - y_9 - y_{10} + y_{11} + y_{12} - y_{13}$$

with

$$y_7 = \frac{\tilde{Q}_2}{\tilde{p}_2 A_{22}}\,, \qquad y_8 = \frac{s_3}{A_{22}} \frac{\sin q_2 + \bar{s}}{\tilde{p}_2}\,, \qquad y_9 = \frac{\bar{Q}_2 + s_3 s}{\tilde{p}_2 \cdot A_{22}}\,,$$

$$y_{10} = \frac{\tilde{A}_{21}\tilde{Q}_1}{\tilde{p}_2 A_{22} A_{11}}\,, \qquad y_{11} = \frac{\tilde{A}_{21}\bar{Q}_1}{\tilde{p}_2 A_{22} A_{11}}\,,$$

$$y_{12} = \frac{\tilde{A}_{21} Q_1}{\tilde{p}_2 A_{22} A_{11}}\,, \qquad y_{13} = \frac{\bar{A}_{21} Q_1}{\tilde{p}_2 A_{22} A_{11}}\,.$$

Further on, we introduce the following state-variables

$$y_{14} = \tilde{p}_2 \frac{\cos q_2 + \bar{c}}{\sin q_2 + \bar{s}}\,, \qquad y_{15} = \bar{p}_2 \frac{\cos q_2 + \bar{c}}{\sin q_2 + \bar{s}}\,, \qquad y_{16} = \tilde{p}_2 \frac{\bar{c}}{\sin q_2 + \bar{s}}\,,$$

$$y_{17} = \frac{\tilde{p}_2 \bar{c}}{\sin q_2 + \bar{s}}\,, \qquad y_{18} = \tilde{p}_2 \frac{\sin q_2 + \bar{s}}{\cos q_2 + \bar{c}}\,, \qquad y_{19} = \bar{p}_2 \frac{\sin q_2 + \bar{s}}{\cos q_2 + \bar{c}}\,,$$

$$y_{20} = \tilde{p}_2 \frac{\bar{s}}{\cos q_2 + \bar{c}}, \qquad y_{21} = \bar{p}_2 \frac{\bar{s}}{\cos q_2 + \bar{c}},$$

$$y_{22} = \tilde{p}_2 \frac{2s_2(\sin q_2 + \bar{s})(\cos q_2 + \bar{c})}{s_1 + s_2 \sin^2 q_2}, \qquad y_{23} = \bar{p}_2 \frac{2s_2(\sin q_2 + \bar{s})(\cos q_2 + \bar{c})}{s_1 + s_2 \sin^2 q_2},$$

$$y_{24} = \tilde{p}_2 \frac{2s_2(\sin q_2 + \bar{s})\,\bar{c}}{s_1 + s_2 \sin^2 q_2}, \qquad y_{25} = \bar{p}_2 \frac{2s_2(\sin q^2 + \bar{s})\,\bar{c}}{s_1 + s_2 \sin^2 q_2},$$

$$y_{26} = \tilde{p}_2 \frac{2s_2\bar{s}(\cos q_2 + \bar{c})}{s_1 + s_2 \sin^2 q_2}, \qquad y_{27} = \bar{p}_2 \frac{2s_2\bar{s}(\cos q_2 + \bar{c})}{s_1 + s_2 \sin^2 q_2},$$

$$y_{28} = \tilde{p}_2 \frac{2s_2\overline{sc}}{s_1 + s_2 \sin^2 q_2}, \qquad y_{29} = \bar{p}_2 \frac{2s_2\overline{sc}}{s_1 + s_2 \sin^2 q_2}.$$

For the purpose of rational simulation, we introduce the following notations:

$$z_1 = y_1 - y_2\,, \qquad z_3 = y_5 - y_4\,, \qquad z_5 = y_5 - y_6\,,$$
$$z_7 = y_7 + y_8 - y_9 - y_{10} + y_{11} + y_{12} - y_{13}\,, \qquad z_{14} = y_{14} - y_{15} - y_{16} + y_{17}\,,$$
$$z_{18} = y_{18} - y_{19} - y_{20} + y_{21}\,, \qquad z_{12} = y_{22} - y_{23} - y_{24} + y_{25} - y_{26} + y_{27} + y_{28} - y_{29}$$

and get the following regular Lotka-Volterra representation

$$F\tilde{q}_1 = z_1\,, \qquad F\tilde{q}_2 = z_3\,, \qquad F\tilde{p}_1 = z_5\,, \qquad F\tilde{p}_2 = z_7\,,$$
$$Fy_1 = -z_1 + z_5\,, \qquad Fy_2 = -z_1\,, \qquad Fy_3 = -z_3 + z_7\,, \qquad Fy_4 = -z_3\,,$$
$$Fy_5 = -z_5 - z_{22} + F\widetilde{Q}_1\,, \qquad Fy_6 = -z_5 - z_{22}\,, \qquad Fy_7 = -z_7 + F\widetilde{Q}_2\,,$$
$$Fy_8 = -z_7 + z_{14}\,, \qquad Fy_9 = -z_7\,, \qquad Fy_{10} = -z_7 - z_{18} - z_{22} + F\widetilde{Q}_1\,,$$
$$Fy_{11} = -z_7 - z_{18} - z_{22}\,, \qquad Fy_{12} = -z_7 - z_{22} + F\widetilde{Q}_1\,, \qquad Fy_{13} = -z_7 - z_{22}\,,$$
$$Fy_{14} = z_7 - z_{14} - z_{18}\,, \qquad Fy_{15} = -z_{14} - z_{18}\,, \qquad Fy_{16} = z_7 - z_{14}\,,$$
$$Fy_{17} = -z_{14}\,, \qquad Fy_{18} = z_7 + z_{14} + z_{18}\,, \qquad Fy_{19} = z_{14} + z_{18}\,,$$
$$Fy_{20} = z_7 + z_{18}\,, \qquad Fy_{21} = z_{18}\,, \qquad Fy_{22} = z_7 + z_{14} - z_{18} - z_{22}\,,$$
$$Fy_{23} = z_{14} - z_{18} - z_{22}\,, \qquad Fy_{24} = z_7 + z_{14} - z_{22}\,, \qquad Fy_{25} = z_{14} - z_{22}\,,$$
$$Fy_{26} = z_7 - z_{18} - z_2\,, \qquad Fy_{27} = -z_{18} - z_{22}\,, \qquad Fy_{28} = z_7 - z_{22}\,,$$
$$Fy_{29} = -z_{22}\,.$$

This system was simulated on a PDP 11-40, and compared with the original mechanical system, with the result that the values of the mechanical state-variables were approximately equal to each other. Because of the relatively high dimension of the regular Lotka-Volterra representation, the computation time was about twice the necessary time in computations with the original system of equations. Obviously the joint-robot is not a suitable object for the Lotka-Volterra approach.

# 10. Simulation Studies of Evolutionary Systems with a Microcomputer[1)]

In this last chapter we will demonstrate that many interesting studies of the dynamic properties of evolutionary systems can be performed on a relatively small microcomputer. We address many problems in connection with the application of the Volterra approach on rather different systems. All computations are performed on the microcomputer ZX 81 with an additional memory of only 16 K bytes. In all examples, we formulate the problem, then we solve it with a computer program in BASIC and then give some graphics of the results thus received.

*1. Lotka-Volterra predator-prey model of the second order*

We first simulate the simple Lotka-Volterra predator-prey model with

$Fy = A - Kx\,,$ $y$ is the prey,

$Fx = Ly - B\,,$ $x$ is the predator.

We simulate this behaviour with and without a perturbation of the reproduction rate $A$ of the prey in order to check the robustness of the trajectory.

```
10 PRINT "USUAL PRED./PREY"
15 PRINT "SIGMA"
16 INPUT SIGMA
20 PRINT "A"
25 INPUT A
30 PRINT "K"
35 INPUT K
40 PRINT "L"
45 INPUT L
50 PRINT "B"
55 INPUT B
56 PRINT "Y0,X0"
57 INPUT Y0
58 INPUT X0
60 PRINT "D"
65 INPUT D
70 PRINT "M"
75 INPUT M
76 PRINT "S"
80 INPUT S
81 LET Y=Y0
82 LET X=X0
90 FOR N=1 TO M
94 LET Y=Y+DY(A-KX)
95 LET X=X+DX(-B+LY)
110 PLOT 5+XS,5+SY
112 Let A=A+SIGMA(2RND-1)
115 NEXT N
125 GOTO 90
```

1) In all examples we have computed on the personal computer ZX 81, we used for the integration the Euler forward method. From this simple method we can easily get an understanding of the qualitative path of trajectories. If yon want more accurate results from the quantitative point of view the trapez-rule or a Runge-Kutta method should be applied.

We simulate this system for the following parameter set

A = 2 , K = 1 , L = 0.5 , B = 1 ,

D = 0.1, Y0 = 3 , X0 = 5 ,

first for SIGMA = 0, and then for SIGMA = 0.1. We get the following results.

SIGMA = 0

| $x$ | $y$ | $x$ | $y$ |
|---|---|---|---|
| 5.00 | 2.00 | 0.49 | 1.17 |
| 5.03 | 2.10 | 0.48 | 1.34 |
| 4.89 | 1.46 | 0.47 | 1.55 |
| 4.66 | 1.04 | 0.46 | 1.79 |
| 4.37 | 0.76 | 0.46 | 2.07 |
| 4.06 | 0.58 | 0.47 | 2.38 |
| 3.75 | 0.46 | 0.49 | 2.74 |
| 3.44 | 0.38 | 0.52 | 3.16 |
| 3.16 | 0.33 | 0.56 | 3.62 |
| 2.87 | 0.29 | 0.62 | 4.15 |
| 2.64 | 0.26 | 0.71 | 4.72 |
| 2.40 | 0.25 | 0.82 | 5.33 |
| 2.19 | 0.24 | 0.99 | 5.96 |
| 2.00 | 0.23 | 1.21 | 6.56 |
| 1.82 | 0.23 | 1.52 | 7.08 |
| 1.66 | 0.24 | 1.93 | 7.41 |
| 1.52 | 0.24 | 2.46 | 7.47 |
| 1.38 | 0.26 | 3.09 | 7.12 |
| 1.26 | 0.27 | 3.76 | 6.35 |
| 1.16 | 0.29 | 4.37 | 5.22 |
| 1.06 | 0.32 | 4.81 | 3.99 |
| 0.97 | 0.35 | 5.01 | 2.87 |
| 0.89 | 0.38 | 5.02 | 2.00 |
| 0.82 | 0.42 | | |
| 0.76 | 0.47 | | |
| 0.70 | 0.53 | | |
| 0.65 | 0.60 | | |
| 0.61 | 0.68 | | |
| 0.57 | 0.78 | | |
| 0.54 | 0.89 | | |
| 0.52 | 1.02 | | |

SIGMA = 0.1

| $x$ | $y$ | $x$ | $y$ |
|---|---|---|---|
| 5.00 | 2.00 | 1.12 | 5.62 |
| 5.03 | 2.10 | 1.36 | 6.30 |
| 4.89 | 1.46 | 1.69 | 6.93 |
| 4.65 | 1.04 | 2.16 | 7.43 |
| 4.36 | 0.77 | 2.76 | 7.60 |
| 4.06 | 0.59 | 3.49 | 7.26 |
| 3.75 | 0.48 | 4.25 | 6.39 |
| 3.45 | 0.40 | 4.93 | 5.17 |
| 3.16 | 0.35 | 5.39 | 3.89 |
| 2.90 | 0.32 | 5.59 | 2.73 |
| 2.66 | 0.30 | 5.54 | 1.84 |
| 2.43 | 0.29 | 5.33 | 1.24 |
| 2.22 | 0.28 | | |
| 2.03 | 0.28 | | |
| 1.86 | 0.29 | | |
| 1.70 | 0.29 | | |
| 1.55 | 0.31 | | |
| 1.42 | 0.32 | | |
| 1.31 | 0.35 | | |
| 1.20 | 0.37 | | |
| 1.10 | 0.41 | | |
| 1.02 | 0.46 | | |
| 0.94 | 0.51 | | |
| 0.88 | 0.57 | | |
| 0.82 | 0.65 | | |
| 0.77 | 0.75 | | |
| 0.72 | 0.86 | | |
| 0.69 | 0.99 | | |
| 0.66 | 1.16 | | |
| 0.64 | 1.36 | | |
| 0.62 | 1.62 | | |
| 0.62 | 1.91 | | |
| 0.63 | 2.25 | | |
| 0.65 | 2.66 | | |
| 0.69 | 3.13 | | |
| 0.74 | 3.67 | | |
| 0.83 | 4.26 | | |
| 0.95 | 4.92 | | |

We intend now to simulate the same behaviour with the Riccati representation.

$$Fv = x = e^{-Bt} u^L , \qquad Fu = y = e^{At}/v^K .$$

The corresponding BASIC program is

```
10 PRINT "PP-RICCATI"
15 PRINT "SIGMA"
16 INPUT SIGMA
20 PRINT "A"
25 INPUT A
30 PRINT "K"
35 INPUT K,
40 PRINT "L"
45 INPUT L
50 PRINT "B"
55 INPUT B
60 PRINT "D"
65 INPUT D
70 PRINT "M"
75 INPUT M
76 PRINT "S"
80 INPUT S
81 PRINT "Y0;X0"
82 INPUT Y0
83 INPUT X0
84 LET U=X0 (1/L)
85 LET V=1/(Y0 (1/K))
86 LET T=1
90 FOR N = 1 TOM
91 LET T=T+1
94 LET Y = EXP(A.T.D)/V K
95 LET U = U+D.Y.U
99 LET X = EXP(-B.T.D). U L
100 LET V = V + D.X.V
105 PRINT "Y="; Y
106 PRINT TAB 8; "X="; X
110 STOP
112 LET A = A + SIGMA .(2.RND -1)
115 NEXT N
125 GOTO 90
```

We now simulate a system with the same parameter condition as before, but with increased accuracy $D = 0.025$, and print only every fourth value. With lower accuracy good reproduction of the values computed formerly is impossible.

We get the following results

| $x$ | $y$ | $x$ | $y$ |
|---|---|---|---|
| 5.00 | 2.00 | 0.64 | 1.25 |
| 5.00 | 2.10 | 0.62 | 1.44 |
| 4.93 | 1.60 | 0.60 | 1.65 |

| $x$ | $y$ | $x$ | $y$ |
|---|---|---|---|
| 4.78 | 1.24 | 0.59 | 1.90 |
| 4.56 | 0.97 | 0.60 | 2.19 |
| 4.30 | 0.77 | 0.60 | 2.52 |
| 4.02 | 0.63 | 0.63 | 2.90 |
| 3.75 | 0.53 | 0.66 | 3.33 |
| 3.47 | 0.46 | 0.71 | 3.80 |
| 3.21 | 0.41 | 0.78 | 4.32 |
| 2.96 | 0.37 | 0.88 | 4.86 |
| 2.61 | 0.33 | 1.02 | 5.42 |
| 2.40 | 0.32 | 1.20 | 5.95 |
| 2.21 | 0.31 | 1.46 | 6.40 |
| 2.03 | 0.30 | 1.79 | 6.70 |
| 1.82 | 0.31 | 2.21 | 6.77 |
| 1.68 | 0.32 | 2.72 | 6.55 |
| 1.54 | 0.33 | 3.29 | 6.03 |
| 1.42 | 0.35 | 3.86 | 5.26 |
| 1.31 | 0.37 | 4.37 | 4.36 |
| 1.21 | 0.40 | 4.74 | 3.47 |
| 1.11 | 0.44 | 4.94 | 2.69 |
| 1.03 | 0.48 | 5.00 | 2.05 |
| 0.96 | 0.53 | | |
| 0.89 | 0.59 | | |
| 0.83 | 0.66 | | |
| 0.78 | 0.74 | | |
| 0.73 | 0.84 | | |
| 0.70 | 0.96 | | |
| 0.66 | 1.09 | | |

If we continue the computation, we necessarily run into overflow because of the monotony of the Riccati state-variables $u$ and $v$. This phenomenon can easily be repaired by rescaling, putting

$$t = t^* + \Delta .$$

*2. Exponential tower, hypercycle and a hypercycle above an exponential tower*

Here we compare the behaviour of an exponential tower which is cut off at a certain order with an input signal equal to 1 with the behaviour of a hypercycle of order 1, and we study the behaviour of hypercycles of low order and the behaviour of exponential chains of lower order closed by a certain hypercycle.

We give here a BASIC program which allows us to study all three questions.

```
10 PRINT "EXPONENTIAL-TOWER, HYPERCYCLE"
20 PRINT "D"
25 INPUT D
40 PRINT "M"
45 INPUT M
50 PRINT "N= ORDER"
```

```
 55 INPUT N
 60 DIM X(N)
 65 DIM K(N)
 70 FOR R=1 TO N
 75 PRINT "X(R)"
 76 INPUT X(R)
 77 PRINT "K(R)"
 78 INPUT K(R)
 80 NEXT R
 81 PRINT "LOG"
 82 INPUT LOG
 83 PRINT "S0"
 84 INPUT S0
 85 PRINT "BACK"
 86 INPUT BACK
 90 FOR T = 1 TO M
120 FOR R= 1 TO N-1
130 LET X(R) = X(R) + D.K(R).X(R).X(R+1)
135 NEXT R
136 IF LOG = 1 THEN LET X(N) = X(N) + D.X(N).X(BACK.K(N)
140 IF LOG  = 0 THEN LET X(N) = X(N) + D.X(N).K(N)
145 PRINT X(S0)
165 NEXT T
```

The following is a comparison of an exponential tower with order $N$ and all $x(R)_0 =$ $= 1$ and all $K(R) = K$ with the hypercycle of order $N = 1$ with the same $K$.

a) $K = 1$. We choose $D = 0.01$, but print only every second value

| $N = 1$ | $N = 2$ | $N = 3$ | $N = 4$ | $N = 5$ | Hypercycle $N = 1$ |
|---|---|---|---|---|---|
| 1.00 | 1.00 | 1.00 | 1.00 | 1.00 | 1.00 |
| 1.02 | 1.02 | 1.02 | 1.02 | 1.02 | 1.02 |
| 1.04 | 1.04 | 1.04 | 1.04 | 1.04 | 1.04 |
| 1.06 | 1.06 | 1.06 | 1.06 | 1.06 | 1.06 |
| 1.08 | 1.09 | 1.09 | 1.09 | 1.09 | 1.09 |
| 1.10 | 1.11 | 1.11 | 1.11 | 1.11 | 1.12 |
| 1.12 | 1.13 | 1.13 | 1.13 | 1.13 | 1.14 |
| 1.15 | 1.16 | 1.16 | 1.16 | 1.16 | 1.16 |
| 1.17 | 1.19 | 1.19 | 1.19 | 1.19 | 1.19 |
| 1.19 | 1.22 | 1.22 | 1.22 | 1.22 | 1.22 |
| 1.22 | 1.24 | 1.25 | 1.25 | 1.25 | 1.25 |
| 1.24 | 1.28 | 1.28 | 1.28 | 1.28 | 1.28 |
| 1.27 | 1.31 | 1.31 | 1.31 | 1.31 | 1.31 |
| 1.30 | 1.34 | 1.35 | 1.35 | 1.35 | 1.35 |
| 1.32 | 1.38 | 1.38 | 1.38 | 1.38 | 1.38 |
| 1.35 | 1.41 | 1.42 | 1.42 | 1.42 | 1.42 |
| 1.37 | 1.45 | 1.46 | 1.46 | 1.46 | 1.46 |
| 1.40 | 1.49 | 1.50 | 1.51 | 1.51 | 1.51 |
| 1.43 | 1.53 | 1.55 | 1.55 | 1.55 | 1.55 |

| $N = 1$ | $N = 2$ | $N = 3$ | $N = 4$ | $N = 5$ | Hypercycle $N = 1$ |
|---|---|---|---|---|---|
| 1 | | | | | |
| 1.46 | 1.58 | 1.60 | 1.60 | 1.60 | 1.60 |
| 1.49 | 1.63 | 1.65 | 1.65 | 1.65 | 1.65 |
| 1.52 | 1.67 | 1.70 | 1.71 | 1.71 | 1.71 |
| 1.55 | 1.73 | 1.76 | 1.77 | 1.77 | 1.77 |
| 1.58 | 1.78 | 1.82 | 1.83 | 1.83 | 1.83 |
| 1.61 | 1.84 | 1.89 | 1.90 | 1.90 | 1.90 |
| 1.64 | 1.90 | 1.96 | 1.97 | 1.97 | 1.97 |
| 1.68 | 1.96 | 2.04 | 2.05 | 2.05 | 2.05 |
| 1.71 | 2.03 | 2.12 | 2.14 | 2.14 | 2.14 |
| 1.75 | 2.10 | 2.21 | 2.23 | 2.23 | 2.23 |
| 1.78 | 2.17 | 2.30 | 2.33 | 2.33 | 2.33 |
| 1.82 | 2.25 | 2.40 | 2.44 | 2.44 | 2.45 |
| 1.85 | 2.33 | 2.51 | 2.58 | 2.57 | 2.57 |
| 1.89 | 2.42 | 2.63 | 2.70 | 2.70 | 2.70 |
| 1.93 | 2.51 | 2.76 | 2.83 | 2.85 | 2.85 |
| 1.98 | 2.61 | 2.90 | 3.00 | 3.02 | 3.02 |
| 2.03 | 2.72 | 3.06 | 3.17 | 3.20 | 3.21 |
| 2.07 | 2.83 | 3.23 | 3.37 | 3.41 | 3.42 |
| 2.11 | 2.94 | 3.42 | 3.60 | 3.65 | 3.66 |
| 2.15 | 3.07 | 3.62 | 4.14 | 3.92 | 3.94 |
| 2.19 | 3.20 | 3.85 | 4.47 | 4.24 | 4.27 |
| 2.22 | 3.34 | 4.10 | 4.85 | 4.60 | 4.65 |
| 2.26 | 3.49 | 4.39 | 5.29 | 5.03 | 5.10 |
| 2.31 | 3.66 | 4.70 | 5.81 | 5.54 | 5.65 |
| 2.35 | 3.83 | 5.06 | 6.43 | 6.16 | 6.32 |
| 2.40 | 4.01 | 5.45 | 7.16 | 6.92 | 7.17 |
| 2.45 | 4.21 | 5.91 | 8.05 | 7.86 | 8.28 |
| 2.50 | 4.41 | 6.42 | 9.15 | 8.06 | 9.77 |
| 2.55 | 4.64 | 7.01 | 10.50 | 10.63 | 11.87 |
| 2.60 | 4.88 | 7.68 | 12.22 | 12.73 | 15.04 |
| 2.65 | 5.14 | 8.46 | 14.43 | 15.67 | 20.30 |
| 2.70 | 5.42 | 9.36 | 17.34 | 19.91 | 30.39 |
| | 5.71 | | | | 55.32 |
| | 6.04 | | | | |
| | 6.38 | | | | |
| | 6.76 | | | | |
| | 7.16 | | | | |
| | 7.60 | | | | |
| | 8.07 | | | | |
| | 8.58 | | | | |
| | 9.14 | | | | |
| | 9.74 | | | | |
| | 10.40 | | | | |
| | 11.11 | | | | |
| | 11.89 | | | | |

b) $K = -1$. We choose $D = 0.1$

| $N = 1$ | $N = 2$ | $N = 3$ | $N = 4$ | $N = 5$ | Hypercycle $K = -1$ |
|---|---|---|---|---|---|
| 1.00 | 1.00 | 1.00 | 1.00 | 1.00 | 1.00 |
| 0.90 | 0.90 | 0.90 | 0.90 | 0.90 | 0.90 |
| 0.81 | 0.82 | 0.82 | 0.82 | 0.82 | 0.82 |
| 0.73 | 0.75 | 0.75 | 0.75 | 0.75 | 0.75 |
| 0.66 | 0.70 | 0.70 | 0.70 | 0.70 | 0.70 |
| 0.59 | 0.65 | 0.65 | 0.65 | 0.65 | 0.65 |
| 0.53 | 0.61 | 0.60 | 0.61 | 0.61 | 0.61 |
| 0.48 | 0.58 | 0.57 | 0.57 | 0.57 | 0.57 |
| 0.43 | 0.55 | 0.53 | 0.54 | 0.54 | 0.54 |
| 0.39 | 0.53 | 0.51 | 0.51 | 0.51 | 0.51 |
| 0.35 | 0.51 | 0.48 | 0.48 | 0.48 | 0.48 |
| 0.31 | 0.49 | 0.45 | 0.46 | 0.46 | 0.46 |
| 0.28 | 0.48 | 0.43 | 0.44 | 0.44 | 0.44 |
| 0.25 | 0.46 | 0.41 | 0.42 | 0.42 | 0.42 |
| 0.23 | 0.45 | 0.39 | 0.40 | 0.40 | 0.40 |
| 0.21 | 0.44 | 0.37 | 0.39 | 0.38 | 0.38 |
| 0.19 | 0.43 | 0.36 | 0.37 | 0.37 | 0.37 |
| 0.17 | 0.42 | 0.34 | 0.36 | 0.36 | 0.36 |
| 0.15 | 0.42 | 0.33 | 0.35 | 0.34 | 0.34 |
| 0.14 | 0.41 | 0.31 | 0.33 | 0.33 | 0.33 |
| 0.12 | 0.40 | 0.30 | 0.32 | 0.32 | 0.32 |
| 0.11 | 0.40 | 0.29 | 0.31 | 0.31 | 0.31 |
| 0.10 | 0.39 | 0.28 | 0.31 | 0.30 | 0.30 |
| 0.09 | 0.39 | 0.27 | 0.30 | 0.29 | 0.29 |
| 0.08 | 0.38 | 0.26 | 0.29 | 0.28 | 0.28 |
| 0.07 | 0.38 | 0.25 | 0.28 | 0.27 | 0.28 |
| 0.06 | 0.38 | 0.24 | 0.27 | 0.27 | 0.27 |

Obviously the even order chains are monotonously approaching the hypercycle trajectory from above, whereas the odd order chains are monotonously approaching the hypercycle trajectory from below.

— The behavioural types of the hypercycle of order $N = 2$

| $K_1 = K_2 = 1$, $D = 0.05$ | $K_1 = 3$, $K_2 = 1$ | $K_1 = 1$, $K_2 = 3$ |
|---|---|---|
| 1.00 | 1.00 | 1.00 |
| 1.05 | 1.15 | 1.11 |
| 1.17 | 1.56 | 1.19 |
| 1.24 | 1.84 | 1.28 |
| 1.31 | 2.20 | 1.40 |
| 1.40 | 2.70 | 1.56 |
| 1.50 | 3.37 | 1.78 |
| 1.61 | 4.36 | 2.10 |

| $K_1 = K_2 = 1$ | $K_1 = 3,\ K_2 = 1$ | $K_1 = 1,\ K_2 = 3$ |
|---|---|---|
| 1.75 | 5.92 | 2.60 |
| 1.90 | 8.67 | 3.45 |
| 2.09 | 14.42 | 5.17 |
| 2.32 | 30.90 | 9.75 |
| 2.60 | 120.77 | 31.01 |
| 2.96 | 2592.84 | 413.08 |
| 3.42 | | |
| 4.05 | | |
| 4.95 | | |
| 6.32 | | |
| 8.62 | | |
| 13.11 | | |
| 24.42 | | |
| 71.17 | | |
| 692.42 | | |

$K_1 = 1,\ K_2 = -1$

| | | |
|---|---|---|
| 1.00 | 1.54 | 1.82 |
| 1.05 | 1.57 | 1.84 |
| 1.10 | 1.60 | 1.85 |
| 1.15 | 1.63 | 1.86 |
| 1.20 | 1.66 | 1.87 |
| 1.24 | 1.69 | 1.88 |
| 1.29 | 1.71 | 1.89 |
| 1.34 | 1.73 | 1.90 |
| 1.38 | 1.75 | 1.91 |
| 1.42 | 1.77 | 1.91 |
| 1.46 | 1.79 | 1.92 |
| 1.50 | 1.81 | 1.92 |

$K_1 = -1,\ K_2 = 1$

| | | |
|---|---|---|
| 1.00 | 0.43 | 0.15 |
| 0.95 | 0.39 | 0.14 |
| 0.85 | 0.36 | 0.12 |
| 0.80 | 0.33 | 0.11 |
| 0.75 | 0.31 | 0.10 |
| 0.71 | 0.28 | 0.09 |
| 0.66 | 0.26 | 0.08 |
| 0.62 | 0.24 | 0.08 |
| 0.58 | 0.22 | 0.07 |
| 0.54 | 0.20 | 0.06 |
| 0.50 | 0.18 | 0.06 |
| 0.46 | 0.16 | 0.05 |

$K_1 = -3,\ K_2 = -1$

| | |
|---|---|
| 1.00 | 0.12 |
| 0.85 | 0.10 |
| 0.73 | 0.09 |
| 0.63 | 0.08 |
| 0.54 | 0.07 |
| 0.47 | 0.07 |
| 0.41 | 0.06 |
| 0.36 | 0.05 |
| 0.32 | 0.05 |
| 0.28 | 0.04 |
| 0.25 | 0.04 |
| 0.22 | 0.03 |
| 0.19 | 0.03 |

$K_1 = -1,\ K_2 = -3$

| | |
|---|---|
| 1.00 | 0.69 |
| 0.95 | 0.69 |
| 0.91 | 0.69 |
| 0.88 | 0.68 |
| 0.85 | 0.68 |
| 0.82 | 0.68 |
| 0.80 | 0.68 |
| 0.79 | 0.67 |
| 0.77 | 0.67 |
| 0.76 | 0.67 |
| 0.75 | 0.67 |
| 0.74 | 0.67 |
| 0.73 | 0.67 |

| $K_1 = -3,\ K_2 = -1$ | | $K_1 = -1,\ K_2 = -3$ | |
|---|---|---|---|
| 0.17 | 0.03 | 0.72 | 0.67 |
| 0.15 | 0.02 | 0.71 | 0.67 |
| 0.13 | 0.02 | 0.70 | 0.67 |

— The behavioural types of a chain of order $N = 1$, closed with a hypercycle of order $N = 1$.

| $K_1 = 2,\ K_2 = 1$ | | $K_1 = 2,\ K_2 = -1$ | | $K_1 = -2,\ K_2 = 1$ | | $K_1 = -2,\ K_2 = -1$ | |
|---|---|---|---|---|---|---|---|
| 1.00 | 12.84 | 1.00 | 3.14 | 1.00 | 0.09 | 1.00 | 0.29 |
| 1.10 | 17.71 | 1.10 | 3.31 | 0.90 | 0.07 | 0.90 | 0.28 |
| 1.21 | 25.71 | 1.20 | 3.48 | 0.81 | 0.04 | 0.81 | 0.26 |
| 1.35 | 39.92 | 1.31 | 3.66 | 0.72 | 0.03 | 0.74 | 0.25 |
| 1.51 | 68.11 | 1.43 | 3.84 | 0.63 | 0.01 | 0.68 | 0.23 |
| 1.69 | 133.17 | 1.54 | 4.04 | 0.55 | 0.00 | 0.62 | 0.22 |
| 1.91 | 321.15 | 1.67 | 4.23 | 0.48 | | 0.57 | 0.21 |
| 2.18 | 1094.39 | 1.79 | 4.43 | 0.41 | | 0.53 | 0.20 |
| 2.50 | 6901.64 | 1.93 | 4.63 | 0.35 | | 0.49 | 0.19 |
| 2.91 | 140691.26 | 2.06 | 4.84 | 0.30 | | 0.45 | 0.18 |
| 3.42 | | 2.20 | 5.05 | 0.24 | | 0.42 | 0.18 |
| 4.06 | | 2.35 | 5.26 | 0.20 | | 0.40 | 0.17 |
| 4.90 | | 2.50 | 5.48 | 0.16 | | 0.37 | 0.16 |
| 6.01 | | 2.65 | 5.71 | 0.12 | | 0.35 | 0.16 |
| 7.53 | | 2.81 | 5.94 | | | 0.33 | 0.15 |
| 9.68 | | 2.97 | 6.17 | | | 0.31 | 0.14 |

### 3. *Autocatalytic Lotka-Volterra system*

We shall simulate this system in the Riccati representation

$$du/dt = K_1 v^{E_1}\,,$$

$$dv/dt = K_2 u^{E_2}$$

which corresponds to a Lotka-Volterra system of the form

$$Fx = E_1 K_2 y - K_1 x\,,$$

$$Fy = E_2 K_1 x - K_2 y\,,$$

$y$ shall be the prey and
$x$ shall be the predator.

We get

$$x = v^{E_1}/u\,, \qquad u = (xy^{E_1})^{1/(E_1E_2-1)}\,,$$

$$y = u^{E_2}/v\,, \qquad v = (x^{E_2}y)^{1/(E_1E_2-1)}\,.$$

Obviously, the Riccati system has the following first integral

$$\frac{u^{(1+E_2)}}{(1+E_2)\,K_1} - \frac{v^{(1+E_1)}}{(1+E_1)\,K_2} = \text{const}\,.$$

The autocatalytic Lotka-Volterra system plays an important role in the Lotka-Volterra representation of the EVOLON model from the hyperlogistic differential equation

$$Fx_0 = Kx_1 ,$$

$$Fx_1 = (k - 1) Kx_1 - wlKx_2 , \qquad dx_0/dt = Kx_0^k(B - x_0^w)^l ,$$

$$Fx_2 = (k + w - 1) Kx_1 - w(l - 1) x_2 .$$

Therefore we get the following correspondences

$$K_1 = w(l - 1) K , \qquad E_1 = - (k + w - 1)/(k - 1) ,$$

$$K_2 = -K(k - 1) , \qquad E_2 = -l/(l - 1) .$$

For this application, only the following four cases are of interest

| | I | II | III | IV |
|---|---|---|---|---|
| | $k > 1, l > 1$ | $k < 1, l < 1$ | $k > 1, l < 1$ | $k < 1, l > 1$ |
| $K_1$ | + | − | − | + |
| $K_2$ | − | + | − | + |
| $E_1$ | − | + | − | + |
| $E_2$ | − | + | + | − |

We will first run the program with a test case, and then run it with four cases corresponding to the relationship to the hyperlogistic EVOLON model.

```
10 PRINT "RICCATI 2ND ORDER"
21 PRINT "LOG"
22 INPUT LOG
23 IF LOG=2 THEN GOTO 200
30 PRINT "K1"
35 INPUT K1
40 PRINT "K2"
45 INPUT K2
50 PRINT "E1"
55 INPUT E1
60 PRINT "E2"
65 INPUT E2
70 PRINT "D"
75 INPUT D
80 PRINT "M"
85 INPUT M
90 PRINT "S"
95 INPUT S
96 PRINT "LOG2"
97 INPUT LOG2
101 PRINT "Y0,X0"
102 INPUT Y0
103 INPUT XO
104 LET U=(X0.Y0^E1)^(1/(E1E2-1))
105 LET V=(Y0.X0^E2)^(1/(E1E2-1))
110 FOR N=1 TO M
111 LET CON=U^(1+E2)/(K1(1+E2))
         - V^(1+E1)/(K2(1 + E1))
115 LET U=U+K1.D.U**E1
120 LET V=V+K2.D.V**E2
125 LET X=(V**E1)/U
130 LET Y=(U**E2)/V
135 IF LOG2=1 THEN GOTO 165
145 PLOT 10+X.S,10 + Y.S
150 NEXT N
155 CLS
160 GOTO 110
165 PRINT Y
166 PRINT TAB 7; "X= " ; X
175 GOTO 150
200 PRINT "K"
201 INPUT K
205 PRINT "W"
206 INPUT W
```

```
207 PRINT "KK, L"
208 INPUT KK
209 INPUT L
210 LET K1 = W.(L−1).K
211 LET K2 = −K.(KK−1)
212 LET E1 = −(KK+W−1)/(KK−1)
213 LET E2 = −L/(L−1)
214 GOTO 70
```

b) Test case

We put LOG = 1 and K1 = 1, K2 = −2, E1 = −3, E2 = −2

D = 0.05, LOG2 = 1 Y0 = 3, X0 = 5

| Y | X | Y | X |
|---|---|---|---|
| 3.00 | 5.00 | 0.06 | 2.77 |
| 2.38 | 7.58 | 0.05 | 2.47 |
| 1.43 | 8.21 | 0.04 | 2.22 |
| 0.77 | 7.28 | 0.03 | 2.01 |
| 0.43 | 6.06 | 0.03 | 1.85 |
| 0.26 | 5.03 | 0.02 | 1.70 |
| 0.17 | 4.23 | 0.02 | 1.58 |
| 0.12 | 3.61 | 0.02 | 1.47 |
| 0.08 | 2.77 | 0.01 | 1.38 |

both species are, in this case, extincted

b) EVOLON cases

Here we put K = 1, W = 1 , LOG = 2 and LOG2 = 1

| KK = 2, L = 2 | | KK = 0.8, L = 0.7 | | KK = 2, L = 0.7 | | KK = 0.8, L = 2 | |
|---|---|---|---|---|---|---|---|
| Y | X | Y | X | Y | X | Y | X |
| 3.00 | 5.00 | 3.00 | 5.00 | 3.00 | 5.00 | 3.00 | 5.00 |
| 2.12 | 4.89 | 2.44 | 5.97 | 2.86 | 7.06 | 1.88 | 4.32 |
| 1.47 | 4.53 | 1.92 | 7.08 | 2.47 | 9.97 | 0.89 | 3.26 |
| 1.03 | 4.08 | 1.46 | 8.40 | 1.85 | 14.00 | 0.65 | 2.88 |
| 0.74 | 3.65 | 1.05 | 10.03 | 1.13 | 19.77 | 0.49 | 2.57 |
| 0.54 | 3.25 | 0.71 | 12.12 | 0.51 | 29.56 | 0.39 | 2.31 |
| 0.41 | 2.91 | 0.44 | 15.13 | 0.13 | 53.81 | 0.31 | 2.10 |
| 0.32 | 2.62 | 0.24 | 19.76 | 0.00 | 279.0 | 0.25 | 1.92 |
| 0.25 | 2.38 | 0.11 | 28.21 | | | 0.21 | 1.76 |
| 0.20 | 2.17 | 0.03 | 48.96 | | | 0.18 | 1.63 |
| 0.17 | 1.99 | 0.00 | 184.47 | | | 0.15 | 1.52 |
| 0.14 | 1.83 | | | | | 0.13 | 1.42 |
| 0.12 | 1.70 | | | | | 0.11 | 1.33 |
| 0.10 | 1.58 | | | | | 0.10 | 1.25 |
| 0.09 | 1.48 | | | | | 0.09 | 1.18 |
| 0.08 | 1.39 | | | | | 0.08 | 1.12 |
| 0.07 | 1.31 | | | | | 0.07 | 1.06 |
| 0.06 | 1.23 | | | | | 0.06 | 1.01 |

| KK = 2 L = 2 | | KK = 0.8, L = 0.7 | | KK = 2 L = 0.7 | | KK = 0.8, L = 2 | |
|---|---|---|---|---|---|---|---|
| Y | X | Y | X | Y | X | Y | X |
| 0.05 | 1.17 | | | | | 0.06 | 0.97 |
| 0.04 | 1.05 | | | | | 0.05 | 0.92 |
| 0.03 | 0.96 | | | | | 0.05 | 0.88 |
| 0.03 | 0.88 | | | | | 0.04 | 0.84 |

*4. Volterra representation of the hyperlogistic EVOLON model*

We will now extend the program in example 3 in such a manner that we can use it for the simulation of the hyperlogistic EVOLON model. We know from 4.1. that the following relations hold

$$dx_0/dt = Kx_0^k(B - x_0^w)^l\,,$$

$$Fx_0 = Kx_1 \quad \text{with} \quad x_1 = x_0^{k-1}(B - x^w)^l\,,$$

$$Fx_1 = K(k-1)\,x_1 - Kwlx_2 \quad \text{with} \quad x_2 = x_0k + w - 1(B - x^w)^l\,,$$

$$Fx_2 = K(k + w - 1)\,x_1 - Kw(l-1)\,x_2\,.$$

With the program in example 3, we can already deal with an autocatalytic Lotka-Volterra system of the variables $x_1$, $x_2$. From the given parameters $k$, $w$, $l$ and $B$, and the initial value $x_0(0)$, we can now determine the corresponding initial values $x_1(0)$ and $x_2(0)$, which we need in the program of the example 3. For the determination of $x_0(t)$ it is necessary, in addition to integrate the first equation. For our purpose, we must insert the following additional statements into the program of example 3.

```
 98 PRINT "LOG3"
 99 INPUT LOG3
100 IF LOG3 = 1 THEN GOTO 180
131 IF LOG3 = 1 THEN LET XOO= XOO + D.K.XOO.Y
133 IF LOG3 = 1 THEN PRINT TAB 4; "XOO="; XOO
180 PRINT "XOO"
181 INPUT XOO
182 PRINT "B"
183 INPUT B
184 LET YO= XOO^(KK-1)(B - XOO^W)^L
185 LET XO= XOO^(KK+W-1)(B - XOO^W)^(L-1)
186 GOTO 104
```

Now we repeat the same EVOLON examples from example 3, putting the initial value XOO = 0.5 in all four cases. For the logical variables we choose the following configuration: LOG = 2, LOG2 = 1 and LOG3 = 1. For D we choose again D = 0.05. For the saturation value, we choose B = 5.

| KK = 2, L = 2 | | | KK = 0.8, L = 0.7 | | |
|---|---|---|---|---|---|
| XOO | Y | X | XOO | Y | X |
| 0.92 | 16.61 | 3.57 | 0.58 | 3.15 | 0.42 |
| 2.28 | 29.77 | 18.76 | 0.67 | 3.01 | 0.48 |
| 3.77 | 13.13 | 26.56 | 0.76 | 2.87 | 0.54 |

KK = 2, L = 2

| X00 | Y | X |
|---|---|---|
| 4.29 | 2.76 | 14.77 |
| 4.50 | 0.96 | 9.32 |
| 4.60 | 0.45 | 6.65 |
| 4.66 | 0.26 | 5.12 |
| 4.70 | 0.17 | 4.15 |
| 4.73 | 0.11 | 3.47 |
| 4.74 | 0.08 | 2.98 |
| 4.76 | 0.06 | 2.61 |
| 4.77 | 0.05 | 2.32 |
| 4.78 | 0.04 | 2.09 |
| 4.79 | 0.03 | 1.90 |
| 4.80 | 0.03 | 1.74 |
| 4.80 | 0.02 | 1.60 |
| 4.81 | 0.02 | 1.49 |
| 4.81 | 0.02 | 1.39 |

KK = 0.8, L = 0.7

| X00 | Y | X |
|---|---|---|
| 0.87 | 2.74 | 0.61 |
| 0.98 | 2.61 | 0.68 |
| 1.10 | 2.49 | 0.76 |
| 1.23 | 2.37 | 0.85 |
| 1.37 | 2.25 | 0.94 |
| 1.51 | 2.13 | 1.04 |
| 1.67 | 2.01 | 1.14 |
| 1.82 | 1.89 | 1.25 |
| 1.99 | 1.78 | 1.37 |
| 2.15 | 1.67 | 1.50 |
| 2.32 | 1.55 | 1.63 |
| 2.49 | 1.45 | 1.78 |
| 2.65 | 1.34 | 1.93 |
| 2.82 | 1.24 | 2.08 |
| 2.98 | 1.13 | 2.25 |
| 3.13 | 1.04 | 2.43 |
| 3.28 | 0.94 | 2.62 |
| 3.42 | 0.85 | 2.82 |
| 3.55 | 0.76 | 3.04 |
| 3.67 | 0.68 | 3.27 |
| 3.78 | 0.60 | 3.52 |
| 3.88 | 0.53 | 3.80 |
| 3.97 | 0.46 | 4.10 |
| 4.05 | 0.39 | 4.44 |
| 4.11 | 0.33 | 4.82 |
| 4.17 | 0.28 | 5.25 |

KK = 2, L = 0.7

| X00 | Y | X |
|---|---|---|
| 0.54 | 1.53 | 0.19 |
| 0.58 | 1.65 | 0.22 |
| 0.63 | 1.78 | 0.26 |
| 0.70 | 1.94 | 0.31 |
| 0.77 | 2.12 | 0.38 |
| 0.86 | 2.34 | 0.48 |
| 0.97 | 2.60 | 0.82 |
| 1.11 | 2.91 | 0.82 |
| 1.30 | 3.29 | 1.13 |
| 1.54 | 3.76 | 1.62 |
| 1.87 | 4.31 | 2.45 |
| 2.33 | 4.93 | 3.95 |
| 2.97 | 5.43 | 6.80 |
| 3.76 | 5.36 | 12.16 |
| 4.52 | 4.02 | 21.47 |

KK = 0.8, L = 2

| X00 | Y | X |
|---|---|---|
| 0.89 | 15.43 | 4.47 |
| 1.30 | 9.34 | 5.41 |
| 1.65 | 5.47 | 5.33 |
| 1.93 | 3.30 | 4.82 |
| 2.13 | 2.10 | 4.22 |
| 2.28 | 1.41 | 3.69 |
| 2.39 | 0.99 | 3.24 |
| 2.48 | 0.73 | 2.87 |
| 2.55 | 0.56 | 2.57 |
| 2.60 | 0.43 | 2.32 |
| 2.65 | 0.35 | 2.69 |
| 2.69 | 0.28 | 1.93 |
| 2.72 | 0.24 | 1.77 |
| 2.75 | 0.20 | 1.64 |
| 2.77 | 0.17 | 1.53 |

| KK = 2, L = 0.7 | | | KK = 0.8, L = 2 | | |
|---|---|---|---|---|---|
| X00 | Y | X | X00 | Y | X |
| 4.92 | 1.77 | 37.53 | 2.79 | 0.15 | 1.43 |
| 4.98 | 0.26 | 88.13 | 2.81 | 0.13 | 1.34 |

*5. Determination of limit cycles by simulation of the hypercycle in replicator representation*

In 7.2. we have shown, that the replicator representation of a hypercycle

$$Fx_i = (K_i x_{i+1} \bmod N - \Phi)$$

with the flow function $\Phi = \sum x_k K_k x_{k+1}$ can be simulated in the following way

$$Fw_i = K_i w_{i+1}/w \quad \text{with} \quad i \bmod N\,,$$

$$w = \sum w_j\,.$$

The following BASIC program is based on this possibility.

```
10 PRINT "HYPERCYCLUS"
20 PRINT "D"
25 INPUT D
30 PRINT "K = SCALE"
35 INPUT K
40 PRINT "M"
45 INPUT M
46 PRINT "S"
47 INPUT S
50 PRINT "N = ORDER"
55 INPUT N
60 DIM W(N)
65 DIM K(N)
70 FOR R = 1 TO N
75 PRINT "W(R)"
76 INPUT W(R)
77 PRINT "K(R)"
78 INPUT K(R)
80 NEXT R
90 FOR T = 1 TO M
95 LET W = 0
100 FOR R = 1 TO N
105 LET W = W + W(R)
110 NEXT R
120 FOR R = 1 TO N - 1
125 IF W(R) < K THEN GOSUB 200
130 LET W(R) = W(R) + D.K(R).W(R).
    W(R + 1)/W
135 NEXT R
140 IF W(N) K < THEN GOSUB 200
145 LET W(N) = W(N) + D.K(N).W(N).
    W(1)/W
150 PLOT 10 + S.W(1)/W,10 + S.W(2)/W
151 PRINT W(1)/W
165 NEXT T
166 CLS
167 GOTO 90
170 STOP
200 FOR F = 1 TO N
205 LET W(F) = W(F)/K
210 NEXT F
212 LET W = W/K
215 RETURN
```

N = 2

| W | X1 | X2 | W | X1 | X2 |
|---|---|---|---|---|---|
| 6.00 | .689 | .356 | 15.44 | .583 | .467 |
| 6.27 | .682 | .364 | 16.22 | .580 | .470 |
| 6.56 | .674 | .372 | 17.03 | .577 | .477 |
| 6.86 | .668 | .379 | 18.80 | .571 | .480 |
| 7.18 | .661 | .386 | 19.75 | .568 | .482 |
| 7.52 | .655 | .393 | 20.76 | .566 | .485 |
| 7.87 | .648 | .399 | 21.81 | .563 | .488 |
| 8.25 | .642 | .406 | 22.92 | .561 | .490 |
| 8.65 | .637 | .412 | 24.09 | .559 | .492 |
| 9.07 | .631 | .417 | 25.31 | .556 | .495 |
| 9.51 | .626 | .423 | 26.61 | .554 | .497 |
| 9.97 | .621 | .428 | 27.96 | .552 | .499 |
| 10.46 | .616 | .433 | 29.39 | .550 | .500 |
| 10.98 | .611 | .438 | 30.89 | .549 | .502 |
| 11.52 | .607 | .443 | 32.47 | .547 | .504 |
| 12.09 | .602 | .447 | 34.13 | .545 | .506 |
| 12.69 | .598 | .452 | 35.87 | .544 | .509 |
| 13.33 | .594 | .456 | 37.70 | .542 | .509 |
| 14.00 | .590 | .460 | 39.63 | .541 | .510 |
| 14.70 | .587 | .463 | 41.66 | .540 | .512 |

N = 3 with K(R) = 1
for R = 1, 2, 3

| W | X1 | X2 | W | X1 | X2 | W | X1 | X2 |
|---|---|---|---|---|---|---|---|---|
| 8 | .130 | .394 | 12.00 | .427 | .254 | 11.00 | .467 | .278 |
| 8.24 | .131 | .401 | 12.29 | .427 | .252 | 11.26 | .469 | .276 |
| 8.49 | .132 | .408 | 12.59 | .427 | .250 | 11.52 | .470 | .275 |
| 8.74 | .133 | .415 | 12.90 | .427 | .248 | 11.79 | .472 | .273 |
| 9.01 | .134 | .422 | 13.22 | .427 | .246 | 12.06 | .474 | .272 |
| some further values | | | some further values (every 7th) | | | | | |
| 10.79 | .144 | .457 | 14.93 | .425 | .236 | 13.54 | .480 | .263 |
| 13.34 | .159 | .485 | 17.71 | .418 | .221 | 16.31 | .486 | .246 |
| 17.04 | .181 | .503 | 21.56 | .406 | .205 | 19.22 | .484 | .231 |
| 21.13 | .204 | .507 | 25.63 | .390 | .192 | 22.70 | .477 | .215 |
| 26.26 | .230 | .502 | 30.49 | .372 | .179 | 27.52 | .462 | .196 |
| 33.76 | .260 | .489 | 37.19 | .349 | .167 | 32.63 | .443 | .181 |
| 42.17 | .287 | .471 | 44.27 | .327 | .158 | 38.77 | .419 | .166 |
| 2.64 | .311 | .455 | 54.00 | .302 | .149 | 47.26 | .389 | .152 |
| 3.42 | .336 | .428 | 64.25 | .281 | .144 | 56.25 | .361 | .141 |
| 4.30 | .353 | .408 | 76.43 | .260 | .140 | 66.94 | .333 | .131 |
| 5.40 | .367 | .390 | interrupt | | | 73.92 | .317 | .127 |
| 7.03 | .377 | .371 | | | | interrupt | | |

| W | X1 | X2 | W | X1 | X2 | W | X1 | X2 |
|---|---|---|---|---|---|---|---|---|
| 8.56 | .381 | .360 | | | | | | |
| 11.14 | .383 | .347 | | | | | | |
| 14.51 | .382 | .337 | | | | | | |
| 18.29 | .378 | .331 | | | | | | |
| 23.05 | .374 | .326 | | | | | | |
| 30.04 | .368 | .324 | | | | | | |
| 37.88 | .362 | .323 | | | | | | |
| 47.76 | .356 | .324 | | | | | | |
| 3.11 | .350 | .326 | | | | | | |
| 3.92 | .345 | .328 | | | | | | |

N = 4 W(R) = 5, 3, 2, 1 K(R) = 1, 1, 1, 1

| W | X1 | X2 |
|---|---|---|
| 11.00 | .467 | .278 |
| 11.26 | .469 | .276 |
| 11.52 | .470 | .275 |
| 11.79 | .472 | .273 |
| 12.06 | .474 | .272 |

only every 7th value

| W | X1 | X2 |
|---|---|---|
| 13.54 | .480 | .263 |
| 16.31 | .486 | .246 |
| 19.22 | .484 | .231 |
| 22.70 | .477 | .215 |
| 27.52 | .462 | .196 |
| 32.63 | .443 | .181 |
| 38.77 | .419 | .166 |
| 47.26 | .389 | .152 |
| 56.25 | .361 | .141 |
| 66.94 | .333 | .131 |
| 73.92 | .317 | .127 |
| 2.72 | .302 | .123 |
| 3.24 | .275 | .118 |
| 3.85 | .250 | .114 |
| 4.69 | .225 | .113 |
| 5.57 | .204 | .114 |
| 6.63 | .186 | .118 |
| 8.08 | .167 | .125 |
| 9.61 | .153 | .135 |
| 11.42 | .142 | .148 |
| 13.91 | .132 | .167 |
| 16.51 | .125 | .187 |
| 19.61 | .120 | .211 |
| 23.87 | .117 | .243 |
| 28.36 | .117 | .273 |
| 33.71 | .120 | .304 |

| W | X1 | X2 |
|---|---|---|
| 3.33 | .202 | .433 |
| 4.05 | .231 | .433 |
| 4.82 | .259 | .425 |
| 5.73 | .289 | .412 |
| 6.98 | .323 | .392 |
| 8.31 | .351 | .371 |
| 9.89 | .376 | .347 |
| 12.06 | .398 | .319 |
| 14.35 | .412 | .295 |
| 17.07 | .420 | .271 |
| 19.80 | .421 | .252 |

nearly period

N = 4 W(R) = 5, 3, 2, 1

| W | X1 | X2 |
|---|---|---|
| 41.08 | .125 | .339 |
| 48.86 | .133 | .367 |
| 58.09 | .145 | .392 |
| 70.77 | .161 | .414 |
| 2.80 | .180 | .427 |

*6. Is a chain consisting of hyperlogistic modules hyperlogistic again?*[1])

Here, we build up chains with hyperlogistic behaviour as a basic behaviour of the modules, meaning

$$dx_i/dt = K_i x_i^{k_i}(B_i - x_i^{w_i})^{l_i}\, x_{i+1}$$

for $i = 1, 2, \ldots, N$ and put $x_{N+1} \equiv 1$. We will also take into account the possibility that $x_{i+1}$ is a feedback signal from a certain level of the chain. Therefore, we can also construct hypercycles of these basic modules.

We are mainly interested in the behaviour of $x_1(t)$ under the influence of the chain above. We shall study how the slope and the delay of the signal $x_1(t)$ as an EVOLON will be influenced by the other modules.

```
10 PRINT "HYPERLOG-CHAIN"
11 PRINT "N = ORDER"
12 INPUT N
13 DIM X(N + 1)
14 DIM K(N)
15 DIM E(N)
16 DIM B(N)
17 DIM W(N)
18 DIM L(N)
19 FOR R = 1 TO N
20 PRINT "X(";R;")"
21 INPUT X(R)
22 PRINT "E(";R;")"
23 INPUT E(R)
24 PRINT "B(";R;")"
25 INPUT B(R)
25 PRINT "W(";R;")"
28 INPUT W(R)
29 PRINT "L(";R;")"
30 INPUT L(R)
31 PRINT "K(";R;")"
32 INPUT K(R)
33 NEXT R
34 LET X(N + 1) = 1
35 PRINT "M"
36 INPUT M
37 PRINT "S"
38 INPUT S
39 PRINT "LOG"
40 INPUT LOG
41 PRINT "BACK"
42 INPUT BACK
43 PRINT "PRINT"
44 INPUT PRINT
45 PRINT "EPS"
46 INPUT EPS
47 PRINT "D"
48 INPUT D
49 CLS
50 FOR F = 1 TO M
55 FOR R = 1 TO N - 1
60 IF B(R) - X(R)^W(R) < EPS THEN
   GOSUB 90
65 LET X(R) = X(R) + D.K(R).(B(R) -
   X(R)^W(R))^L(R).X(R)^E(R).X(R + 1)
66 NEXT R
```

[1]) The idea for studying chains with hyperlogistic modules originates from several problem discussions with Dr. P. Allen, Prigogine-Institute, Bruxelles, who studies nets with nodes, all of which logistic modules.

```
67 LET R = N
68 IF B(N) – X(N)^W(N) < EPS THEN GOSUB 90
70 LET X(N) = X(N) + D.K(N).(B(N) – X(N)^W(N))^L(N).X(N)^E(N).X(BACK)
75 IF LOG = 1 THEN PLOT S.X(PRINT), S.X(PRINT + 1)
76 IF LOG = 1 THEN GOTO 80
78 PLOT F, S. X(PRINT)
79 PLOT F, S. X(PRINT + 1)
80 PRINT "X(D) ="; X(PRINT)
81 NEXT F
82 CLS
83 GOTO 50
90 IF B(R) – X(R).^W(R) < EPS THEN LET X(R) = B(R)^(1/W(R)) – EPS
91 RETURN
```

We now study for different values of N, the trajectories for X(PRINT) and X(PRINT + 1), and put K(R) = E(R) = W(R) = L(R) = X(R) = 1 and B(R) = 8 for all R = 1, 2, ... , N. We let D = 0.01 and EPS = 0.1.

| N = 2 | | N = 3 | | N = 4 | | N = 5 | |
|---|---|---|---|---|---|---|---|
| X1 | X2 | X1 | X2 | X1 | X2 | X1 | X2 |
| 1.07 | 1.07 | 1.07 | 1.07 | 1.07 | 1.07 | 1.07 | 1.07 |
| 1.15 | 1.14 | 1.15 | 1.15 | 1.15 | 1.15 | 1.15 | 1.15 |
| 1.24 | 1.22 | 1.24 | 1.24 | 1.24 | 1.24 | 1.24 | 1.24 |
| 1.34 | 1.31 | 1.34 | 1.34 | 1.34 | 1.34 | 1.34 | 1.34 |
| 1.46 | 1.39 | 1.46 | 1.46 | 1.46 | 1.46 | 1.46 | 1.46 |
| 1.59 | 1.48 | 1.60 | 1.60 | 1.60 | 1.60 | 1.60 | 1.60 |
| 1.74 | 1.58 | 1.77 | 1.74 | 1.77 | 1.77 | 1.77 | 1.77 |
| 1.92 | 1.68 | 1.96 | 1.92 | 1.96 | 1.96 | 1.96 | 1.96 |
| 2.11 | 1.79 | 2.19 | 2.11 | 2.20 | 2.19 | 2.20 | 2.20 |
| 2.33 | 1.90 | 2.45 | 2.33 | 2.47 | 2.45 | 2.48 | 2.48 |
| 2.59 | 2.02 | 2.77 | 2.59 | 2.81 | 2.77 | 2.81 | 2.81 |
| 2.87 | 2.14 | 3.15 | 2.87 | 3.21 | 3.15 | 3.22 | 3.21 |
| 3.18 | 2.26 | 3.58 | 3.18 | 3.70 | 3.58 | 3.72 | 3.70 |
| 3.53 | 2.39 | 4.09 | 3.53 | 4.27 | 4.09 | 4.31 | 4.27 |
| 3.91 | 2.53 | 4.65 | 3.91 | 4.92 | 4.65 | 4.99 | 4.92 |
| 4.31 | 2.66 | 5.26 | 4.31 | 5.62 | 5.26 | 5.73 | 5.62 |
| 4.73 | 2.81 | 5.88 | 4.73 | 6.33 | 5.88 | 6.46 | 6.33 |
| 5.17 | 2.95 | 6.47 | 5.17 | 6.95 | 6.47 | 7.09 | 6.95 |
| 5.60 | 3.10 | 6.98 | 5.60 | 7.42 | 6.98 | 7.54 | 7.42 |
| 6.02 | 3.25 | 7.38 | 6.02 | 7.72 | 7.38 | 7.80 | 7.72 |
| 6.41 | 3.41 | 7.66 | 6.41 | 7.88 | 7.66 | 7.92 | 7.88 |
| 6.75 | 3.56 | 7.82 | 6.75 | 7.95 | 7.82 | 8.00 | 7.95 |
| 7.05 | 3.72 | 7.92 | 7.05 | 8.00 | 7.92 | 8.00 | 8.00 |
| 7.30 | 3.88 | 8.00 | 7.30 | 8.00 | 8.00 | | |
| 7.50 | 4.04 | 8.00 | 7.50 | | | | |
| 7.65 | 4.20 | | | | | | |

| N = 2 | | N = 3 | | N = 4 | | N = 5 | |
|---|---|---|---|---|---|---|---|
| X1 | X2 | X1 | X2 | X1 | X2 | X1 | X2 |
| 7.76 | 4.36 | | | | | | |
| 7.84 | 4.52 | | | | | | |
| 7.90 | 4.68 | | | | | | |
| 7.94 | 4.83 | | | | | | |
| 7.99 | 4.99 | | | | | | |
| 7.99 | 5.14 | | | | | | |

The same experiment, but with E(R′) = 2 for R = 1, ... N

| N = 2 | | N = 3 | | N = 4 (D = 0.005) | | N = 5 (D = 0.0025) | |
|---|---|---|---|---|---|---|---|
| X1 | X2 | X1 | X2 | X1 | X2 | X1 | X2 |
| 1.07 | 1.07 | 1.07 | 1.07 | 1.04 | 1.04 | 1.02 | 1.02 |
| 1.15 | 1.15 | 1.15 | 1.15 | 1.07 | 1.07 | 1.04 | 1.04 |
| 1.26 | 1.24 | 1.26 | 1.26 | 1.12 | 1.12 | 1.06 | 1.06 |
| 1.39 | 1.34 | 1.40 | 1.40 | 1.16 | 1.16 | 1.08 | 1.08 |
| 1.56 | 1.46 | 1.57 | 1.56 | 1.23 | 1.22 | 1.10 | 1.10 |
| 1.80 | 1.60 | 1.82 | 1.80 | 1.28 | 1.28 | 1.12 | 1.12 |
| 2.12 | 1.77 | 2.19 | 2.12 | 1.35 | 1.35 | 1.14 | 1.14 |
| 2.58 | 1.96 | 2.78 | 2.58 | 1.43 | 1.43 | 1.17 | 1.17 |
| 3.29 | 2.20 | 3.83 | 3.29 | 1.53 | 1.53 | 1.20 | 1.20 |
| 4.41 | 2.48 | 5.84 | 4.41 | 1.64 | 1.64 | 1.23 | 1.23 |
| 6.14 | 2.81 | 9.09 | 6.14 | 1.78 | 1.78 | 1.26 | 1.26 |
| 8.11 | 3.23 | 8.03 | 8.11 | 1.96 | 1.95 | 1.29 | 1.29 |
| 8.01 | 3.72 | | | 2.19 | 2.17 | 1.33 | 1.33 |
| 8.01 | 4.32 | | | 2.49 | 2.46 | 1.37 | 1.37 |
| 8.02 | 5.00 | | | 2.91 | 2.84 | 1.41 | 1.41 |
| 8.03 | 7.13 | | | 3.52 | 3.37 | 1.45 | 1.45 |
| 8.04 | 7.57 | | | 4.46 | 4.14 | 1.50 | 1.50 |
| 8.04 | 7.82 | | | 5.92 | 5.25 | 1.56 | 1.56 |
| 8.04 | 7.93 | | | 7.83 | 6.73 | 1.62 | 1.62 |
| 8.04 | 8.00 | | | 8.18 | 8.06 | 1.69 | 1.69 |
| | | | | | | 1.77 | 1.76 |
| | | | | | | 1.85 | 1.85 |
| | | | | | | 1.95 | 1.95 |
| | | | | | | 2.06 | 2.06 |
| | | | | | | 2.19 | 2.19 |
| | | | | | | 2.34 | 2.34 |
| | | | | | | 2.52 | 2.52 |
| | | | | | | 2.74 | 2.73 |
| | | | | | | 3.01 | 2.99 |
| | | | | | | 3.35 | 3.31 |
| | | | | | | 3.78 | 3.72 |

| N = 2 | | N = 3 | | N = 4 (D = 0.005) | | N = 5 (D = 0.0025) | |
|---|---|---|---|---|---|---|---|
| X1 | X2 | X1 | X2 | X1 | X2 | X1 | X2 |
| | | | | | | 4.35 | 4.24 |
| | | | | | | 5.08 | 4.90 |
| | | | | | | 6.00 | 5.71 |
| | | | | | | 7.03 | 6.64 |
| | | | | | | 7.83 | 7.49 |
| | | | | | | 8.03 | 7.95 |
| | | | | | | 8.00 | 8.00 |

*7. A signal-generator for the Tschebyscheff polynomials*

The Tschebyscheff polynomials are given by the following explicit formulas

$$T_n(t) = x_0 = \cos\big(n \arccos(2\,t - 1)\big)\,.$$

We have the following Lotka-Volterra equations

$$Fx_0 = nx_1\,,$$
$$Fx_1 = Fy_0 - nx_1 - nx_2 \quad \text{with} \quad y_0 = 1/squr\big(t(1-t)\big),$$
$$Fx_0 = Fy_0 + nx_1 + nx_2$$

and the Riccati representation

$$du_0/dt = u_0u_1\,, \qquad x_0 = u_1\,,$$
$$du_1/dt = nu_2y_0\,, \qquad x_1 = u_2y_0/u_1\,,$$
$$du_2/dt = -nu_1y_0\,, \qquad x_2 = u_1y_0/u_2\,.$$

The help state-variables are

$$x_1 = \frac{\sin\big(n \operatorname{arc} \cos(2t-1)\big)\,y_0}{\cos\big(n \operatorname{arc} \cos(2t-1)\big)}\,, \qquad x_2 = \frac{\cos\big(n \operatorname{arc} \cos(2t-1)\big)\,y_0}{\sin\big(n \operatorname{arc} \cos(2t-1)\big)}\,.$$

We get the following BASIC program,

```
 10 PRINT "TSCHEBYSCHEFF"
 11 PRINT "R0"
 12 INPUT R0
 13 PRINT "M"
 14 INPUT M
 15 PRINT "D"
 16 INPUT D
 17 PRINT "S"
 18 INPUT S
 20 PRINT "N=ORDER"
 30 INPUT N
 31 PRINT "LOG"
 32 INPUT LOG
 35 IF LOG= 1 THEN GOTO 219
120 PRINT "SIGMA"
```

```
130 INPUT SIGMA
133 LET U1 = cos (N.arccos (2 R0 – 1)
134 LET U2 = sin (N.arccos (2R0 – 1)
140 FOR R=1 TO M
150 LET T= R0 + R.D
155 GOSUB 200
165 LET U1 = U1 + D.N.Y0.U2
170 LET U2 = U2 – D.N.U1.Y0
175 PRINT U1
180 NEXT R
181 CLS
182 GOTO 140
183 STOP
200 LET Y0 = 1/squr(T.(1 – T))
210 RETURN
219 FOR R=1 TO M
220 LET F= cos (N.arccos(2.(R0+D.R)–1))
221 LET G = sin (N.arccos(2.(R0+D.R)–1))
225 PRINT F
230 NEXT R
```

R0 = 0.02 D = 0.01

N = 1

| X0-model | X0 orig. |
|---|---|
| –0.94 | –0.94 |
| –0.93 | –0.92 |
| –0.91 | –0.90 |
| –0.89 | –0.88 |
| –0.87 | –0.86 |
| –0.85 | –0.84 |
| –0.84 | –0.82 |
| –0.82 | –0.80 |
| –0.80 | –0.78 |
| –0.78 | –0.76 |
| –0.76 | –0.74 |
| –0.74 | –0.72 |
| –0.72 | –0.70 |
| –0.70 | –0.68 |
| –0.69 | –0.66 |
| –0.67 | –0.64 |
| –0.65 | –0.62 |

N = 1

| XO-model | XO-orig. |
|---|---|
| –0.63 | –0.60 |
| –0.61 | –0.58 |
| –0.59 | –0.56 |
| –0.57 | –0.54 |
| –0.55 | –0.52 |
| –0.53 | –0.50 |
| –0.51 | –0.48 |
| –0.49 | –0.46 |
| –0.47 | –0.44 |
| –0.45 | –0.42 |
| –0.43 | –0.40 |
| –0.41 | –0.38 |
| –0.39 | –0.36 |
| –0.37 | –0.34 |
| –0.36 | –0.32 |
| –0.34 | –0.30 |

# 11. References


[1] Volterra, V.: Leçons sur la théorie mathématique de la lutte pour la vie, Gauthier-Villars, Paris, 1931

[2] Svireshev, J. M., Logofet, D. O.: Stability in biological communities (Russian), Edition "Nauka", Moscow, 1978

[3] May, R. M.: Stability and complexity in model ecosystems, Princeton Univ. Press, Princeton, New Jersey, 1973

[4] Malthus, T. R.: An essay on the principle of population, Johnson, London, 1897

[5] Maynard Smith, J.: Models in Ecology, Cambridge University Press, 1974

[6] Williamson, M.: The analysis of biological populations. Arnold, London, 1972

[7] Basykin, A. D.: The Volterra system and the Michaelis-Menthon equation, in "Voprosy matematiceskoi genetiki" (Russian), Novosibirsk, Soobschtsch. AN SSSR, 1974, pp. 103—143

[8] Odum, H. T.: Environment, power and society, Wiley (Interscience), New York, 1971

[9] Eigen, M., Schuster, P.: The hypercycle, Springer-Verlag, Berlin-Heidelberg-New York 1979

[10] Verhulst, P. E.: Nuov., Mem. Acad. Roy. Bruxelles 18, 1 and 20, 1, 1847

[11] Schuster, P., Sigmund, K., Hofbauer, J., Wolff, R.: Selfregulation of behaviour in animal societies, Biological Cybernetics I in **40**, 1—8, II in **40**, 9—15, III in **40**, 17—25, Springer-Verlag, Berlin-Heidelberg-New York 1981

[12] Maynard Smith, J.: On evolution, Edinburgh, University Press, 1972

[13] Schuster, P., Sigmund, K.: Coyness, philandering and stable strategies. Animal Behaviour **29** (1981) 186—192

[14] Hofbauer, J., Schuster, P., Sigmund, K.: Game dynamics in Mendelian populations, Biol. Cybernetics **43** (1982) 51—57

[15] Peschel, M., Mende, W., Ahlbehrendt, N., Voigt, M., Grote, U.: Dynamic problems of evolution, **WP-80-1**, IIASA, Laxenburg, 1980.

[16] Hofbauer, J., Schuster, P., Sigmund, K.: A note on evolutionary stable strategies and game dynamics, Jour. theor. Biol. **81** (1979) 609—612

[17] Mende, W., Peschel, M.: Structure-building phenomena in systems with power-product forces, In: Haken, H.: Chaos and order in nature, 196—206, Springer-Verlag, Berlin-Heidelberg-New York 1981

[18] Mende, W., Peschel, M.: Strukturentwurf für instationäre und nichtlineare Systeme, msr **10** (1981)

[19] Peschel, M., Mende, W.: Probleme der mathematischen Modellierung von Evolutionsprozessen, msr **12** (1981)

[20] Peschel, M., Mende, W.: Volterra-Modelle für nichtlineare und instationäre Systeme und deren Äquivalenztransformationen, msr **2** (1982)

[21] Peschel, M., Mende, W.: Parameteridentifikation von Potenz-Produkt-Triebkraftmodellen, msr **5** (1982)

[22] Peschel, M., Mende, W., Grauer, M.: An ecological approach to systems analysis based on the Volterra equations, Collaborative Paper, IIASA, **CP-82-20**, 1982

[23] Peschel, M.: Modellbildung für Signale und Systeme, VEB Verlag Technik, Berlin, 1978

[24] Peschel, M., Wunsch, G.: Methoden und Prinzipien der Systemtheorie, VEB Verlag Technik, Berlin, 1972

[25] Solodownikow, W. W., Semjonow, W. W., Peschel, M., Nedo, D.: Berechnung von Regelsystemen auf Digitalrechnern, Anwendung von Spektral- und Interpolationsmethoden, VEB Verlag Technik, Berlin, 1978, Oldenbourg-Verlag, München, 1978 und Verlag Maschinostrojenija, Moscow, 1979

[26] Peschel, M.: Moderne Anwendungen algebraischer Methoden, VEB Verlag Technik, 2. Auflage, Berlin, 1972

[27] Peschel, M.: Ingenieurtechnische Entscheidung — Modellbildung und Steuerung mit Hilfe der Polyoptimierung, VEB Verlag Technik, Berlin, 1980

[28] Arneodo, A., Coullet, P., Tresser, C.: Occurrence of strange attractors in three-dimensional Volterra equations, Physics Letters, **79 A**, 4 (1981) 259—263

[29] Hofbauer, J., Schuster, P., Sigmund K.: Competition and cooperation in catalytic selfreplication, Journ. Math. Biology **11** (1981) 155—168

[30] Hofbauer, J.: On the occurrence of limit cycles in the Volterra-Lotka equation, Nonlinear Analysis, Theory, Methods and Applications, **5** (1981) 9, 1003—1007

[31] Robert van der Vaart, H.: Conditions for periodic solutions of Volterra differential systems, Bulletin of Math. Biology, **40** (1978 133—160

[32] Takeuchi, Y., Adachi, N., Tojumaru, H.: Global stability of ecosystems of the generalized Volterra type, Math., Biosci., **42** (1978) 1—2, 119—136

[33] Grossberg. St.: Decisions, patterns and oscillations in nonlinear competitive systems with applications to Volterra-Lotka systems, J. theor. Biol. **73** (1978), 101—130

[34] Takeuchi, Y.: Existence of stable equilibrium points for dynamical systems of Volterra type, Jour. of math. analysis and applications **19** (1981) 141—162

[35] Logofet, D. O., Svireshev, J. M.: Stability in models of interacting populations (Russian), Problemy Kibernetiki, **32** (1977) 187—202

[36] Feinberg, M.: Chemical oscillations, multiple equilibria and reaction network structure, in 'Dynamics and modelling of reactive systems', edited by W. Stuart, W. H. Ray and C. Couley, Academic Press, New York, 1980

[37] Narendra, S. Goel, Samaresh, C. Maitra, Ellitt, W. Montroll: On the Volterra and other nonlinear models of interacting populations, Reviews of Modern Physics **43** (1971) 2, part I, 231—275

[38] Gompertz, B.: Phil. Trans. Roy. Soc. **115** (1825) 513

[39] Lotka, A. J.: Analytical note on certain rhythmic relations in organic systems, Proceedings of the National Acad. of Sciences **6** (1920) 410

[40] Haken, H.: Dynamics of synergetic systems, Springer-Verlag, Berlin-Heidelberg-New York, 1980

[41] Engels, F.: Dialektik der Natur, Dietz-Verlag, Berlin, 1952

[42] Lenin, W. I.: Materialismus und Empiriokritizismus, Dietz-Verlag, Berlin, 1949

[43] Parisi, E., Filosa, S., Monroy, A.: Spatial-temporal coordination of mitotic activity in developing sea urchin embryos, In: Haken, H.: Chaos and order in nature, Springer-Verlag, Berlin–Heidelberg–New York, 1981

[44] Zadeh, L. A.: Fuzzy sets, Information and Control 8 (1965) 338

[45] Richardt, J.: Eine unscharfe Modellierung für die Lösung quadratischer Zuordnungsprobleme, In: Berichte des Seminars 'Unscharfe Modellbildung und Steuerung', Teil III (1980), 35—51, TH Karl-Marx-Stadt

[46] Odum, E. P.: Fundamentals of ecology, 4th edition, Saunders, New York, 1980

[47] Peschel, M., Mende, W., Köhler, L., Eckardt, G.: New developments in systems theory with consequences to the technique of measurements, Plenary Paper 9. World Congress of IMEKO, Preprints, Berlin-West, 1982

[48] Nicolis, G., Prigogine, J.: Self-Organization in nonequilibrium systems, Wiley, New York, 1977

[49] Mende, W.: Zur Energieflußanalyse der ökologischen Grundsituation, In: Forschungsbeiträge zur Ökologie und zum Umweltschutz, Greifswald 1980

[50] Mende, W., Peschel, M.: Problems of fuzzy modelling, control and forecasting of time-series and some aspects of evolution, IFAC Symposium on Control Mechanisms in Bio- and Ecosystems, Plenary Paper, Leipzig, 1977

[51] Prigogine, J.: Vom Sein zum Werden, Piper, München, 1979

[52] Prigogine, J., Stengers, I.: Dialog mit der Natur, Piper, München, 1981

[53] Pearson, K.: Contributions to the mathematical theory of evolution III (1893), 71—110, University College, London

[54] Pearson, K.: Contributions to the mathematical theory of evolution X, Scew variation in homogenous materials (1896) 343—412, University College, London

[55] Kaltofen, H.: Quantitative Aspekte des Zusammenwirkens mehrerer Wachstumsfaktoren, Tagungsberichte der Akademie der Landwirtschaftswissenschaften der DDR, Berlin **139** (1975) 71—112

[56] Liebig, J. v.: Die Naturgesetze des Feldbaus, Verlag Vieweg und Sohn, Braunschweig, 1862

[57] Mitscherlich, E. A.: Das Gesetz des Minimums und das Gesetz des abnehmenden Bodenertrages, Landwirtschaftliche Jahresberichte **38** (1909) 537—552

[58] Mitscherlich, E. A., Dühring, F.: Das Liebigsche Gesetz vom Minimum und das Wirkungsgesetz der Wachstumsfaktoren. Schriften der Königsberger gelehrten Gesellschaft, Naturwissenschaftliche Klasse **3** (1926) 1—30

[59] Peschel, M.: Präferenz, Referenz, Konvexität und Effizienz-Grundbegriffe für Entscheidungsprozesse mit mehrfacher Zielsetzung, Wissenschaftliche Zeitschrift der TH Ilmenau **26** (1980) 6, 135—145

[60] Kalman, R., Falb, P., Arbib, M.: Topics in mathematical systems theory, McGraw Hill Book Company, New York, 1969

[61] Naslin, P.: Dynamik linearer und nichtlinearer Systeme, VEB Verlag Technik, Berlin, 1967

[62] Göldner, K.: Mathematische Grundlagen für Regelungstechniker, VEB Fachbuchverlag, Leipzig, 1968

[63] Suchovitzki, S. I., Avdejevna, L. I.: Linear and convex programming (Russian), Edition "Nauka", Moscow, 1964

[64] Weidlich, W., Haag, G.: Dynamics of interacting groups in society with application to the migration of population, In: Haken H.: Dynamics of synergetic systems, 235—243, Springer-Verlag, Berlin-Heidelberg-New York, 1980

[65] Gerskova, M. I.: Stepwise parabolic equation for the growth of chicken embryos (Russian), In: 'Analysis of the growth dynamic of biological objects (Russian)', 5—15, Edition 'Nauka', Moscow, 1978

[66] Lotka, A. J.: Elements of physical biology, Williams Baltimore, and Wilkins Comp., 1925

[67] Bertalanffy, L. v.: General systems: Foundations, development, applications, Braziller, New York, 1968

[68] Whittaker, R. H.: Communities and ecosystems. Mc Millan Publ. Co, 2nd. ed. New York, 1975

[69] Darwin, Ch.: Architecture and distribution of coral reefs, London, 1842

[70] Kolmogoroff, A.: Sur la notion de la moyenne, In: Atti della Reale Academii nazionale dei Lincei Rend. adunance solenni Ser. **6** (1930) 12, 388—391
[71] Naur, P.: Revised report on the algorithmic language ALGOL 60, Comm. **ACM6** (1963) 1, 1—17
[72] Schmalhausen, I. I.: Fundamentals of an evolution process from the point of view of cybernetics (Russian), In: Problemy Kibernetiki **4**, 121—150, Moscow, 1960
[73] Bongard, M. M.: On the notion of 'useful information' (Russian), In: Problemy Kibernetiki **9**, 71—102, Moscow, 1963
[74] Gössel, M.: Angewandte Automatentheorie I und II, WTB Mathematik-Physik, Akademie-Verlag, Berlin, 1972
[75] Grossman, M.: The first nonlinear system of differential and integral calculus, University of Lowell, 1979
[76] Darwin, Ch.: The formation of vegetable mould through the action of worms, with observation of their habits, London, 1881
[77] Großmann, A., Fiedler, K., Grauert, B.: Beitrag zur Modellierung heterogener Reaktionen, Zeitschrift für physik. Chemie, Leipzig **261** (1980) 2, 265—270
[78] Mende, W.: Phänomenologische Betrachtungen zur Wachstumsdynamik, Diss. A, AdW d. DDR, Forschungsbereich Geo- und Kosmoswissenschaften, 1984
[79] Broda, E.: Bioenergetics of growth, London, 1947
[80] Kriegel, U., Mende, W., Grauer, M.: An evolutionary analysis of world energy consumption and world population, **CP-83-34**, IIASA, Collaborative Paper, Laxenburg, 1983
[81] Peschel, M., Mende, W., Grauer, M.: An Ecological approach to systems analysis based on the Volterra equations, **CP-82-20**, IIASA, Collaborative Paper, Laxenburg, 1982
[82] Peschel, M., Mende, W.: Ecological approach to systems analysis based on Volterra equations, Cybernetics and Systems: An International Journal (1982) **13**: 175—186
[83] Peschel, M., Mende, W.: A unified modelling concept for nonlinear systems with Lotka-Volterra Equations, Systems Analysis, Modelling and Simulation **1** (1984), 17—26
[84] Peschel, M., Mende, W., Grauer, M.: Qualitative analysis of nonlinear systems by the Lotka-Volterra approach. **CP-83-60**, IIASA, Collaborative Paper, Laxenburg, 1983
[85] Scudo, F. M., Ziegler, J. R.: The golden age of theoretical ecology: 1923—1940, vol. 22 in Lecture Notes in Biomathematics, Springer-Verlag, Berlin-Heidelberg-New York, 1978
[86] Yodzis, P.: Competition for space and the structure of ecological communities, vol. 25 of Lecture Notes in Biomathematics, Springer-Verlag, Berlin-Heidelberg-New York, 1978
[87] Thomas, R. (editor): Kinetic logic — A Boolean approach to the analysis of complex regulatory systems. Proceedings, vol. 29 in Lecture Notes in Biomathematics, Springer-Verlag, Berlin-Heidelberg-New York, 1979
[88] Getz, W. M. (editor): Mathematical modelling in biology and ecology vol. 33 in Lecture Notes in Biomathematics, Springer-Verlag Berlin–Heidelberg–New York, 1980
[89] Mesarovič, M. D., Pestel, E.: Menschheit am Wendepunkt, Deutsche Verlagsanstalt, Stuttgart, 1976
[90] Holling, C. S. (editor): Adaptive environmental assessment and management, Wiley, IIASA International Series on Applied Systems Analysis, Vol. 3, John Wiley, Chichester, England, 1978
[91] Meadows, D., Richardson, J., Bruckmann, G: Groping in the dark: The first decade of global modelling, John Wiley, Chichester, England, 1982

[92] IIASA Energy Systems Program Group, Leader W. Häfele, Energy in a finite world, Ballinger Publishing Company, Cambridge, Mass., 1981
[93] Keyfitz, N.: Population of the world and its regions: 1975—2050, **WP-77-7**, IIASA, Working Paper, Laxenburg, 1977
[94] Lance, G. N.: Numerical methods for high speed computers, Iliffe & Son Ltd., London, 1960, p. 56
[95] James, F.,Ross, M.: MINUIT-system for function minimization and analysis of parameter errors and correlations, Computer Physics Communication, **10** (1975) 343—367
[96] The global 2000 report to the president, Washington, 1980
[97] Neefe, M.: Statistisches Jahrbuch deutscher Städte, Korn, Breslau, 1890
[98] Albrecht, K. F., Mende, W., Grauer, M.: Hyperlogistic growth study of European cities, IIASA-Collaborative (in preparation)
[99] Bunke, H.: Gewöhnliche Differentialgleichungen mit zufälligen Parametern, Akademie-Verlag, Berlin, 1972
[100] Chow, S. N., Hale, J. K.: Methods of bifurcation theory. Springer-Verlag, Berlin-Heidelberg-New York, 1982
[101] Collatz, L.: Numerische Behandlung von Differentialgleichungen. Springer-Verlag, Berlin-Heidelberg-New York, 1982
[102] Curry, H. J.: An algorithm for finding closed orbits. Lecture Notes in Math. 819, Springer-Verlag, Berlin-Heidelberg-New York, 1976
[103] Ebeling, W., Feistel, R.: Physik der Selbstorganisation. Akademie-Verlag, Berlin, 1983
[104] Feigenbaum, M. J.: J. Stat. Phys. **19** (1978), 19.
[105] Gückenheimer, J., Moser, J., Newhouse, Sh. E.: Dynamical systems. Birkhäuser, Basel, 1980
[106] Jansen, W., Kriegel, U.: DEAN– a differential equations analyzer (in Preparation)
[107] Kliemann, W.: Qualitative Theorie Nichtlinearer systeme. Diss., Universität Bremen, 1980
[108] Mandelbrot, B. B.: Fractals-form, change and dimension, Freeman, San Francisco, 1977
[109] Pugachev, W. S.: Theory of random functions. (Russian) Fismat., Moscov, 1960
[110] Peschel, M., Mende, W.: Leben wir in einer Volterra-Welt? Akademie-Verlag, Berlin, 1983
[111] Schuster, P., Sigmund, K.: A mathematical model of the hypercycle. In: H. Haken: Dynamics of synergetic systems, Springer-Verlag Berlin-Heidelberg-New York, 1980
[112] Allee, W. C.: Cooperation among animals with human implications, Schumann New York, 1951
[113] Hegel, G. W. F.: Geschichte der Logik (Herausgeber: G. Laffon), 2. Teil, Akademie-Verlag, Berlin, 1975
[114] Bossel, H.: Umweltdynamik. Verlag te-wi, München, 1985
[115] Keyfitz, N.: Population of the World and its Regions: 1975—2050, **WP-77-7**, IIASA, Laxenburg, 1977
[116] Peschel, M., Mende, W., Voigt, H. M.: Anwendung der Polyoptimierung auf Evolutionsprozesse. Österreichische Studiengesellschaft für Kybernetik, Rep.Nr. 16, Wien, 1979
[117] Pych, J. A.: Equilibrium and stability in models of population dynamics. (Russian) Nauka, Moscov, 1983
[118] Hofbauer, J., Sigmund, K.: Evolutionstheorie und dynamische Systeme, Paul Parey, Berlin, 1984

[119] Peschel, M.: Einige Gedanken zur Systembeschreibung, msr **15** (1972) H. 12
[120] Peschel, M.: Untersuchungen über den Begriff der Konvexität, Diss. A, Math. nat. wiss. Fakultät, Humboldt-Universität, Berlin, 1966
[121] Haken, H.: Synergetics. An introduction. Vol. 1, Springer Series in Synergetics. Springer-Verlag, Berlin-Heidelberg-New York-Tokyo, 1978
[122] Allee, W. C., Emerson, W. C., Park, O., Park, Th., Schmidt, K.: Principle of animal ecology. Saunders, Philadelphia, 1949
[123] Straube, B.: Anwendung der Theorie unscharfer Mengen bei der Modellierung realer Systeme, Diss. B, AdW der DDR, Berlin, 1983
[124] Marcus-Roberts, H., Thompson, M.: Life, science, models. Springer-Verlag New York-Heidelberg-Berlin, 1983
[125] Braun, M. Coleman, S. C., Drew, A. D.: Differential equation models. Springer-Verlag, New York-Heidelberg-Berlin, 1983

# 12. Index

Bei Fragen zur Produktsicherheit wenden Sie sich bitte an:
If you have any questions regarding product safety,
please contact:

Walter de Gruyter GmbH
Genthiner Straße 13
10785 Berlin
productsafety@degruyterbrill.com